Signs of the end times
How to discern the times and be
prepared for what lies ahead of you
Copyright ©2021
Contact address: Oteng Montshiti
P O Box M1139
Kanye
Botswana

E-mail address:
otengmontshiti@gmail.com
Contact number: (+267) 74 644 954

Contents
Part 1
Signs of the end times... 8
Part 2
How to get prepared... 112

Acknowledgements

Writing a book is a challenging task which requires time. I would like thank my family especially my lovely wife for supporting me.

INTRODUCTION

When the issue known as the end of times is mentioned people's minds are filled up with various forms of pictures. Some think of nuclear annihilation, while other segments of people imagine the atmosphere being depleted and being affected by various diseases like cancer and so on. While another group of people views the issue of the end of the age as a myth or fairy tale story which was created by certain people to instil fear in the lives of others so that they can submit to their authority.

In other words, it is an issue that has given birth to confusion and fear in the hearts of my folks but it mustn't bring confusion and pandemonium. The best book which or person who has written or given precise prophetic words while walking on earth was Jesus Christ the son of the living God. One thing is certain the word of God has and will continue to fulfil itself before our eyes. In the past, many people tried to destroy it but it has stood the test of time. All the biblical prophecies were fulfilled at the exact time and season. You can't add or subtract anything in the word of God. In other, words it is alive, breathing, and carries the very nature of God. And he is going to make sure that whatever it says comes to pass or is fulfilled. One thing is certain everything happening under heaven or the sun isn't an end product of guesswork but if you are familiar with the word of God you know why certain events happen. God has inspired me to share the knowledge that he revealed to me with the entire human race. And I know your life will never be the same again. This book has been divided into two, part one discusses the issue of the signs of the end times and the second part discusses the way out or how you can be prepared for what lies ahead of you.

Sgns of the end times
Chapter 1
Destruction of the temple

Matthews 24: 1 And Jesus went out, and departed from the temple: and his disciples came to *him* for to shew him the buildings of the temple. **2** And Jesus said unto them, See ye not all these things? verily I say unto you, There shall not be left here one stone upon another, that shall not be thrown down. One day, Jesus Christ was sitting on Mount Olives when his followers gathered around and showed him a magnificent temple. He glanced at it and began to release prophetic words which left them shocked to the core. Prophecy simply means speaking God's plan or purpose concerning the future. And remember God does nothing outside the provision of his word. He and the word are one. The Bible says, God does nothing without revealing it to His servants the prophets, and remember Jesus Christ was a prophet. He told them that the temple was going to be destroyed after God had struck the shepherd and Jesus Christ is the shepherd and they were the sheep- John 10: 10. Around 30 AD and 33 AD Jesus Christ was falsely accused, tried, crucified, buried, and on the third day rose from the dead. Ultimately he ascended to heaven and sat at the right hand of God the father. Let me turn back the hands of time back a bit, one day the Pharisees and Sadducees came before Jesus Christ with a mission of trapping him and ask him to perform miracles in the book of Matthew 16. He told them point blank that he wouldn't perform them for them and told them that the only sign or miracle available for them was the sign or miracle of Jonah. Jonah was one of the prophets of God in the Old Testament. One day, God gave him the assignment to go and warn the inhabitants of Nineveh about God's impending judgment over their nation. Instead of obeying God's instructions, he disobeyed. Then, he travelled on a ship that was going to Tarshis away from the presence of God. Guess what happened, God sent a storm across his path, when others asked him to pray to God and the storm didn't stop. He told them that he had rejected God's assignment and asked them to throw him into the raging ocean. As he landed into the huge body of a rumbling sea a big fish swallowed him up and he stayed in its belly for three days and nights and after three days it spits him out. From there he went to Nineveh and preached for forty days, people repented including their leader and God almighty spared their lives. In the book of Matthew 16, he told them that the best miracle or sign that they could get was repentance. Repentance means turning away from your wicked thoughts, ways, and behaviour and walk with the Lord. However, they didn't believe in him. They couldn't realize him as the true saviour of the world. They depended on their effort for their salvation. To make matters worse, they expected the Messiah to come as a king to deliver them from the oppressive hands of the Roman Empire. They quoted all manner of scriptures before him. The person the prophets of the Old Testament prophesied about was before their eyes. That's to say, he was the fulfilled of prophecies that were released by prophets like Prophet Zachariah and many others. Remember, the word of God says, God has given us the power of choice between life and death. Jesus Christ came to give life in abundance and the gift of eternal life. While Satan's mission is to kill, steal and destroy life. Jewish rejected the will of God which was to embrace and believe in his only begotten son. Guess what happened, around 70 AD the temple and Jerusalem were destroyed to fulfil the prophetic words of our Lord Jesus Christ. Romans invaded Jerusalem under Titus, Jewish were captured and taken away as captives.

Chapter 2
Persecution of the chosen people or Christians

Luke 21:12 But before all these, they shall lay their hands on you, and persecute *you*, delivering *you* up to the synagogues, and into prisons, being brought before kings and rulers for my name's sake.

After the temple, Jerusalem had been reduced to a rumble, the Jewish were captured and held as captives. Remember, when Jesus Christ was still on earth the Roman Empire was at the highest reigning power. In other words, it was a superpower. The Jewish and followers of Jesus Christ were severely persecuted, left, right and centre as he had prophesied. During the reign of Kaiser Nero, Christians were persecuted. They were killed in a gruesome manner. Some were crucified while others suffered other manner of terrible afflictions. For example, Apostle John the revelatory wrote the book of Revelation around 90 AD on the island of Patmos without sight. It was the darkest period in the history of Christians. Other followers of Jesus Christ like Apostle Paul and Silas faced severe persecutions. They were thrown into jail but we thank God for his intervention. At one point people beat the hell out of Apostle Paul and left him half dead and, God revived him. That's also the fulfilment of the scriptures in the book of 2nd Timothy 3, which says that those who love and obey God will face many tribulations or challenges. Today, the persecution of Christians has escalated. In other countries which are ruled using religions and where there is no religious tolerance, they are imprisoned. Terrorist groups have targeted them and asked them to denounce their faith. If they refuse gruesome videos will go viral on social media like Facebook, Twitter, and so forth showing Christians being beheaded in public places to humiliate them. But we thank God most of them have refused to denounce Jesus Christ and a reward of the crown of glory awaits them in heaven. In other countries, especially communist nations Christians are not allowed to carry their Bibles around preaching the gospel. If you are caught doing that you are in hot soup, and the Bible is prohibited also.

If you want to send it to somebody in such nations you must snuggle it inside. Indeed, Christians are facing persecution.

<u>Chapter 3</u>
<u>The rise of false prophets or doctrines</u>

Matt 24:11 And many false prophets shall rise, and shall deceive many.

1Timothy 4:1 Now the Spirit explicitly says that in the later times some will desert the faith and occupy themselves with deceiving spirits and demonic teachings, Another prophetic word that Jesus Christ released was the rise of false prophets or (False doctrines). I was jaw dropped when one of the servants of God told us that false prophets were with them at the Bible College. They had put aside the garments they usually wear when they are consulting people at their shrines. They told him that they had lost clients due to the rise of the true churches of God because people had abandoned them and seek the face of God. So they couldn't sit down somewhere curl their knees and bury their heads between thighs. They had to come up with a strategy to counterattack the floor crossing. They had decided to go to Bible Colleges to study the doctrine of God and graduate.

After graduating they would

open their churches (cast demons, heal, bless people, and so on). Do you see how false prophets have infiltrated the church of God? From there they twist scriptures to lure people to their churches. During the day, they carry the Bible but at night they visit theirs shines for empowerment.

These kinds of people are very dangerous because they are not leading people to heaven but to hell. That's what is called the doctrines of Demons. They went on to tell him that in their churches they don't condemn sin because people would move out of their church or they would lose members. Instead, they preach what people want to hear or entertains them. In other words, they preach what supports their weaknesses.

False prophets have been there throughout history. In 1995, the world watched in horror when few people were killed and many others were wounded when sarin gas was released underway in Tokyo, Japan. The leader of that cultic group said he was disappointed at the low death toll. He had brainwashed people with wrong teaching or false doctrine about Armageddon. His followers were brainwashed to the level where they couldn't form a single thought of their own. They committed such atrocity as a sign of devotion to their leader.

When the millennium bug was making massive rounds or waves around the world, many false prophets stepped on the scene and prophesied that when the hands of the clock struck midnight that's the end of the world or the return of Jesus Christ. I remember, my family were outside glancing up at the starry sky, some people's eyes glued to their watches with the help of torches. The hands kept on ticking until the hands hit midnight. The sense of silence descended upon us as we waited for the howling of the trumpet of God or sign up in the sky but nothing happened. Remember, the word of God says, nobody knows the day or hour of the return of the son of God. It shall come like a thief in the night.

In recent years we have witnessed the rise of false prophets who portray the state of holiness on the outside but inside they are wolves in sheepskins. They go to certain places or shrines to get empowered and perform false signs and wonders. To make matters worse, they are sleeping with young girls in their congregations and encourage them to abort children when they fall pregnant to cover their tracks. Some of them are involved in criminal activities like money laundering, rape, and obtaining by false pretences. After committing such atrocities their congregations continue to support them because they are have been brainwashed. They quote scriptures like, "touch not my anointed and harm not my prophets,"

To make matters worse, I watched with shock on social media a certain pastor allowing people to kneel as he swung the door of the car open and stepped out. This isn't biblical, it simply idol worshipping. We must worship God alone and give all the adoration he deserves. What is happening is one of the signs of the end time. We are not supposed to worship leaders. I am not saying we should disrespect them but we should honour them and worship God alone.

<u>Chapter 4</u>
<u>Wars and rumors of wars</u>

Matt 24:6 And ye shall hear of wars and rumours of wars: see that ye be not troubled: for all *these things* must come to pass, but the end is not yet. Numerous wars have been fought throughout history and Southern African wasn't an exception. Around the 1800s intertribal wars were fought in Southern Africa, they are known as the Mfecane wars which is a Sesotho word meaning to "crush". During that time many people lost their lives, others were displaced and small tribes were incorporated into bigger tribes and ultimately lost their cultural identities. Mfecane started as a spark of fire that leaped up into the air and in no time turned into a raging inferno. People were using spears and shields.

In 1914, the First World War broke and many innocent souls were killed, nations were reduced to ashes. Nations that participated in that war used weapons like rifles, bombs, and so forth. Then, two decades later in 1939, another war broke in Europe and its destruction was beyond assessment. More sophisticated weapons were used like tanks, and the first atomic bomb was thrown at Hiroshima and Nagasaki in Japan on the 6th August 1945, many people were killed and other people are still suffering from the effects of that bomb. It was the first atomic bomb to be used and the ceasefire was declared. Then, the United Nations was established with a mandate of promoting international peace.

From 1933 to 1945 over six million Jewish was killed during the holocaust. From 1959 up to 1964, a war broke between Vietnam and

United States. It was one of the terrible wars and both nations suffered a large amount of human loss which is something that isn't good because human life is precious. I
n 1967 another war broke when Israel was surrounded by enemies but within seven days she emerged victorious and captured Jerusalem.

Another horrible war was between Iran and Iraq from 1980 to 1988 many innocent people died. When it came to an end and the whole world thought it was breathing a sigh of relief in 1990, the Iraq regime captured and occupied Kuwait.

An ultimatum was made and remained defiant and war broke which was known as the Gulf War 1 between Iraq and the coalition forces. Iraqi soldiers were killed as they driving out of Kuwait and it is called the highway of death. Many people especially civilians were gruesomely killed. Infrastructures like telecommunications, roads, and buildings were destroyed. In 2002 another way broke again between Iraq and coalition forces because they believed that Iraq had stockpiles of weapons of mass destruction (WMD) something which has not been established up to today. Iraq today is a heaven of terrorist groups, lawlessness has set in.

Chapter 5
Pestilences or pandemics

Matt 24:7 For nation shall rise against nation, and kingdom against kingdom: and there shall be famines, and pestilences, and earthquakes, in divers places.

Throughout history, there have been outbreaks of deadly diseases which had claimed people's lives. In 1918, there was an outbreak of deadly influenza known as the Spanish flu. It is regarded as the deadliest pandemic in all history because it killed over 50 million people across the world. Fortunately, scientists were able to contain it but the damage had already been done. Polio killed young children and we thank God because it was contained too. Today, children receive polio vaccination the world over. Then, as the wheel of history was turning another pandemic surfaced its ugly head know as HIV/AIDS. It has killed millions of people around the world since 1980. It spread mostly where there is ignorance. I remember, in my country especially my village where people are very deep in tradition and believe in superstition people brushed aside knowledge from medical practitioners aside and sail in the ocean of ignorance or follow tradition. They started pointing accusing at one another. Some said, "He or she has been bewitched by his or her so and so." While another segment said, "she or he attended so and so weeding that's where he or she was poisoned." As they were busy pointing fingers like that the health of the victim of HIV/AIDS would deteriorate and ultimately succumbed to the scourge. False prophets at that time robbed people of their hard-earned cash as they cast their bones and offer them herbal medicines.

However, we must thank God for releasing his anointed knowledge in the scientific world. New drugs have been discovered which can give the victim of HIV/AIDS a chance to live a longer life and healthier. The graph of death rates has been flattened a bit but ultimately the best way is to abstain from sex until marriage. Obedience to the word of God is the master key to all issues of life.

There was an outbreak of swine flu and people were vaccinated in recent years. In 2019, what started as a whispering sound in the streets of Wuhan in China has become an ear-splitting sound? The Covid-19 has grabbed the whole world by the throat, economies of nations have suffocated, and people have succumbed to Covid 19 because of lack of air. Life has changed forever, we no longer shake hands, as usual, we wear masks, keep social distance, and sanitizing our hands. The destruction of Covid 19 is beyond the scales of assessment. We are still feeling its effects today. Millions of people have died and may their souls rest in peace.

We thank God for the discovery in the scientific community,

because of advanced technology vaccines have been manufactured and available for members of the public within a year.

Chapter 6
Lovers of pleasure more than God

2 Timothy 3: 4, ...loving pleasure rather than loving God The word of God is categorically clear that in the last days, people will be lovers of pleasure more than God, I remember, a certain sister who approached a certain servant of God asking for permission to go out and watched a particular artist. The servant of God was shocked to the core. The question was, what she was lacking in the house of God or his presence. But we thank God she

changed her mind after she was rebuked.

Today, if you take a look around at music concerts conducted by famous people, the stadia are filled to the rafters while the church of God is all most empty. In these places, there are lots of sinful or wrong things happening there. People rape one another, others sell their bodies, drug uses is on the rise, and so on. That's shows that indeed people are lovers of pleasure more than God. One day, I was strolling down the streets of Kanye herding home from a tuck shop when I met a certain sister.

We exchanged pleasantries and walked together and I asked her;

"Where is your husband?"

"We have divorced a long time back" she answered.

"Why did you do that?" I asked again.

"Because I wanted freedom, I wanted to go wherever I want, come home when I want, associate with other people freely" she replied beaming with confidence. I tried to reason with her but she rejected my suggestions. Then, we parted ways. Do you see how true the word of God is? Today, people want freedom but real freedom is found in God.

People don't want to be corrected. They want to spend time on the wrong things. They only find delight in using the issue of divorce as an exit road to freedom.

Today, some women and men don't want to tie the knot.

<u>Chapter 7</u>
<u>Pride</u>

₂**Tim 3:2** For men shall be lovers of their own selves, covetous, boasters, proud, blasphemers, disobedient to parents, unthankful, unholy,

Pride simply means somebody who has no teachable spirit. In other, words it is somebody who doesn't allow himself or her to be corrected and rebuked by the word of God. Everybody must go to school and acquire knowledge or skills for survival but the anointed knowledge must be made a top priority in your life.

In recent years pride has saturated people's hearts. They can't be taught the things of the Kingdom of God with humility. Everybody has knowledge of his own to present before God which is wrong. They don't want to submit to the Lordship of Jesus Christ however, we should not rely on our strength but the unlimited power of God?

. Out of pride, people now debate with Biblical concepts like the creation of the world and mankind. Philosophers have coined a term known as evolution. It says, man originated from apes and evolved or changed over time from one form to another to adapt to the ever-changing environment. While the word of God makes it categorically clear that God formed man in his image and he used lump from the dust to form his outer shell which is the physical body, which he uses to interact with the outside world.

These theories are the door or entry point for the spirit of pride to enter your life and oppose everything that concerns God. Worldly knowledge can lead to arrogance and pride that's why wisdom from God is needed. Intellectual knowledge must submit to the absolute truth which is knowledge inspired by the Holy Spirit. When divine knowledge is substituted by carnal knowledge its leads to pride, confusion, arrogance, and destruction. That's why the word of God says, pride comes before down falling.

Chapter 8
Gospel shall be preached in all the world

Matthew 24:14 And this gospel of the kingdom shall be preached in all the world for a witness unto all nations; and then shall the end come.

Today, the word of God is preached at a tremendous speed. It penetrates the barriers established by the dictatorship regime of our time. The gospel started in Jerusalem as a spark of fire and diffused beyond her borders, swept across the world at an alarming speed. While Jesus Christ was on the earth and his followers they were under the limitation of distance and time. That's to say, they could be in one place at a time. However, it can now be preached through radios, which have wide coverage and a larger audience. It can be preached through internet facilities like YouTube and social media like Facebook and Twitter etc. The world is slowly but surely moving from traditional radio to online which can be watched or listened to in the palm of your hand. People who live inside rogue nations can't watch Christian channels in the comfort of their homes. Today, various institutions offer free online Bible training courses. Something which fulfils the prophetic words of Daniel 9. Even, conference, seminars and others can be conducted online across the world through social media like Facebook, Twitter and so forth. Indeed the world has turned into a global village where people can interact with one another, discuss biblical topics at the comfort of their home while sipping a steaming delicious cup of cocoa, tea, or coffee.

Chapter 9
Iniquity shall abound

Matthew 24:12 And because iniquity shall abound, the love of many shall wax cold.

Just like the days of Noah so shall be the days of our lives. During the era of Noah, God gave him a mammoth task of constructing an ark because God had revealed to him that man had embraced sin which separated him from his creator. God regretted creating man in his image. Sin was overflowing. Iniquity is like pouring some water in a cup when it is full it overflows.   If you take a glance around you, iniquity is everywhere. You access the internet pornographic materials are everywhere. To make matters worse, it is available for public consumption. Young children's minds are poisoned, homes are broken because watching pornographic materials lead to masturbation and it leads to detachment from real spouses. When you are detached from her you are simply venturing down the path of divorce.

Men and women live together as husbands and wives when they aren't legally married. It is known as cohabitation. They can have children outside the institution of marriage. To make matters worse, their conscience is dead. That's to say, it can't tell them what they are doing is wrong. When your consciousness is dead you are indeed a dead person walking upon the earth because God uses your conscience to talk with you.

The world is full of wickedness. People kill one another. Passion killings are on the rise. Our minds are cluttered with wrong or ungodly things. Everywhere, you look it is evil, women are selling the wonderfully made bodies in the streets (sex workers). People are demon-conscious rather than God-conscious. Indeed the world will not be the same again.

Chapter 10
Earthquakes and natural disasters

Matthew24: 7, there shall be famines,...and earthquakes, in divers places. As the world is racing towards the end of ages we have heard and experienced earthquakes in various places across the world. We have heard about tornados, volcanic eruptions which have killed people and destroying infrastructures. In 79 CE Mountain Vesuvius in Italy destroying the ancient city of Pompeii and thousands of people died. Volcanic eruptions have hit nations like DRC in

Africa.

In recent years the world has watched with horror as tornados hit one place after another. Natural disasters have hit the nation of Japan, Indonesia, the United States, China, and Russia, and so on. The issue of climate change has been topping various and powerful organizations' agendas upon the surface of the earth. The word of God says as the coming of Jesus Christ comes closer these signs shall intensify like birth pains. There are eclipses of the sun and moon. The moon has turned blood red in the past.

Famines and drought has hit many nations across the world throughout history and they continue even today.

Chapter 11
People will faint with fear

Luke 21:23 But woe unto them that are with child, and to them that give suck, in those days! for there shall be great distress in the land, and wrath upon this people.

After the first atomic bomb was dropped on Japan on the 6th August 1945 life changed forever. Since then, people have been filled with fear of nuclear annihilation especially if it could fall into the wrong hands like rogue nations or terrorist organizations. Rogue nations where dictatorship regimes are in power can access materials that manufacture nuclear bombs like uranium and use such weapons against their opponents or even the whole world. Today, such nations have told the whole world that they have them and they can use them whenever they want.

I remember during the first Gulf war we lived in constant fear that may be the Iraqi dictatorship regime was going to use weapons of mass destruction. We watched the raging war on television, listened to the radio, and so on but we thank God because they were not used. Indeed the world today is living in fear as the fulfilment of the prophetic words of the word of God. In the area of economics, nations are living in constant fear because the economies of nations across the world are not stable. Around 1930, there was an economic depression known as the Great depression. The Corona Virus has crippled the economies of nations and the outbreak of its variants has filled people with fear after killing many people across the world. The outbreak of pandemics also makes people live in constant fear because they makes the future uncertain.

Since, 9 September 2001, the terrorist attack in the United States of America has changed life forever. War against terrorism was declared and terrorist organizations especially in the Middle East were targeted. Of recent, in Mozambique, they have emerged there killing innocent people in a gruesome manner. This has sent chilling waves of fear across Southern Africa because nobody knows when they will strike next.

Indeed, as the word of God as prophesied people are living and fainting because of fear.

They are walking in the street with fear because criminals' attack them, eating in fear because there are evil people who can poison others, and so on. At family level husbands are living in constant fear because as their wives are climbing the corporate ladder she is likely to divorce him especially when he is being paid lower than her.

Chapter 12
Covetousness
What is covetousness?

2 Timothy 3: 2 for men shall be lovers of their own selves, covetous It means the craving or desiring to have more in life. It also means to desire what other people have. The opposition of covetousness is contentment. It is a very dangerous spirit because it makes you lose focus on the Lord and you will start to focus on earthly materials like power to get wealth, control other people, and get what doesn't belongs to you at all costs.

In the Bible, there is a story of King David who was walking on the balcony of his palace one day when he saw Bathsheba bathing instead of looking away he started to lust after her. Then, called her and slept with her. To make matters worse, she was a married woman. In the time of war, to try to conceal his evil deeds he planned her husband's death by placing him at the frontline but his plan failed. Ultimately, he managed to kill him and he married her something which didn't please the Lord and he was punished severely by God. You see, how covertness is very dangerous?

It can lead to other atrocities.

Some people lack contentment in life, they want to have more without questioning the source of their blessings. There is a sister who doesn't want her siblings to have anything. When they have the money she usually goes to them and borrows it while she has a huge amount of money in her accounts. That's covetousness. If it isn't checked it can lead to destructive behaviour. As human beings, we must be content with what we have and thank and celebrate God for that. Do you know that people can even sell their children or offer them as a sacrifice because they want more power, money, and influence, or fame? They can do that to get more out of life which is very dangerous. God Almighty wants us to thank him and be content with what we have. To make matters worse, some nations are very greedy they can attack other nations to grab their mineral resource and enrich themselves. As they are doing that they kill innocent people, reduce magnificent nations to ashes.

Chapter 13
Nation shall rise against nation and kingdom against kingdom

Matthew 24:7 For nation shall rise against nation, and kingdom against kingdom Jesus Christ prophecies never miss or can be aborted. If there is anything that has the power to fulfil whatever it has said or released as a prophecy is the word of God. I always tell people that God will always make sure his word comes to pass. As I have said earlier, nations have been fighting or rising against one another. In 70 AD the nation of

Israel was attacked by Romans. Jewish people were captured, tormented. The leader of Germany Hitler killed over 6 million Jewish during the holocaust.

The nation of Israel for example today is surrounded by enemies who want to destroy her. However, we thank God because she has one of the most powerful armies in the world. The level at which nations hate one another is worrisome. Some rogue nations have acquired materials and manufactured weapons which can travel across continents and hit the target with precision.

They are known as continental missiles. They have threatened to destroy any nation they regard as their enemy.

In spiritual things, the kingdom of darkness has risen against the kingdom of God. In this world, there are only two kingdoms namely the kingdom of darkness and light or God. Throughout history Satan, the enemy of your soul has been engaged in a war against the children of God. He fights them using various methods like doubts, fear, and so forth. He has divided children of God and make them fight one another within the Christian faith.

That's to say, churches are fighting one another. Do you know why? Because he knows that a nation that fights within can't win any battle. There are times when he uses his agents to attack. However, we thank God for his protection, deliverance, and victory through the blood of Jesus Christ.

Chapter 14
Prevalence of lack of forgiveness

Lack of forgiveness has reached an alarming proposition and the cause of divorce is lack of forgiveness. People no longer rely on the principles of God but they rely entirely on their emotions or human strength. Marriages are no longer regarded as holy institutions but people take them for granted. When a husband is unfaithful instead of praying for him a wife will simply jump into a wagon of divorce. No matter how hard you try to reconcile them they will reject your advice and continue with his or her intention.

Leaders, can't forgive and forget. They fight one another and this has negative effects on the economies of their nations and the masses. One wise man once said lack of forgiveness is like somebody who eats some poison and expects it to kill his enemies. It destroys or widens the crack of division among nations and families, it is difficult to forgive and forget but one thing is certain it is the right thing to do that. So the right thing has to be done for unity for all.

Today, a lack of forgiveness has saturated people's hearts. I remember a certain sister who told me that her elder sister has done her wrong. When I asked her to give her a call, talk, and reconcile with her. She told me point blank, that she will never forgive her in her life time.

Chapter 15
False accusers

2Timothy 3:3 Without natural affection, trucebreakers, false accusers, incontinent, fierce, despisers of those that are good,

The word of God is telling the truth because false accusers are everywhere. In politics, people accuse one another wrongly because they want to be the favourite leaders of political organizations. People don't climb the ladder because they are competent but they backstab other people to get where they want to be. In the Christian world, people who aren't genuinely born again are busy using social media to accuse one another. They can manufacture a story and destroy fellow brethren's image or accuse him of crimes he or she has never committed. There are a certain group of people who like to use other people to climb the ladder of success especially the so-called men or women of God. They try to make names for themselves at the expense of other people.

Chapter 16
Knowledge shall increase

Daniel 12:4 But thou, O Daniel, shut up the words, and seal the book, *even* to the time of the end: many shall run to and fro, and knowledge shall be increased. The word of God says, in the last days' knowledge shall increase tremendously. They say, knowledge is power but it can be destructive especially if it draws you away from God or his presence. Since the beginning, God has never created man as an independent being rather as a dependent creature. That's to say, a man was created to rely or depend on God for his provision and guidance.

Now back to our discussion the word of God says, in the last days' knowledge shall increase in all spheres of life.

Military

In the olden days, people used to fight using spears, shields and maneuvered around on foot. Soldiers would throw a long spear at their enemies and kill them or he would absorb it using a shield. During the intertribal wars that were fought in Southern Africa around the 1800s, King Shaka of the Zulu nation introduced new methods of fighting.

He introduced short stabbing spears which didn't disarm soldiers, long shields that covered the whole body to replace short ones which didn't cover the whole body and his soldiers fought barefooted to maximize their speed on the battlefields. His soldiers were allowed to marry at the age of 30 so that they could focus on their military lives. People who were 30 years old and above were regarded as mature beings. As the years and centuries went by more sophisticated weapons were manufactured which could hit targets with utmost precision.

Today, some weapons can be guided using sophisticated and powerful computers. There are continental missiles that can travel across continents and they are very precise. On the 6th August 1945, the first atomic bomb was dropped down at Hiroshima and Nagasaki killing many people and injuring others. Today, people are living with the effects of those bombs. Do you see, how knowledge and technology have increased tremendously in recent years?

Medicine

In the past, it took scientists many years to discover cures and vaccines but today it can take less than a year. The knowledge in the scientific world has increased tremendously. They can manufacture millions of vaccines in a day using complicated equipment. In the past, people used herbs to treat diseases and other ailments, but today medicine comes in liquid, tablets, powdered form and they are scientifically proven.

Therefore, they cure diseases precisely with minimum or no side effects. Today diseases that used to kill millions of people across the world like TB can be treated easily, HIV/AIDS therapy is a breakthrough in the scientific world. There are various Covid-19 vaccines today, to curb and flatten the curve of infection. We thank God for releasing anointed knowledge in the scientific world.

Space travel

In 1964, three United States of America astronauts landed safely on the moon. They collected some rocks and brought them back to earth. It was one of the greatest achievements of mankind and it answered many questions that had been rolling and turning in his mind. It answered the following questions;

What is that thing traveling across the sky at night?

Is there life on the moon?

If there is life who lives there?

Today, man can travel to space at will. He has also visited planet Mars which according to scientific findings there is the potential of life there. There are satellites also at Mars. Man has also discovered that the sun doesn't move what orbit around the sun at high speed are the planets. And their positions can be determined with precision because of sophisticated equipment.

<u>Weather forecasting</u>

Scientific equipment which is sophisticated in nature can predict the weather with utmost precision today. If you want to visit somebody in South Africa in Durban you don't have to live by guesswork. You can just flip through television channels and watch news bulletins and the weather conditions will be revealed to you.

Today, because of technological advancement you just need a smartphone, laptop access Facebook and the weather conditions in Durban will be revealed to you. Meteorologists can even measure the speed at which the tornado is traveling with the highest precision. They can even determine its path. That's is to say, places where it will pass through.

Big brother is watching you

One day, as the sun sunk behind the western skies, casting shadows on the streets of Gaborone I was walking along one of the tarred roads enjoying the cool breeze as it whipped my face at peace and harmony with nature. The stillness of the day was occasionally interrupted by the rumbling of vehicles on the tarmac as they travel along the streets of Gaborone. Suddenly, t

, he atmosphere was disturbed by the stumping of footsteps howling of a siren and police van came to a skidding halt as a troop of policemen hopped out from the back of the van carrying guns.

"Where are they?" their senior asked.

"They are gone," a tall, coffee-coloured police officer answered.

The darkness was now hugging the city and we were astonished. We were told that somebody had stolen a driver's wallet by the traffic lights. We were amazed,

"Did the driver called you?" I

asked them. "No, I have been watching them at the police headquarters, called other members of disciplined forces for back up," another police officer answered still glancing around.

"How? "

"This is a surveillance camera, "he said as he lifted his eyes towards the top of the traffic lights.

"Wow, that's amazing. How are you going to find them because it's already dark,"

"Don't worry. These cameras can reveal pictures clearly at night'" he answered as he waved at his men, hooped onto the van, and speed off. The following day the culprits were arrested easily. Do you see, how the government agencies are keeping on us without our knowledge?

In more developed nations surveillance cameras are everywhere. If you commit a crime in the street you are in deep trouble. In prisons, governments are investing heavily to mount cameras to monitor jailed persons. Since September 11, 2001, the intelligence community is working harder than ever before. Today, there is sophisticated equipment that is

used to tap, decode information as it travels along the telephone lines. These technological machines can detect your location and can be used to target leaders of opposition parties in certain countries. In other words, intelligent agents can infiltrate the terrorists' communication network, crackdown their communication codes because most of them use codes when they transfer information from one person to another. So that somebody who is outside their network can't understand it. However, because of advanced technology they can infiltrate their network.

Well-known criminals, influential people telephones can be tapped by intelligence agents without their knowledge and listen to their private calls. If a well-known syndicate of arm robbers is planning to rob a certain bank, they can be intercepted easily because big brother is watching them. They can even use sophisticated software to identify a well-known criminal strolling around the street, give the criminal investigation Department or intelligence agency a warning signal, and they will keep an eye on him or her as walk down the streets like that.

The language barrier has been broken

When you go to school you go there to open up the line of communication between you and other people from around the world. I am a Motswana before I went to school I couldn't understand and an English person speaking to me and respond accordingly. As I became learned that barrier was removed, which is a good thing because it makes interaction easier. Today, technology has broken the language barrier even further. You don't need somebody to translate Spanish for you online if you aren't Spanish speaking person especially on social media like Facebook. If somebody sends you a message in Spanish and you have programed your smartphone in English, you mustn't worry. You just click translation and the message will be translated from Spanish to English.

Agriculture

In the past, people used to plant plants that took three to four months to reach the maturity stage. Sometimes, even longer. Fruit trees were bearing small fruits but today plants that took 3 to 4 months in the past can be harvested in two months with better or maximized output. They are known as hybrid seeds.

In the past, people used to scatter kraal manure around the farm to improve the quality of the soil. Something which was sometimes very destructive to crops but toady fertilizers have been manufactured which can be applied to improve the quality of the soil and maximize yields. Soil samples can be collected, sent to a laboratory to be tested, and find out how it can be improved using scientific equipment.

Air and road travel

In the olden days, people used to travel from one place to another on foot. It was an industrious task. Today, it takes few hours to travel from Gaborone to New York in the USA. Indeed the world is racing at a huge speed. There are trains and cars which don't use diesel they use electricity.

Electricity

In the past, people used to gather wood to make fires and warm their homes. When rain fell upon them disaster sat in. However, today, the story has changed forever, houses can be warmed up with heaters and other electrical appliances. In the streets where I grew up during the night, it would be pitch black especially when there was no moon. But today, the street lights are twisting and turning along the streets making Kanye a glorious village at night. Electric bulbs in homes have turned night into daytime too.

Online buying

In the past information travelled at a snail pace because books took many months to arrive from the USA to Botswana. Goods like furniture etc. took longer to arrive too. But today, you can buy an electronic book online, download and read it on your laptop, smartphone, and so on at the comfort of your home. Audiobooks are available too. Today, you don't need to go around looking for a traditional publishing company you just write a book, edit, proofread, format, and upload it on Amazon without any difficulty. This has helped aspiring authors positively because people have a platform where their talents can be displayed and put food on the table.

Chapter 17
Israel surrounded by enemies

Luke 21:20 And when ye shall see Jerusalem compassed with armies, then know that the desolation thereof is nigh.

When the Jewish people finally returned to their ancestral land, on the 6th May 1948 Israel was declared as an independent and sovereign state. Sovereignty means a nation that can make its decision without external influence or interference. As years went by, the enemy of Israel multiplied within the region and encircled her. In 1967, Israel fought a war in which many countries in the region joined hands and attacked her. However, on the 6th day, she emerged victorious. Throughout history some of her neighbours have worked day and night with one purpose in mind, to destroy her. In recent years, a conflict between Israel and terrorist organizations in the region has reached an alarming rate. At the writing of this book, the world watched with horror when thousands of rockets were thrown towards Israel but we thank God for his protection and the sophisticated weaponswhich were able to detect them and destroy over 90 percent of them.

Indeed, Israel is surrounded by many enemies which are the fulfilment of the prophetic word of God. Her enemies want Jerusalem. However, God has promised in his word that nobody is going to uproot them.

Chapter 18
Idol worship

Exodus 20: 3 Thou shalt have no other gods before me. 4 Thou shalt not make unto thee any graven image, or any likeness *of any thing* that *is* in heaven above, or that *is* in the earth beneath, or that *is* in the water under the earth: 5 Thou shalt not bow down thyself to them, nor serve them: for I the LORD thy God *am* a jealous God, visiting the iniquity of the fathers upon the children unto the third and fourth *generation* of them that hate me; In recent has been on the rise. Idol is anything that takes priority in your life. In simple words, it's anything that replaces God in your life. Do you know that your wife can be an idol? Yes, if you love her more than God she becomes your idol. Similarly, if you love your children more than God they become your idol. Your car, country, house, and so on can be your idol if you love it more than God.

Your pastor, prophets, evangelists, teacher of the word of God, the apostle can be your idol if you love them more than God. A line of years idol-worshipping distinction must be drawn between God and his servants. As children of God, we mustn't worship them. Worship and glory must be given to God, the only person who deserves it. Certain people do things that aren't scriptural whereby they kiss servants of God's shoes, kneel forming a straight line when they appear. This is idol worshipping something which must be given to God alone, To make matters worse, some people worship trees, cows, goats, and mountains while other worship snakes of the caves and rivers. They go there on a periodical basis to be empowered spiritually. After they will prophesy and do all manner of things that aren't biblical like sleeping with young girls as part of their covenant with the forces of the dark side. Some people visit the graveyards on a periodical basis and consult or fellowship with the dead, something which God doesn't approve and it defiles their spirit. There are places of worship and they worship images of Mary the mother of Jesus Christ and so forth. Others have all manner of images around their churches and worship them which is wrong.

Chapter 19
Blasphemy

2Tim 3:2 For men shall be lovers of their own selves, covetous, boasters, proud, blasphemers, disobedient to parents, unthankful, unholy,

There is a certain segment of people around the world who have cultivated a habit of ridiculing the word of God. They say the Bible wasn't written through the inspiration of the Holy Spirit instead it's introduced to colonize nations. Something, which is wrong because when the word of God spread across the world it's a fulfilment of the prophetic words of Jesus Christ.

Today, some people have a tendency of attacking servants Of God calling them all manner of names. They call them robbers, thieves, and so on. Speaking against servants of God isn't something new. When Jesus Christ was walking upon the surface of the earth, Pharisees said he used Beelzebub to cast out demons. The question is can Satan cast himself out? Highly impossible. It was blasphemy against God almighty because God and Jesus Christ are one.

Some people attacked the word of God calling it a novel something which is wrong. They call it a collection of novels to be exact which isn't true. The work of God has been under attack and under demined. The return of the Lord has been ridiculed throughout history. Some people say it will not come to pass. However, one thing is certain Jesus Christ is coming soon and we must be ready whether people like it or not.   Some people have told people that they can't worship God they can't see or touch. Therefore, when they worship them they are worshipping

God. That's blasphemy in its purest form because God has no peer. People must learn to respect God as the ultimate ruler who has no weakness and they mustn't be drunk with the spirit of blasphemy. Because what they are doing can attract judgment from God.

Chapter 20
Disobedience of children

2Tim 3:2 For men shall be lovers of their own selves, covetous, boasters, proud, blasphemers, **disobedient to parents**, unthankful, unholy,

Disobedience of children has reached skyrocketing proportion. When parents try to discipline them they usually run away from home. And by so doing they become vulnerable to predators which are roaming our streets looking for somebody to devour or attack. In other, words they end up falling into the wrong hands. They end up meeting the wrong people who entice with things like exclusive phones, cars, and so forth to have their carnal knowledge. Something which must be an issue of concern to all of us because if children are disobedient to their parents then the future of nations is not certain.

Disobedience to parents by children has destroyed their lives because some of them are doing drugs and so forth. Others are involved in criminal activities like rape, robbery, and the like. Something which should give us as parents' headaches. Disobedient children don't like a peaceful atmosphere.

They incite violence and another manner of social ills. Teachers are living in constant fear because of such children because they can assault them or stab them with sharp instruments and sustain serious injuries. Indeed the word of God is being fulfilled before our eyes.

Part 2
Chapter 1
How to get prepared?
Be born again

John 3:3 Jesus answered and said unto him, Verily, verily, I say unto thee, Except a man be born again, he cannot see the kingdom of God.

The first thing that you must do as an individual is to accept Jesus Christ as your Lord and saviour. It simply means to decide to follow the Lord and his ways. Without salvation, it is highly impossible to be upright before God. The righteousness that I am talking about isn't attained by personal efforts or works but the gift of righteousness which is found in Jesus Christ. Not whereby go to the river and bath and claim that you are saved. Not whereby you distribute food hampers to the needy, widow, widower, and orphans and claim that you are trying to live an upright life before God. Some people claim that prayer and fasting every week can make them right before God. Other people splash their millions on the poor through charity organizations while some purchase vaccines for the developing nations and think that it a guarantee to enter into heaven. It is not wrong to do that however when you do that without being saved you are doing that all in vain. Because salvation comes through Jesus Christ. After you have been born again the gift of rightness is imparted into your spirit. The Bible is very clear, Jesus Christ said, I am the way the truth, life, and nobody comes to the father except through me. There is only one gate that leads to heaven and it's Jesus Christ of Nazareth. Don't be deceived apostles, prophets, evangelists, teachers of the word and pastors are too, small to save you.

No matter how powerful your nation might be in the world, it is too small to save your soul. Your material possession can't save you. No matter how beautiful or handsome some of you may be your physical appearance is too small to save you. Your talent no matter how powerful it's it can't save you. The level of your education can't save you. Only Jesus Christ of Nazareth has the power to rescue and save you. When you are born again your spirit man is connected to the spirit of God. And it is going to help you to walk in a Godly path and fulfil divine assignment on earth. The Holy Spirit is going to give you the gift of discernment to differentiate the error from the truth you, your family, nation, and what the world at large are going through or where God is leading you, church, families, and nations. In other words, you will not live a life of guesswork. You will live, walk-in sync with the heavens. Confusion is a sign of lack of being led by God because God isn't an author of confusion.

Do you see how salvation is very important? As human

beings as we are racing towards the end of ages, we must seek salvation. Because if you aren't born again no matter how you try to place reasons before people, the truth is you aren't a covenant child of God. In the Bible, God calls his children sheep, John 10: 10

If you aren't born again you can't read and understand the things of the spirit because they are only understood through revelation, not carnality.

Chapter 2
Build your foundation the word of God

Then, you have to build your foundation on the word of God and the word of God is Jesus Christ. When you are grounded on the word of God when situations come you are going to remain standing. But if you aren't grounded on it when challenges come or signs of the end times come you are going to backslide or fall from grace. The only tool at your disposal as a covenant child of God that is going to strengthen you as you sail through the storms or sings of the end time is the word of God like never before and declare it, after meditating upon it. The women or men of God in the past are your example, to follow them maintained your ground on the word of God even when the environment around speaks the opposite.

The Kingdom of God is like a man who builds his house on a rock and storms come across it and remained standing. While another man builds his house on sandy soil and when storms break out his house collapses. That shows that you won't be shaken when challenges come across your path when you are loaded with scriptures. However, if you don't have the word in your spirit you aren't going to survive.

Remember, when Jesus Christ came face to face with the Devil, when he was tempted to turn stones into bread he quoted scriptures. He didn't use his mental faculty to answer him. He dug deep into his spirit and released words that state that a man couldn't live by bread alone but by every word that proceeds from the mouth of God.

As you are passing through signs of the end times and intensify the only thing that is going to sustain you is the word of God. As we are racing towards the end of ages Satan is going or has already raised an army in the unseen world against the body of Christ (Church of God). The only weapon at our disposal is the anointed word of God, not carnality. David conquered Goliath using a stone which is the word of God because Jesus Christ said he was the chief corner 'stone. Without the word of God, you won't make it. You will be like a soldier on the battlefield without a loaded gun. Instead of being an overcomer,you are going to be a victim on the battlefield.

On the battlefield to win a war you need a superior strategy to that of your opponent. You sit down first, read or study it. Military strategists explain everything to you. That's to say, ways you are going to use to conquer your enemies. In Christianity, the Bible is the strategy that you must follow and execute to emerge victoriously. While the Holy Spirit is the strategist who explains it to you. For example, during the era of Prophet Joshua when the walls of Jericho stared at them defiantly. They saw a man and when they ask who he was he told them he was the captain of the heavenly host and he was to guide or help them to capture Jericho. That's the Holy Spirit and after encircling the walls of Jericho seven times, with trumpets on the 7th day it's miraculously collapsed.

Chapter 3
Winning souls

Proverb 11:30 The fruit of the righteous *is* a tree of life; and he that winneth souls *is* wise.

When one soul is added into the kingdom of God there is joy and celebration in heaven. That shows how God honours the grace of soul winning. When you win soul for God, he declares you are a wise person when you do that. Because in the Kingdom of God you aren't saved alone you are saved to save others. The greatest gift you can give to somebody is the gift of salvation which has no ˋ

price tag.

As we are racing towards the end of ages God is going to empower his church to harvest souls for him like never before. Please, don't waste any time my brother and sister hop onto the wagon of soul winning. Don't let this opportunity pass you by God has placed resources for you at the comfort of your homes. You just need an internet open Facebook page and start sharing the word of God. You can do the same on other social media platforms like Twitter and so forth. Today, because of technology you can reach many places in a minute.

My dear reader, don't say where can I start because the harvest is plenty but labourers are few. Don't fight to preach in the church of God, you can have your altar of soul-winning like Jesus Christ. Wherever he went his tongue was saturated with words of his father's kingdom. For example, he preached at the well, in a boat, at weddings, in public places, and so forth. That's shows that there are many places where you can preach the good news and win souls for God. Don't limit yourself.

Chapter 4

Be in the spirit of prayer

Luke 18:1 And he spake a parable unto them *to this end*, that men ought always to pray, and not to faint; As the clock is ticking towards the end of ages, prayer is a very important tool. Life of prayer is an integral part of Christianity. You can't be a child of God when you are living a life without prayer because you are called to imitate the life of Jesus Christ. He was a man of prayer and instructed his disciples to be in the spirit of prayer to avoid temptations. The same command still extends into your life today, it didn't end at the Garden of Gethsemane.

Prayer is going to help you to avoid temptations as you are sinking deeper and deeper into the perilous times. One wise man of God once said break-in prayer is a break with Jesus Christ. Do you know why? It is because it is a connection or it keeps the lines of heaven open. It also positions you well where you can hear God's voice clearly and fulfil his assignment on earth. The word of God says these signs of the end times shall increase and intensify as the coming of Jesus Christ comes closer and closer.

So as children of God, you must intensify your prayer as well. Prayer must be combined with fasting a. fasting simply means abstaining from food to cultivate or deepen your relationship with God. Remember, when you eat or read the word of God you are feeding your spirit man or inner man but when you are engaged in fasting and prayer you are strengthening him. It also makes you be sensitive to the voice of God and be able to fulfil his will of purpose because you are crucifying the flesh. It also leads you to divine ideas.

For example when I ran out of ideas after praying and fasting ideas flow like water from the tap or fountain into my mind.

Chapter 5
Be content

The dangerous part of lack of contentment is that you can easily be misled. If you can't thank God for the little or whatever he has given you, I fear for you, you are likely to be caught by the demonic web that has encircled the world. Somebody who lacks contentment can do anything that contradicts the scriptures to gain more out of life. That's is to say, he or she can commit unthinkable atrocities to move to another level of success.

People who lack contentment can offer their children, wives, family members as sacrifices to the kingdom of darkness. Some people can kill animals as a sacrifice to win political office or gain power. While others can sacrifice their freedom as assign of devotion to the Devil. They usually remain single for the rest of their lives and sleep with snakes in the caves in exchange for power and influence. And this is driven by a lack of contentment, wanting more power and influence in life.

There are times when some people can do unthinkable

things to rule or have dominion over others. A certain brother had so many goats in his kraal while his neighbour has few. Then, one day out of lack of contentment targeted one of the goats which had strayed into his. In the night, he grabbed, slit its throat, and killed it for meat consumption. Within few days police were everywhere across his paths, he was convicted and sent to prison. Do you see, how lack of contentment is destructive?

In life, you should develop a habit of being fulfilled and

content with whatever God has entrusted you with. You should believe that God at his appointed time he will elevate you to another level of success and glory. He should trust you with the little you have before he can trust you with huge things in life. For example, God must trust you with your family before he can trust with a church. If you can't handle fifty pounds God can't trust you a million. Therefore, you should be content in life.

<u>Chapter 6</u>
<u>Endurance or patience</u>

Matthew 24:13 But he that shall endure unto the end, the same shall be saved.

Throughout history, man has to question the return of Jesus Christ. Some people said since we were young the same story has been told but nothing happened. Then, they have ended up giving up on their faith and follow doctrines that aren't consistent with the scriptures. When the Israelites were delivered from the hands of Egyptians and crossed the Red Sea, Moses went to Mount

Sinai to meet God there. Guess what happened when he took longer people took their jewels, melt them and manufacture a golden cow. They started worshipping it or an idol. When Moses descended from the mountain an unthinkable sight greeted him. People had given up on God and worshipped it, something which angered him and God. They were severely punished for their sin.

Similarly, as the hands of time are ticking towards the end of ages, signs intensifying, people are going to give up on God, questioning his return.

That's to say, they are going to become impatient and follow false doctrines which itch their ears or which support their human weaknesses. Therefore, as a follower of Jesus Christ, you must exercise maximum patience and endurance until he returns.

Giving up on God is giving up on the ultimate plan of God. The Lord plan is to raise us to be in his presence and rule with him as kings over all nations of the world during the millennium rule and live for eternity in the new earth, where there is no tear, sorrow, death, and aging. You see, residing in the presence of the almighty God. Remember, when you are living in the presence of God or glory there is no pain and suffering. Adam and Eve were immune to death, suffering, and pain while living in the Garden of Eden in perfect harmony with the will of God. They started experiencing them after disobeying God and fall from glory.

Endurance is to follow God's plan purpose and fulfil every letter in it. It simply means enduring hardship or overcomes obstacles until you reach where God wants you to be. And God wants you to hold onto his word until he returns or death. Challenges will intensify as the return of the Lord looms on the horizon but if you endure God will reward you at the end.

Chapter 7
Nobody knows the day or hour

1Thessalonians 5:2 For yourselves know perfectly that the day of the Lord so cometh as a thief in the night.

Matthew 24:36 But of that day and hour knoweth no *man*, no, not the angels of heaven, but my Father only. In the past, many people tried to predict the return of Jesus Christ but all of them had failed. The word of God categorically clear, it says nobody knows the day of the return of the son of God. If anybody says, he is coming tomorrow or states a specific date in the future, my dear reader runs away from him because he lacks the truth.

If you don't read the word of God on your own and rely on grabbing words from people's mouths, I fear for you, you are going to be deceived. We must be like brethren from Beria who studied the scriptures on their own to verify what Apostle Paul and others had taught them with the word of God. The Bible says, my people are perishing because of lack of knowledge and ignorance leads to spiritual bondage. While the truth sets you free.

The era of false prophets has arrived whereby everybody comes up with his or her revelation of the scriptures. Some people have tried to point people to the return of Jesus Christ starting dates which is something that isn't supported by the scriptures.

The end.

Topics in Current Genetics

10

Series Editor: *Stefan Hohmann*

Available online at

SpringerLink.com

Sirkka Keränen • Jussi Jäntti (Eds.)

Regulatory Mechanisms of Intracellular Membrane Transport

With 23 Figures, 7 of Them in Color; and 5 Tables

 Springer

Dr. SIRKKA KERÄNEN
Dr. JUSSI JÄNTTI
VTT Biotechnology
Tietotie 2
P.O. Box 1500
Espoo
Finland 02044 VTT

The cover illustration depicts pseudohyphal filaments of the ascomycete *Saccharomyces cerevisiae* that enable this organism to forage for nutrients. Pseudohyphal filaments were induced here in a wild-type haploid MATa Σ1278b strain by an unknown readily diffusible factor provided by growth in confrontation with an isogenic petite yeast strain in a sealed petri dish for two weeks and photographed at 100X magnification (provided by Xuewen Pan and Joseph Heitman).

Springer-Verlag Berlin Heidelberg New York

ISSN 1610-2096
ISBN 978-3-642-06095-3
e-ISBN 978-3-540-44476-3

Springer-Verlag is a part of Springer Science + Business Media
springeronline.com

©Springer-Verlag Berlin Heidelberg 2010
Printed in Germany

Data-conversion: PTP-Berlin, Stefan Sossna e.K.
Cover Design: Design & Production, Heidelberg
Printed on acid-free paper - 39/3150-WI - 5 4 3 2 1 0

Table of contents

Introduction: Regulatory processes, an emerging feature in intracellular membrane traffic..1

   Sirkka Keränen and Jussi Jäntti 1

      Vesicle budding...3
      Vesicle targeting – regulation by Rab GTPases 4
      Tethering complexes 4
      SM proteins 5
      Regulation of SNARE assembly by protein phosphorylation 6
      Phosphoinositides in membrane traffic 6
      Regulation of *de novo* membrane formation 6
      References 7

Molecular mechanisms in clathrin-mediated membrane budding...................9

   Brigitte Ritter and Peter S. McPherson...9

      Abstract 9
      1 Cellular sites of clathrin-mediated membrane budding..................9
      2 Formation of endocytic CCVs...10
      3 Machinery for CCV formation at the cell surface 14
         3.1 Role of clathrin 14
      3.2 Role of AP-2..15
         3.3 Role of endocytic accessory proteins...18
         3.4 Interactions of endocytic accessory proteins with AP-2 23
      4 Clathrin trafficking at the TGN 26
         4.1 AP-1 and GGA adaptor complexes..26
         4.2 Structural analysis of the γ-ear and GAE domain 28
         4.3 Enthoprotin (Clint/epsinR): identification of a novel γ-ear/GAE
         domain-binding consensus motif..29
      5 Concluding remarks 30
      Acknowledgments..31
      References...31

Regulating membrane curvature 39

   Joost C. M. Holthuis 39

      Abstract 39
      1 Introduction 39
      2 Topological considerations...40
      3 Extrinsic forces affecting membrane curvature..............................42
         3.1 Cytoskeleton 42
         3.2 Adhesion to curved particles..44
         3.3 Coat assembly...44
      4 Intrinsic forces affecting membrane curvature...............................46

4.1 Transbilayer area asymmetry...46
4.2 Transbilayer curvature asymmetry 52
4.3 Lipid domain-induced curvature...56
5 Concluding remarks 57
Acknowledgements 57
References..58

Functions of Rab GTPases in organelle biogenesis 65
Christopher G. Burd and Ruth N. Collins......................................65
Abstract...65
1 General considerations 65
2 The GTPase cycle 66
2.1 Structural features of Rab 68
2.2 Prenylation and localization of Rab GTPases........................70
3 Regulation of inter-organelle trafficking and organelle transport 71
3.1 Cargo sorting and vesicle budding...71
3.2 Regulation of membrane tethering and fusion........................73
3.3 Rabs and membrane microdomains 75
3.4 Organelle transport 76
3.5 The roles of Rab GTPases in signaling pathways...................77
4 Rab proteins in disease and development....................................78
5 Perspective 78
Acknowledgements 80
References..80

Tethering proteins in membrane traffic...................................89
Sarah Chu and Wei Guo 89
Abstract...89
1 Introduction...89
2 Tethering proteins in various traffic stages 91
2.1 TRAPP..91
2.2 p115..93
2.3 Uso1p..94
2.4 The COG complex 95
2.5 The exocyst...96
2.6 EEA1 98
2.7 HOPS..100
2.8 The GARP/VFT complex 101
3 Postulated functions of the tethering proteins 102
4 Emerging common features of the tethering proteins 103
5 Regulations of the tethering proteins...105
6 Tethering proteins and sorting...105
7 Future studies 106
Acknowledgment 106
References..106

The function of Sec1/Munc18 proteins - Solution of the mystery in sight? 115
   Maria Kauppi, Jussi Jäntti, and Vesa M. Olkkonen 115
     Abstract 115
     1 Introduction 115
     2 Genetic studies indicate an essential positive function for SM proteins in vesicle transport 117
     2 Genetic studies indicate an essential positive function for SM proteins in vesicle transport 118
     3 The structure of SM proteins 120
     4 SM proteins display different binding modes to SNARE molecules 121
     5 Are the data suggesting an inhibitory role for SM proteins physiologically relevant? 123
     6 At what stage of the vesicle tethering-docking-fusion process do SM proteins exert their effects? 125
     7 Evidence for a role of SM proteins in transport and stability of syntaxins 127
     8 Protein phosphorylation regulates interactions of SM proteins with syntaxins 128
     9 Non-syntaxin interaction partners of SM proteins 129
        9.1 Interactions of SM proteins with transport vesicle tethering complexes 129
        9.2 Interaction partners that regulate the SM protein-syntaxin association 131
        9.3 The Mint proteins 133
     10 Do the SM proteins share a conserved function despite the different modes of interaction with SNAREs? 134
     Acknowledgements 135
     References 135

Regulation of SNARE assembly by protein phosphorylation 145
   Adina Weinberger and Jeffrey E. Gerst 145
     Abstract 145
     1 Introduction 145
     2 Phosphorylation and control of the exocytic machinery 147
        2.1 Mammalian syntaxins 147
        2.2 Yeast syntaxins 151
        2.3 *Arabidopsis* syntaxin 152
        2.4 SNAP-25 and SNAP-23 152
        2.5 VAMP/Synaptobrevins 155
        2.6 SM Proteins 155
        2.7 Synaptotagmin 157
     3 Protein phosphorylation in the endocytic pathway 158
        3.1 Yeast Tlg1,2 and Vps45 158
        3.2 VAMP4 159
        3.3 Rabs 4 and 5 159

4 Protein phosphorylation in the early secretory pathway 160
5 Conclusions.. 161
Acknowledgments... 162
References.. 162

Phosphoinositides and membrane traffic in health and disease.................... 171
Anna Godi, Antonella Di Campli, and Maria Antonietta De Matteis............ 171
Abstract... 171
1 PIs metabolism... 171
2 The PI kinases: their subcellular localization and role in the secretory
pathway.. 173
2.1 The PtdIns 3-kinases.. 173
2.2 The PtdIns 4-kinases.. 175
2.3 The PtdIns monophosphate kinases 177
3 The PI phosphatases: localization and function in the secretory
pathway.. 179
3.1 The PI 3-phosphatases 179
3.2 The PI 4-phosphatases 181
3.3 The PI 5-phosphatases 181
4 PI metabolism and disease 182
4.1 PTEN and human cancers.. 182
4.2 Myotubular myopathy and Charcot-Marie-Tooth disease 183
4.3 Lowe syndrome 184
5 Conclusions.. 185
Acknowledgements... 185
References.. 185

Regulation of exocytotic events by centrosome-analogous structures 193
Christof Taxis and Michael Knop... 193
Abstract... 193
1 MTOCs: a brief overview 193
2 Spindle pole bodies and plasma membrane biogenesis in yeast
meiosis 195
2.1 PSM initiation.. 196
2.2 Growth and shaping of the PSM.. 199
2.3 Closure of the PSM.. 200
3 The IMC in *Apicomplexa* 200
4 Centrioles, basal bodies and plasma membrane 204
5 Conclusion 204
Acknowledgement... 205
References.. 205

Index.. 209

List of contributors

Burd, Christopher G.
Department of Cell and Developmental Biology, University of Pennsylvania School of Medicine, 421 Curie Boulevard room 1010, Philadelphia, PA 19104-6158, USA
cburd@mail.med.upenn.edu

Chu, Sarah
University of Pennsylvania, Department of Biology, Philadelphia, PA19104-6018, USA

Collins, Ruth N.
Department of Molecular Medicine, C4-109, VMC, Cornell University, Ithaca NY 14853-6401, USA

De Matteis, Maria Antonietta
Department of Cell Biology and Oncology, Consorzio Mario Negri Sud, Via Nazionale, 66030 Santa Maria Imbaro (Chieti), Italy
demattei@negrisud.it

Di Campli, Antonella
Department of Cell Biology and Oncology, Consorzio Mario Negri Sud, Via Nazionale, 66030 Santa Maria Imbaro (Chieti), Italy

Gerst, Jeffrey E.
Department of Molecular Genetics, Weizmann Institute of Science, Rehovot 76100, Israel
jeffrey.gerst@weizmann.ac.il

Godi, Anna
Department of Cell Biology and Oncology, Consorzio Mario Negri Sud, Via Nazionale, 66030 Santa Maria Imbaro (Chieti), Italy
godi@negrisud.it

Guo, Wei
University of Pennsylvania, Department of Biology, Philadelphia, PA19104-6018, USA
guowei@sas.upenn.edu

Holthuis, Joost C. M.
Department of Membrane Enzymology, Faculty of Chemistry, Padualaan 8, 3584 CH Utrecht, The Netherlands
j.c.holthuis@chem.uu.nl

Jantti, Jussi
VTT Biotechnology, Tietotie 2, Espoo, P.O. Box 1500, FIN-02044 VTT, Finland
jussi.jantti@vtt.fi

Kauppi, Maria
Department of Molecular Medicine, National Public Health Institute, Biomedicum, P.O.Box 104, FIN-00251, Helsinki, Finland
maria.kauppi@ktl.fi

Keränen, Sirkka
VTT Biotechnology, Tietotie 2, Espoo, P.O. Box 1500, FIN-02044 VTT, Finland
sirkka.keranen@vtt.fi

Knop, Michael
EMBL, Cell Biology and Biophysics Programme, Meyerhofstr. 1, 69117 Heidelberg
knop@embl.de

McPherson, Peter S.
Montreal Neurological Institute, McGill University, 3801 University, Montreal, QC, H3A 2B4 Canada
peter.mcpherson@mcgill.ca

Olkkonen, Vesa M.
Department of Molecular Medicine, National Public Health Institute, Biomedicum, P.O.Box 104, FIN-00251, Helsinki, Finland

Ritter, Brigitte
Montreal Neurological Institute, McGill University, 3801 University, Montreal, QC, H3A 2B4 Canada

Taxis, Christof
EMBL, Cell Biology and Biophysics Programme, Meyerhofstr. 1, 69117 Heidelberg

Weinberger, Adina
Department of Molecular Genetics, Weizmann Institute of Science, Rehovot 76100, Israel

Introduction: Regulatory processes, an emerging feature in intracellular membrane traffic

Sirkka Keränen and Jussi Jäntti

The subject of this volume is the molecular mechanism of the intracellular membrane trafficking, a central eukaryotic cell biological process. In the post genomic era, essential molecules involved in intracellular membrane/protein transport are emerging with increasing pace. The present challenge is to compile the molecular networks that govern these processes. Understanding of regulatory processes and participating molecules are likely to reveal global cellular regulatory circuits that couple membrane trafficking with other cellular functions. The part of the membrane transport machinery, which forms stabile protein complexes is rather well known already. However, the regulatory mechanisms that link these more stabile complexes to other cellular functions are only starting to emerge. This book focuses on the regulatory aspects of this process.

Continuous intracellular membrane traffic is needed for growth and maintenance of the compartmental organisation of eukaryotic cells. This trafficking is carried out by an elaborate and highly dynamic tubulo-vesicular network that ensures transport of membrane components, proteins, and lipids from their site of synthesis to their site of function. The biosynthetic pathway, also called protein secretion pathway and exocytotic pathway, leads from endoplasmic reticulum (ER) to cell surface and to the endocytotic compartments via the Golgi complex (Fig. 1). Proteins destined to plasma membrane and to the endocytotic compartments as well as those secreted to cell exterior are first translocated to the ER. In the ER, proteins undergo several types of posttranslational modifications and are folded with the help of a number of chaperones and foldases. The properly folded and oligomerised protein molecules are packaged into transport vehicles in the form of membrane vesicles that are targeted to the next station, the Golgi complex. Fusion of the vesicles with the Golgi membranes releases the proteins into the Golgi lumen, where they may undergo further modifications. The trans-Golgi network is the major sorting compartment in which the proteins are packaged into vesicles that are targeted either to the plasma membrane or to the endocytotic/lysosomal compartments. The endocytotic/lysosomal pathway is used to transport material into the cell to the endosytic compartments and to lysosomes. Basically, the same membrane transport mechanisms operate in all eukaryotic cells from yeast to mammalian neurons and at the different transport steps within the cell. Exocytosis of neurotransmitter-loaded transport vesicles bears striking similarity with molecular mechanisms of transport vesicle fusion in other cell types and even in unicellular eukaryotes like yeast. However, neurotransmitter exocytosis of synaptic plasma membrane associated vesicles is evoked by plasma membrane influx of calcium, which is triggered by membrane depolarisation. In consequence, neuronal exocytosis involves calcium-binding synaptotagmin proteins. For synaptotagmins homologues have not been identified, for example, in

Topics in Current Genetics, Vol. 10
S. Keränen, J. Jäntti (Eds.): Regulatory Mechanisms of Intracellular Membrane Transport
DOI 10.1007/b98736 / Published online: 16 June 2004
© Springer-Verlag Berlin Heidelberg 2004

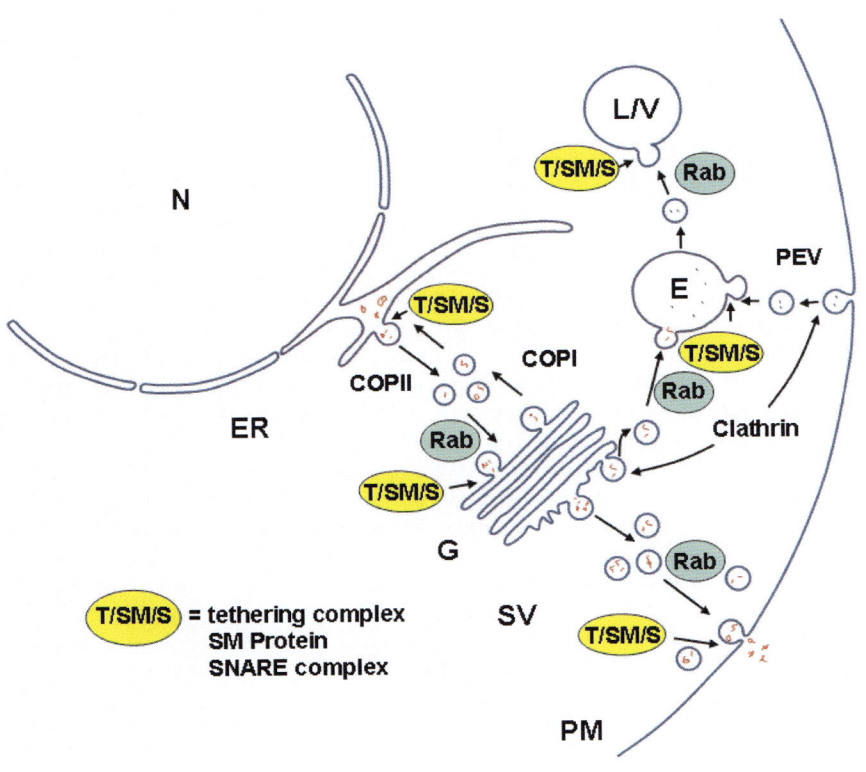

Fig. 1. Schematic presentation of the secretory pathway. Transport of proteins from ER to plasma membrane and the endosomal compartments via the Golgi complex is depicted to occur in vesicles that are budding from the donor membrane and targeted to and fused with the acceptor membrane. The sites of function of the coat complexes (COPI, COPII and Clathrin) as well as those of the Rab proteins, tethering complexes (T), SM proteins (SM), and the SNARE complexes (S) are indicated. ER: endoplasmic reticulum; G: Golgi complex; SV: secretory vesicle; PM: plasma membrane; E: early and late endosomes; L/V: lysosomal-vacuolar compartment; N: nucleus.

yeast cells. Although *in vitro* reconstitution experiments in yeast have shown Ca^{++} dependence of vesicle fusion, the Ca^{++} dependent regulation in non-neuronal cells has not yet been studied in detail.

The basic principle of the biosynthetic or secretory pathway was first proposed by Palade (1975). The outlines of the secretory pathway were elucidated in the late 1970's and early 1980's partially with the help of yeast mutants defective in protein secretion and partially with biochemical and morphological methods in yeast and animal cells. In both systems, *in vitro* reconstitution experiments proved to be excellent tools in dissecting transfer of proteins from one compartment to another. Instrumental for the mapping of different intracellular transport pathways was the pioneering yeast genetic work by P. Novick and R. Schekman for the secretory

pathway and by S. Emr and T. Stevens for the endocytic pathway (Novick and Schekman 1979; Bankaitis et al. 1986; Rothman and Stevens 1986). This, combined with the rigorous biochemical approaches pioneered by Rothman and co-workers in mammalian cells, resulted in a very rapid increase in the number of molecules active in transport and in their functional role in this process. Genetic screens are biased for essential genes, thus, being often unable to reveal the role of functionally redundant genes or non-essential regulatory molecules. Biochemical work, on the other hand, is biased toward isolation and characterisation of sufficiently stable protein complexes that withstand the experimental protocols. Nevertheless, extensive research employing a multitude of elegant genetic and biochemical approaches during the following years, culminated in a model for the transport mechanism, vesicle budding from the donor membrane followed by targeting and fusion with the acceptor membrane, which was introduced by Rothman (1994). Fine-tuning during the past ten years has resulted in a more detailed understanding of the transport pathways and the molecular machineries performing various tasks along it. Subsequent structural work on single components and even protein complexes are now bringing our knowledge of the stable protein complexes to a new level.

Vesicle budding

Transfer of proteins and membrane components from one compartment to another starts with vesicle budding from the donor membrane. A widely held view is that specific coat proteins and adaptors are instrumental in this process by aiding the vesicle budding and selecting the cargo for the vesicles. Clathrin was identified as the first vesicle coat protein. It is involved both in formation of endocytic vesicles at the plasma membrane and in vesicle formation at the trans-Golgi network. As described in the review by B. Ritter and P.S. McPherson, during the past ten years, the relatively simple model of clathrin mediated membrane budding has changed dramatically. In addition to clathrin and its adaptor protein AP2, a large number of accessory proteins have been identified that are involved in the process. The lack of suitable genetic approaches has limited the experimental approaches in mammalian cells in the past. Now, the situation has changed when the loss-of-function approaches have become possible with the introduction of the RNAi techniques. Among other modern methods, comparative genomics in the form of identification of sequence motives, and proteomics approaches have been helpful in identifying new components in this process. Non-clathrin coats, COP I and COP II, play an important role in membrane transport between the ER and the Golgi complex. Unfortunately, the chapter that dealt with these processes, transport from ER and the retrograde transport, was cancelled at such a late stage in the preparation of this book that it was not possible to get a substitution for the chapter. However, some of the other authors in this volume discuss aspects of these processes. In addition, there are some excellent recent reviews on this subject, for example, by Haucke (2003), Barlowe (2003), and Duden (2003).

Although the role of coat complexes in membrane budding is well established, there are numerous examples of membrane budding in different cellular compartments without the coat proteins. The important structural role of the lipids in vesicle formation by regulating the membrane curvature is an important aspect of membrane functionality in transport processes and is described by J.C.M. Holthuis. Certainly, lipid dynamics and their physico-chemical properties are of utmost importance in intracellular membrane traffic. In the future, it will be a great challenge to develop efficient ways to visualise lipids *in vivo* and to study lipid-protein and lipid-lipid interactions. These approaches in combination with proteomics techniques, will open completely new ways of understanding the dynamisc of membrane transport.

Vesicle targeting – regulation by Rab GTPases

Small GTP binding proteins are ubiquitous regulators of different cellular functions. The basic principle of this mode of regulation is cycling of the GTPases between GTP bound active and GDP bound inactive form of the proteins. The rab GTPases have been known to be involved in membrane transport since the 1980's when Salminen and Novick (1987) found that yeast Sec4 protein, essential for protein secretion in yeast, is a member of this protein family. The detailed functions of rab proteins are still poorly understood. Rab proteins function in all intracellular transport steps and regulate vesicle tethering and fusion, cargo selection, and cytoskeleton-dependent organelle transport. A great number of Rab accessory proteins, Rab activators, and downstream effectors have been identified but their role is even less understood than the role of the Rab proteins themselves. C.G. Burd and R.N. Collins described the current state of the art. They discuss the future challenges, which include; analysis of the diversity of specific functions of individual Rab proteins and their auxiliary factors, connection of Rab signalling with other signalling networks, for example, cell growth and differentiation and the roles of the uncharacterized Rab proteins.

Tethering complexes

Targeting of the transport vesicles to the correct intracellular localization is essential for biosynthesis and maintenance of the cell architecture. A commonly held view is that the specificity of targeting is provided by the so-called tethering factors, which can be long coiled-coil proteins or large multiprotein complexes. It is believed that the tethering factors can connect the vesicle from certain distance to the site of the following fusion event involving the SNARE proteins. W. Guo and S. Chu give a detailed account of the present knowledge on various tethering proteins and their proposed functions at different intracellular organelles and at plasma membrane. Most extensively studied tethering complexes are the TRAPP

complex mediating targeting of the ER derived vesicles to the Golgi complex and the Exocyst complex that targets Golgi-derived vesicles to plasma membrane. Both of these complexes were first identified in yeast and were later found also in mammalian cells, and were shown to have a similar function there. The tethering complexes interact with Rab proteins and in some cases have been shown to function as their downstream effectors. There are indications that the tethering proteins may have other functions in addition to the tethering *per se*. Such functions include, for example, cargo sorting and acceleration of the assembly of correctly paired SNARE proteins. It has also been proposed that one Exocyst component, Exo84, is involved in mRNA splicing. Mutations in some Exocyst subcomponents display defects in transport steps other than exocytosis suggesting that tethering components may be participating in tethering or other membrane transport functions at more than one intracellular location. Finally, recent data suggests that at least some components of the Exocyst complex interact with the ER translocation machinery and, thereby, regulate the efficiency of translation of secreted and membrane proteins suggesting a regulatory circuit adapting the translation efficiency with the cells secretory capacity. Clearly, future studies are needed to dissect all the interactions of the tethering complexes with several other cellular functions in addition or in connection to the tethering events.

SM proteins

Similar to the Rab proteins, the Sec1/Munc18 (SM) proteins participate in each targeting/fusion step inside the cell. The SM proteins are implicated in regulation of the SNARE complex formation or function at the target membrane by binding to the target membrane SNAREs. Genetic data suggest a positive role for SM-family proteins in membrane traffic regulation. However, numerous *in vitro* studies have suggested a negative regulatory role for these proteins. Thus, in spite of a multitude of experimental approaches, the SM proteins have resisted attempts to their detailed functional analysis and these controversial results have made it difficult to form a model for SM protein function. The review by M. Kauppi, J. Jäntti, and V. Olkkonen brings together the present knowledge in this field. Structural data available for some SM and SNARE proteins have revealed a remarkable diversity in the mode of SM protein binding to the cognate t-SNARE proteins. There seems to exist at least four different modes of binding between these two proteins. Also, SNARE-independent binding of Munc18 to plasma membrane has been shown to take place in epithelial cells. A model is emerging which takes into account non-SNARE proteins as regulators of the SM protein function. Candidates for such proteins are, for example, the tethering complex components and the Rab proteins as well as other less well-characterised proteins like the Mint proteins in mammalian cells. It is likely that future studies will reveal currently unidentified proteins in association with SM proteins in membrane traffic regulation.

Regulation of SNARE assembly by protein phosphorylation

Protein phosphorylation/dephosphorylation is a widely used regulatory mechanism in biology. Yet, our knowledge on the role of protein kinases and phosphatases in regulation of membrane transport is still in its infancy. It is, nevertheless, emerging as an important controlling mechanism also in this field as described in the review by A. Weinberger and J.E. Gerst. The role of protein phosphorylation along the secretory pathway is perhaps best known for the plasma membrane SNARE complex formation and function and regulation of the membrane fusion. Both the SNARE molecules themselves and SNARE regulators such as the SM proteins are subject to protein phosphorylation. Although SNARE phosphorylation, as such, is necessary for the secretory process, phosphorylation seems to have a negative effect in the SNARE-SNARE or SNARE-SNARE regulator interactions and complex formation. Given the multitude of protein kinases and phosphatases known already and their function in diverse cellular pathways it can be predicted that our understanding on the role of protein phosphorylation in connection to protein secretion and membrane transport and connection of these to the various signalling and regulatory cascades in cell is still in the beginning. A technical challenge is to capture these often low affinity and transient interactions between kinases and phosphatases with their target molecules. Clearly, development of new techniques is on demand.

Phosphoinositides in membrane traffic

Phosphoinositides, the phosphorylated derivatives of phosphatidylinositol, have a central regulatory role in the vesicle formation/coat assembly and signalling complex formation in addition to cytoskeleton remodelling, best known for the endocytotic transport. Typical for this mode of regulation is that the phosphoinositides are rapidly synthesised locally at different organelles by phosphorylation of the inositol ring at different positions and inactivated by dephosphorylation. The function of the various enzymes involved in formation and inactivation of the different phosphoinositides, their role in membrane transport and the human diseases caused by defects in them are dealt with by A. Godi, A. Di Campali, and M.A. De Matteis. Their role in clathrin coat assembly is also dealt with by B. Ritter and P.S. McPherson.

Regulation of *de novo* membrane formation

Last, but not least, M. Knop reviews an interesting aspect of the *de novo* membrane formation that takes place in yeast sporulation and development of *Apicomplexan* inner membrane complex. These processes represent interesting specialised

cases of membrane formation, regulation of which is expected to partly share components of the normal membrane fusion machinery, but also partly to reveal new cellular regulatory cascades linked to cell differentiation.

As the regulatory cascades in the membrane transport field are just now emerging, we felt that it is appropriate to review some of these aspects at the moment. We thank all the authors for excellent contributions to this volume bringing together the present state of the art. We also wish to thank Stefan Hohmann and Ursula Gramm for good collaboration and for swift processing of the manuscripts. With the genome data available, more and more genes and proteins involved in membrane transport can be soon identified in different organisms. With the new research tools available, such as RNAi techniques, the genome wide analyses together with *in vivo* visualisation techniques, we are undoubtedly facing an interesting future in developing our understanding on the regulation of the intracellular membrane trafficking and its connections to the cellular regulatory cascades.

References

Bankaitis VA, Johnson LM, Emr SD (1986) Isolation of yeast mutants defective in protein targeting to the vacuole. Proc Natl Acad Sci USA 83:9075-9079

Barlowe C (2003) Signals for COPII-dependent export from the ER: what's the ticket out? Trends Cell Biol 13:295-300

Duden R (2003) ER-to-Golgi transport: COP I and COP II function. Mol Membr Biol 20:197-207

Haucke V (2003) Vesicle budding: a coat for the COPs. Trends Cell Biol 13:59-60

Novick P, Schekman R (1979) Secretion and cell-surface growth are blocked in a temperature-sensitive mutant of *Saccharomyces cerevisiae*. Proc Natl Acad Sci USA 76:1858-1862

Salminen A, Novick PJ (1987) A ras-like protein is required for a post-Golgi event in yeast secretion. Cell 49:527-538

Palade G (1975) Intracellular aspects of the process of protein synthesis. Science 189:347-358

Rothman JE (1994) Mechanisms of intracellular protein transport. Nature 372:55-63

Rothman JH, Stevens TH (1986) Protein sorting in yeast: mutants defective in vacuole biogenesis mislocalize vacuolar proteins into the late secretory pathway. Cell 47:1041-1051

Keränen, Sirkka
VTT Biotechnology, Tietotie 2, Espoo, P.O. Box 1500, FIN-02044 VTT, Finland
sirkka.keranen@vtt.fi

Jantti, Jussi
VTT Biotechnology, Tietotie 2, Espoo, P.O. Box 1500, FIN-02044 VTT, Finland
jussi.jantti@vtt.fi

Molecular mechanisms in clathrin-mediated membrane budding

Brigitte Ritter and Peter S. McPherson

Abstract

Clathrin-coated vesicles (CCVs) form at the plasma membrane where they select cargo for endocytic entry into cells. CCVs also form at the trans-Golgi network (TGN) where they function in protein transport from the secretory pathway to the endosomal/lysosomal system. It has been long thought that clathrin and the clathrin adaptors 1 (AP-1) and 2 (AP-2) could define a minimal machinery for the formation of CCVs and the selection of cargo. However, it is now clear that although clathrin, AP-1 and AP-2 are key components of these processes, complex protein machineries are necessary to regulate many of the fundamental processes in CCV formation and cargo selection at both the plasma membrane and the TGN.

1 Cellular sites of clathrin-mediated membrane budding

The formation, intracellular transport, and targeted, organelle-specific fusion of membrane vesicles are essential aspects of eukaryotic cell biology. Most if not all transport vesicles are initially encased in proteinaceous coats that assemble from oligomeric complexes recruited from the cytosol (Bonifacino and Lippincott-Schwartz 2003). The protein coats contribute to membrane deformation and aid in the recruitment of appropriate cargo into the nascent vesicles. Upon release from the donor membrane, the vesicles uncoat allowing for transport and fusion with an acceptor compartment. Two of the major sets of coats contain clathrin: One, along with AP-2 regulates the formation of CCVs at the plasma membrane; the other, along with AP-1 regulates CCV formation at the TGN (Brodsky et al. 2001; Robinson 2004).

The AP-2-assisted formation of CCVs at the plasma membrane is the central feature of clathrin-mediated endocytosis (CME). CME is the major route of endocytic entry into cells and, thus, regulates many important physiological processes (Conner and Schmid 2003a). For example, CME allows for the cellular entry of nutrients such as iron and cholesterol. CME of G-protein-coupled and tyrosine kinase receptors, not only regulates their cell surface expression, but is necessary for their ability to couple to specific intracellular signaling pathways (McPherson et al. 2001). Regulated CME of AMPA type glutamate receptors appears to form the basis for long-term depression, an important form of synaptic plasticity (Car-

Topics in Current Genetics, Vol. 10
S. Keränen, J. Jäntti (Eds.): Regulatory Mechanisms of Intracellular Membrane Transport
DOI 10.1007/b98564 / Published online: 25 May 2004
© Springer-Verlag Berlin Heidelberg 2004

roll et al. 2001). Moreover, biochemical and morphological studies have indicated that CME is used for the recycling of synaptic vesicles (Murthy and De Camilli 2003; Blondeau et al. 2004).

The TGN is a second major site of clathrin-mediated membrane budding. One established pathway from the TGN involves AP-1-containing CCVs that function in the transport of cargo from the secretory pathway to the endosomal system and the lysosome (Hinners and Tooze 2003). Important cargo proteins for these vesicles are the mannose-6-phosphate receptors (MPRs) that sort mannose-6-phosphate tagged lysosomal hydrolases. Interestingly, Bonifacino and colleagues have recently described a new class of TGN-derived carriers (Puertollano et al. 2003). These pleiomorphic vesicles and tubules, which are significantly larger than CCVs, also contain clathrin and AP-1 (Puertollano et al. 2003). They undergo rapid microtubule-dependent transport to the peripheral cytoplasm and retain their coats much longer than traditional CCVs. These carriers may be involved in the transport of cargo, including MPRs, to more peripherally localized endosomes (Puertollano et al. 2003). The diversity of clathrin-mediated trafficking pathways and the important cellular processes they control underscore the need to develop a complete understanding of the molecular machineries that account for and regulate these events.

2 Formation of endocytic CCVs

The formation of CCVs has been most extensively studied in relationship to CME occurring at the plasma membrane. The assembly unit of the clathrin coat is the triskelion, which is composed of three copies of the ~190 kDa clathrin heavy chain (CHC), linked through their C-termini (Schmid 1997; Brodsky et al. 2001). Each CHC is associated with either of two ~30 kDa clathrin light chains (CLCs). Under mildly acidic conditions, clathrin triskelions spontaneously self-assemble, forming polyhedral clathrin cages. These cages have a diameter similar to that of endogenous CCVs and it has, thus, been generally assumed that clathrin assembly drives the membrane curvature necessary for the formation of CCVs.

A second major component of clathrin-coated pits (CCPs) and CCVs at the plasma membrane is AP-2 (Brodsky et al. 2001; Fig. 1). AP-2 was originally detected as a protein factor that stimulated coat assembly at physiological pH (Keen et al. 1979), conditions in which spontaneous clathrin assembly is not observed *in vitro*. AP-2 is a heterotetramer (Fig. 1). The N-terminal regions of the two large subunits of AP-2, α- and β2-adaptin, together with the medium sized μ2-adaptin and small σ2-adaptin subunits, form the core structure of the protein complex. The C-terminal regions of α- and β2-adaptins form the ear domains (α-ear and β2-ear) that are linked to the core via flexible hinge regions.

Different functions have been assigned to each of the AP-2 subunits, which give AP-2 its unique role in CME (Fig. 1). The σ2-adaptin subunit appears to be responsible primarily for the structural stability of the core (Collins et al. 2002).

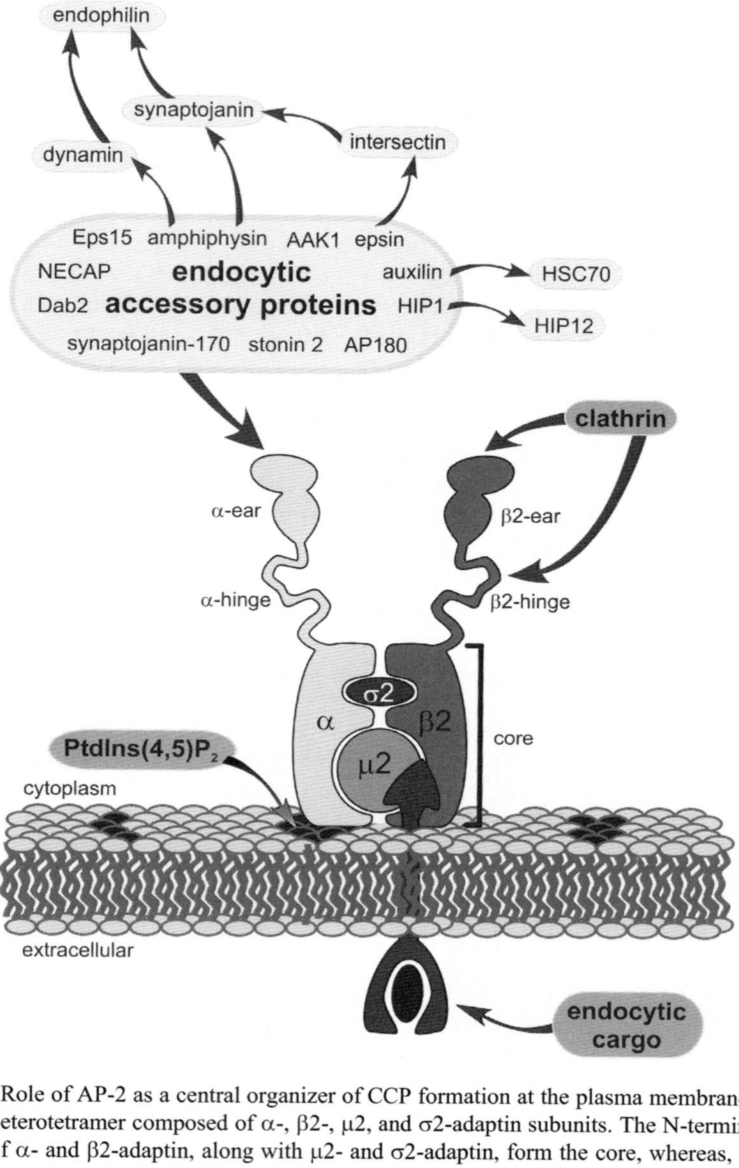

Fig. 1. Role of AP-2 as a central organizer of CCP formation at the plasma membrane. AP-2 is a heterotetramer composed of α-, β2-, μ2, and σ2-adaptin subunits. The N-terminal regions of α- and β2-adaptin, along with μ2- and σ2-adaptin, form the core, whereas, the C-terminal regions of α- and β2-adaptin form the hinge and ears. The large arrows originate from the various interacting partners for AP-2. These include an array of endocytic accessory proteins (contained within the oval) that use three classes of α-ear-binding motifs (discussed in the text) to interact directly with the α-ear. Other endocytic accessory proteins that bind indirectly to the α-ear are indicated outside of the oval and are linked to some of their known binding partners by arrows.

Fig. 2. (overleaf) Standard model for CCV formation at the cell surface. A) AP-2 binds to PtdIns(4,5)P$_2$ at the plasma membrane leading to the surface recruitment of clathrin triskelia. B,C) Clathrin triskelia assemble into a clathrin coat, driving membrane curvature. Endocytic cargo proteins are recruited into the emerging CCPs through interactions with AP-2. D) Dynamin is recruited to deeply invaginated CCPs were it forms an oligomeric ring around the membrane stock connecting the CCP to the surface. E) Dynamin mediates a fission reaction leading to the liberation of a mature CCV. Auxilin binds to the α-ear and recruits HSC70 to the CCVs. F) HSC70 mediates clathrin disassembly and the uncoating of the CCVs.

The N-terminal region of α-adaptin binds to phosphatidylinositol(4,5)P$_2$ (PtdIns(4,5)P$_2$) (Gaidarov and Keen 1999), which facilitates the localization of AP-2 to the plasma membrane (Krauss et al. 2003; Fig. 1). Through cooperative sites in the β2-ear and hinge, AP-2 binds to clathrin triskelia and promotes clathrin assembly (Owen et al. 2000; Fig. 1). Thus, AP-2 can recruit clathrin to the plasma membrane where it stimulates formation of CCPs. The μ2- and β2-adaptin subunits interact with proteins carrying specific endocytic targeting sequences in their cytoplasmic domains, thus, concentrating the cargo in nascent CCPs (Traub 2003; Fig. 1). Through direct or indirect interactions, the α-ear serves as a recruitment platform for numerous endocytic accessory proteins that regulate various aspects of CCV formation (Fig. 1). Based on these features, the AP-2 complex is exquisitely equipped to be a master regulator of the formation of CCPs and CCVs.

Thus, a standard model for CCV formation at the cell surface suggests that AP-2 recruits clathrin to the membrane and stimulates its assembly, leading to membrane deformation and the emergence of a budded CCP (Fig. 2A, B). Cargo proteins that contain appropriate endocytic targeting sequences are concentrated in CCPs through their interactions with AP-2 (Fig. 2B). As cargo is recruited, the pits continue to bud until they obtain a deeply invaginated state (Fig. 2C, D). The α-ear then contributes to the recruitment of endocytic accessory proteins to the emerging vesicles (Fig. 1 and 2D, E). One such protein is the GTPase dynamin, which is recruited to CCPs in part through interactions with amphiphysin, an endocytic adaptor protein that bridges the α-ear and clathrin to dynamin (David et al. 1996; Ramjaun et al. 1997; Owen et al. 1999) (Fig. 1). Dynamin forms oligomeric rings around the membrane stalks that link deeply invaginated CCPs to the plasma membrane (Takei et al. 1995; Hinshaw and Schmid 1995) and it mediates the fission reaction that releases mature CCVs (Sever et al. 2000; Fig. 2D, E). HSC70, which causes clathrin disassembly and promotes the release of the clathrin coat (Schlossman et al. 1984), is recruited to CCVs through its interaction with auxilin, which binds to the α–ear and clathrin (Ungewickell et al. 1995; Fig. 1 and 2E, F).

Thus, AP-2 and clathrin have been seen as the primary components for driving membrane curvature and for selecting appropriate endocytic cargo (Fig. 2). The endocytic accessory proteins have been thought to regulate various aspects of the CCV cycle such as CCV fission and uncoating, but they have not been considered as being directly involved in the formation of membrane curvature or in cargo selection. However, as the understanding of the function of the various endocytic

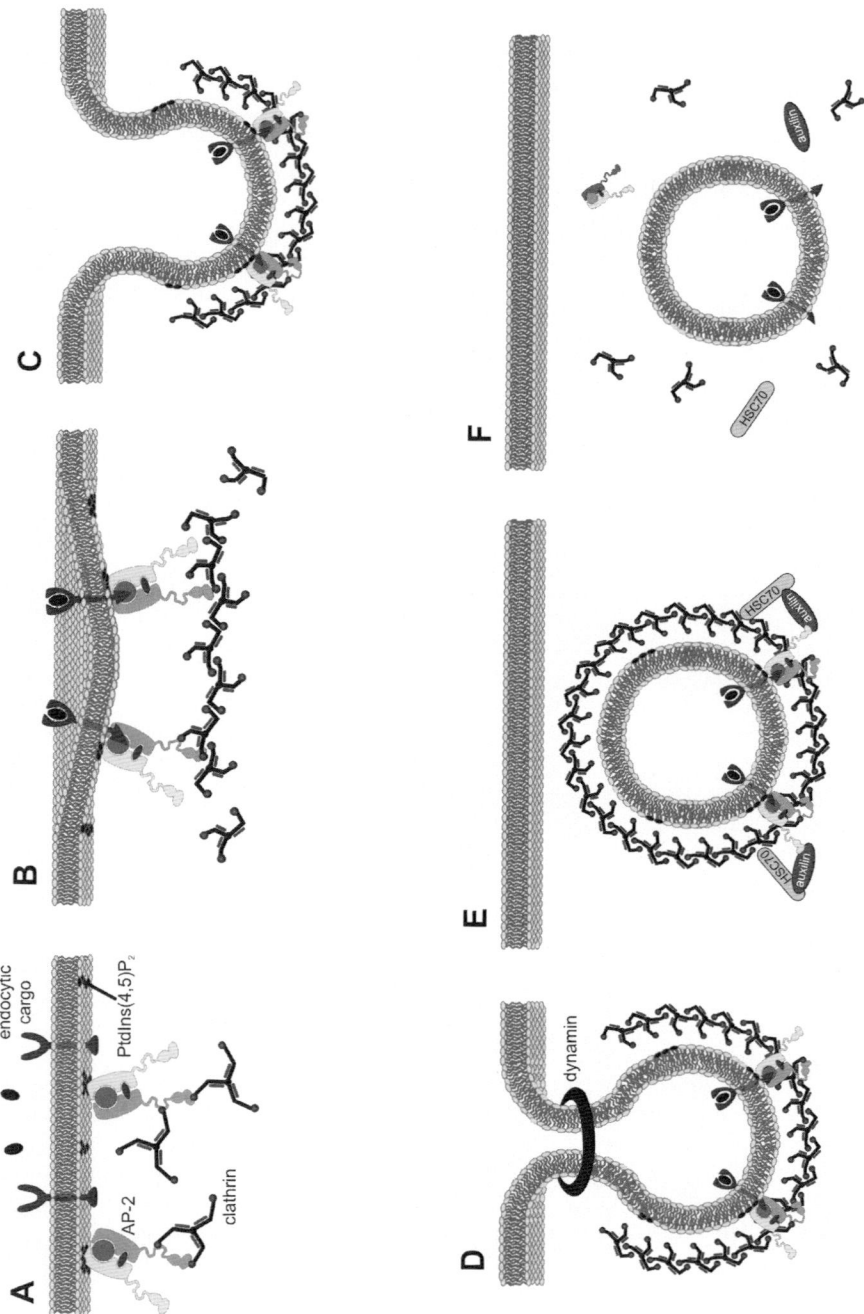

accessory proteins has increased, a more fundamental role for these proteins in membrane budding and cargo recruitment has emerged. In this chapter, we will examine several of these issues.

3 Machinery for CCV formation at the cell surface

3.1 Role of clathrin

Ever since the observation that purified clathrin triskelia can assemble into clathrin cages with the same diameter as mature CCVs, it has been widely held that clathrin is responsible for driving the membrane budding that leads to the formation of CCVs. To better define the role of clathrin in the formation of CCPs and CCVs, several recent studies have utilized clathrin loss-of-function approaches. (Wettey et al. 2002; Hinrichsen et al. 2003; Iversen et al. 2003; Moskowitz et al. 2003; Motley et al. 2003; Huang et al. 2004). The use of RNAi to reduce CHC to near undetectable levels in HeLa cells causes a near complete loss of CCPs and CCVs as detected by EM (Hinrichsen et al. 2003; Motley et al., 2003). Moreover, CHC knock down leads to severe inhibition of the endocytosis of transferrin (Hinrichsen et al. 2003; Motley et al. 2003; Huang et al. 2004), a reliable marker of CME. Similar effects on transferrin endocytosis were also observed when the function of the CHC was perturbed by homologous recombination (Wettey et al. 2002), antisense RNA treatment (Iversen et al. 2003), or chemical inhibition (Moskowitz et al. 2003). These results clearly demonstrate a critical role for clathrin in the formation and endocytic function of CCVs.

Interestingly, in the study by Ungewickell and colleagues, a component of the residual CHC (approximately 20% after 2 days of siRNA exposure) remained on the plasma membrane where it demonstrated a punctate distribution (Hinrichsen et al. 2003). EM analysis of these cells revealed an increase in shallow CCPs at the expense of more deeply invaginated CCPs and CCVs (Hinrichsen et al. 2003). As assembled clathrin transitions from a flat lattice to a spherical CCV, a structural rearrangement of clathrin must occur during the invagination process. Eisenberg and Greene and colleagues have used fluorescence recovery after photobleaching analysis to examine the dynamics of clathrin triskelia marked with GFP-tagged CLC (Wu et al. 2001). Replacement of photobleached CLC in CCPs occurred at the same rate in cells that were blocked for endocytosis as in control cells (Wu et al. 2001). Thus, clathrin triskelia appear to undergo a robust, two-way exchange between the cytosol and assembled CCPs throughout the life cycle of the CCP (Wu et al. 2001). This exchange is likely to be involved in the structural rearrangement of clathrin that occurs as CCPs invaginate. Hinrichsen et al. (2003) have argued that a decrease in cytosolic clathrin concentrations resulting from RNAi treatment leads to a slowing of the clathrin exchange and an increase in the number of shallow CCPs. Thus, a dynamic clathrin-lattice that can rearrange to form a curved clathrin coat appears necessary for the proper formation of CCPs and CCVs. Despite the clear evidence of a critical role for clathrin in endocytosis,

none of these studies reveal if clathrin assembly directly provides the mechanical force for driving membrane curvature. In fact, as discussed in section 3.3.1, clathrin may be insufficient to drive membrane curvature and this process may depend on endocytic accessory proteins that directly influence lipids. Another interesting caveat to the role of clathrin in endocytic function concerns the role of the CLCs. Despite the fact that CHC and CLC have been considered to work as a functional unit within clathrin triskelia, simultaneous knock down of both CLC isoforms by RNAi has no effect on CME (Huang et al. 2004). The functional role of the CLC in CCV formation clearly deserves further attention.

3.2 Role of AP-2

3.2.1 Role of AP-2 in clathrin recruitment

To better understand the function of AP-2, several recent studies have explored the effect of AP-2 knock down or mislocalization on clathrin membrane recruitment, formation of CCPs and CCVs, and the internalization of endocytic cargo (Motley et al. 2003; Hinrichsen et al. 2003; Conner and Schmid 2003b; Huang et al. 2004). The studies of Conner and Schmid stemmed from their identification of the adaptor-associated kinase 1 (AAK1), an endocytic accessory protein that binds to the α-ear and phosphorylates the μ-adaptin subunit (Conner and Schmid 2002). Overexpression of AAK1 caused a dramatic displacement of AP-2 from its normally punctate distribution on the plasma membrane, although the mechanism mediating this effect is unknown. Surprisingly, clathrin recruitment to CCPs did not appear to be altered in these cells, leading to the conclusion that AP-2 is not stoichiometrically required for coat assembly (Conner and Schmid 2003b). A similar conclusion was drawn in a study by Robinson and colleagues, in which the complete elimination of μ2-adaptin by RNAi treatment did not eliminate the formation of morphologically normal CCPs at the plasma membrane. However, in this study, the number of CCPs was reduced to less than 10% of that seen in control cells and α-adaptin, although reduced, was not eliminated, with approximately 10% of the control levels remaining. Moreover, the levels of β2-adaptin were not examined and it, thus, remains possible that residual α- and β-adaptin subunits are functioning to mediate clathrin recruitment. In fact, the study of Hinrichsen et al. (2003) demonstrated that the complete elimination of α-adaptin did indeed lead to a complete loss of membrane-associated clathrin, causing these authors to conclude that AP-2 is critical for clathrin recruitment to the membrane. The nature of the seeming discrepancies between these studies is unclear. One possibility, however, may reside in the fact that clathrin recruitment likely involves a more complex protein machinery than that provided by AP-2 alone. In section 3.3.1, we will describe a number of the components of the endocytic machinery that can bind to both clathrin and to phospholipids, and may, thus, contribute to clathrin membrane recruitment. Moreover, many of these proteins also bind to AP-2 and are likely to

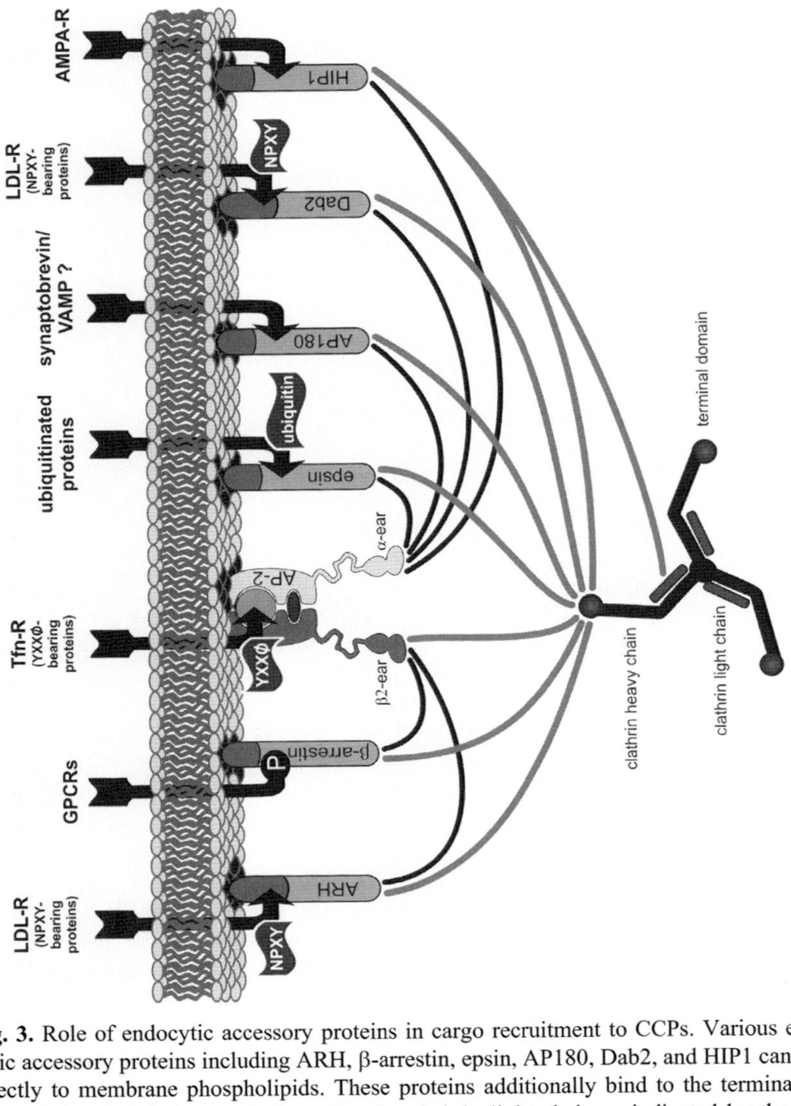

Fig. 3. Role of endocytic accessory proteins in cargo recruitment to CCPs. Various endocytic accessory proteins including ARH, β-arrestin, epsin, AP180, Dab2, and HIP1 can bind directly to membrane phospholipids. These proteins additionally bind to the terminal domain of the clathrin heavy chain and/or the clathrin light chain as indicated by the grey lines. They also interact with AP-2 through the α-ear or the β2-ear as indicated by the black lines. These proteins could, thus, cooperate with AP-2 in recruiting clathrin to the membrane. Additionally, the accessory proteins can bind to various forms of endocytic cargo. These include NPXY motifs in the low-density lipoprotein receptor (LDL-R), YXXØ motifs in the transferrin receptor (Tfn-R), and ubiquitin moieties in ubiquitinated proteins. They can also bind to specific cargo such as G-protein-coupled receptors (GPCRs), synaptobrevin/VAMP, and AMPA-type glutamate receptors (AMPA-R) as indicated.

cooperate with AP-2 in clathrin recruitment. Thus, the phenotype resulting from AP-2 knock down may depend on the background of endocytic accessory proteins that are expressed in the cell type studied. Unraveling the precise role of AP-2 in clathrin recruitment will clearly require further study.

3.2.2 Role of AP-2 in cargo recruitment

Each of the studies referred to in section 3.2.1 also examined the role of AP-2 functional depletion on the endocytosis of specific cargo proteins. In general, the concentration of endocytic cargo into CCPs is dependent on the presentation of internalization motifs within the cytoplasmic region of the cargo molecule. Several of these motifs are recognized by the AP-2 complex (Traub et al. 2003) and the internalization of many types of cargo proteins is thought to depend on AP-2 as mutations in AP-2-binding motifs reduces their internalization (Kirchhausen et al. 1997). The transferrin receptor is the endocytic cargo that is most clearly dependent on AP-2 (Fig. 3). The receptor contains a tyrosine-based endocytic motif that matches the consensus YXXØ (where X is any amino acid and Ø is a hydrophobic amino acid). Tyrosine-based motifs bind to µ2-adaptin (Ohno et al. 1995) and mutation of the motif within the transferrin receptor or the binding site for the motif in µ2-adaptin interferes with transferrin uptake (Nesterov et al. 1999; Olusanya et al. 2001; Fig. 3).

Given these results, it is not surprising that Motley et al. (2003) observed that knock down of µ2-adaptin by RNAi treatment led to a near complete block in transferrin receptor endocytosis. Interestingly, however, there was no effect of µ2-adaptin knock down on endocytosis of EGF receptor and LDL receptor, even though the endocytosis of all receptors was inhibited or greatly reduced in clathrin knock down cells. This led these authors to conclude that AP-2 is not a general endocytic adaptor but is in fact required for the uptake of specific cargo (transferrin receptor) with other cargo proteins using distinct adaptors. Likewise, AAK1 overexpression, which causes AP-2 to mislocalize to the cytosol, led to a block in endocytosis of the transferrin and LPR (low-density lipoprotein-related protein) receptors but had no effect on EGF receptor internalization (Conner and Schmid 2003b). Thus, this study also concludes that AP-2 works as an adaptor for only specific classes of endocytic cargo. In particular, both studies argue that AP-2 does not function as an adaptor for EGF receptor endocytosis. Hinrichsen et al. (2003) also observed a complete block in transferrin receptor endocytosis with normal endocytic uptake of EGF receptor upon siRNA-mediated knock down of α-adaptin. Although, they also observed that EGF receptor endocytosis proceeds normally in clathrin knock down cells. Thus, they concluded that blocking clathrin- and AP-2-dependent endocytosis of the EGF receptor simply caused the receptor to utilize a distinct pathway for internalization. In addition, the occupancy of large numbers of EGF receptors is known to lead to their internalization through clathrin-independent pathways as the clathrin-dependent pathway becomes saturated (Huang et al. 2004). In fact, siRNA-mediated knock down of AP-2 was shown to cause a severe reduction in EGF receptor internalization at low

concentrations of EGF (Huang et al. 2004). Thus, transferrin and EGF receptor appear to use AP-2/clathrin pathways under physiological conditions, whereas, LDL receptor may utilize a distinct adaptor for CME. Earlier studies by Warren et al. (1999) had in fact indicated that LDL receptor does not compete with transferrin receptor or EGF receptor for cellular entry via CME. Thus, AP-2 has a critical role in CME although proteins other than AP-2 may function alone or in cooperation with AP-2 as adaptors for endocytic cargo.

3.3 Role of endocytic accessory proteins

3.3.1 Endocytic accessory proteins in clathrin recruitment, assembly, and membrane curvature

The identification of clathrin and AP-2 and the realization of their roles in CCV formation date back more than twenty years. However, it was only in the last decade, starting with the identification of dynamin, amphiphysin, and the lipid phosphatase synaptojanin, did the concept of a more complex endocytic regulatory machinery first emerge (McPherson et al. 1994, 1996; Takei et al. 1995; David et al. 1996). Since then, a large array of proteins has been identified that regulate various aspects of CCV formation and function (Fig. 1). Many of these proteins bind directly to clathrin and/or AP-2 but traditionally, they have not been viewed as part of the core machinery for CCV formation and selection of endocytic cargo. However, it is now clear that many of these proteins are in fact critical components of the core machinery for CCV formation. For example, recent studies have suggested that several of the endocytic accessory proteins, notably those bearing ENTH domains, function in the recruitment and assembly of clathrin at the plasma membrane and are likely to play a direct mechanical role in generating membrane curvature.

The ENTH (epsin N-terminal homology) domain is an ∼ 150 amino acid module with a compact globular structure composed of eight α-helicies (Hyman et al. 2000) that was first recognized in the N-terminus of the endocytic protein epsin (Rosenthal et al. 1999; Kay et al. 1999; see Legendre-Guillemin et al. 2004 for a recent review). Outside of the ENTH domain, epsin has an unstructured C-terminal domain (Kalthoff et al. 2002a) containing binding motifs for the α-ear of AP-2 (Owen et al. 1999; Traub et al. 1999) and the terminal domain at the C-terminus of the CHC (Hussain et al. 1999; Rosenthal et al. 1999; Drake and Traub 2001; Fig. 3). Structurally, the ENTH domain is highly related to the ANTH (AP180 N-terminal homology) domain (Legendre-Guillemin et al. 2004). The ANTH domain has been best characterized in AP180, a brain-specific protein that functions in the recycling of synaptic vesicles (Morgan et al. 2000). Like epsin, AP180 binds to the CHC (Morgan et al. 2000) and the α-ear (Hao et al. 1999; Brett et al. 2002; Fig. 3). The huntingtin interacting protein 1 (HIP1) is also an ANTH domain-bearing protein that binds to CHC, CLCs, and the α-ear (Fig. 3) and is concentrated on CCPs and CCVs (Metzler et al. 2001; Mishra et al. 2001; Waelter et al. 2001; Legendre-Guillemin et al. 2002).

Interestingly, E/ANTH domains bind to inositol phospholipids and in particular, the epsin ENTH domain and the AP180 and HIP1 ANTH domains display a preference for PtdIns(4,5)P$_2$ (Ford et al. 2001; Itoh et al. 2001; Mishra et al. 2001; Fig. 3). Since all three proteins also bind CHC, they could contribute to clathrin recruitment by linking the CHC to plasma membrane-enriched PtdIns(4,5)P$_2$ (Fig. 3). The proteins have also been demonstrated to stimulate the assembly of soluble clathrin triskelia into clathrin cages with the assembly activity located outside of the E/ANTH domain (Ahle and Ungewickell 1986; Morgan et al. 2000; Kalthoff et al. 2002a; Legendre-Guillemin et al. 2002). In fact, *in vitro* they can each recruit clathrin to PtdIns(4,5)P$_2$ monolayers where they stimulate its assembly into clathrin lattices (Mishra et al. 2001; Ford et al. 2002). Thus, the proteins have the potential to function independent of AP-2 in clathrin membrane recruitment and assembly (Fig. 3). This could provide a mechanism to explain why clathrin recruitment can still be observed in cells in which AP-2 has been knocked down. However, as AP180, HIP1 and epsin each bind strongly to AP-2, it seems most likely that under normal circumstances, the proteins would cooperate with AP-2 by binding to the α-ear and generating a multivalent scaffold for clathrin recruitment (Fig. 3).

Although both AP180 and epsin can stimulate the assembly of a clathrin lattice on PtdIns(4,5)P$_2$ monolayers, clathrin patches formed by epsin are invaginated, whereas, AP180-induced lattices are flat (Ford et al. 2002). This intriguing observation suggests that clathrin assembly may not be sufficient to drive membrane curvature and that the ENTH domain may play an active role in this process. Support for this idea has come from the structural analysis of ENTH and ANTH domains. The extreme N-terminus of the epsin ENTH domain contains a small stretch of unstructured amino acids. Upon binding to PtdIns(4,5)P$_2$, this region forms an α-helix referred to as α0, which is not generated in the ANTH domain (Ford et al. 2001, 2002; Itoh et al. 2001). Basic residues on the inner face of α0 contribute to PtdIns(4,5)P$_2$ binding and their mutation abrogates lipid interactions. On its outer surface, α0 has a series of hydrophobic residues and McMahon and colleagues speculated that α0 may insert into the cytosolic leaflet of the bilayer following PtdIns(4,5)P$_2$ binding (Ford et al. 2002). If true, this mechanism could facilitate the formation of membrane curvature by mechanically pushing the lipid head groups apart (Legendre-Guillemin et al. 2004). In fact, addition of the epsin ENTH domain to phospholipid monolayers produces changes in surface pressure indicating protein insertion into the membrane (Stahelin et al. 2003). This only occurs in the presence of PtdIns(4,5)P$_2$ and mutations of the hydrophobic residues in α0 block the effect (Stahelin et al. 2003). Moreover, no membrane insertion is seen with the ANTH domain of AP180. Thus, the ENTH domain can perform mechanical work on membranes and the physical deformation of the membrane by the actions of the ENTH domain of epsin could be an important mechanical force contributing to membrane curvature. In such a model, a curved clathrin lattice could provide a mold to define the shape of the deformed membrane without contributing directly to the mechanical deformation of the membrane. The clathrin lattice could also be used to stabilize the deformed membrane to prevent it from col-

lapsing back into the plane of the membrane. Clearly, the mechanisms that drive membrane curvature and the role of clathrin in this process deserve further attention.

3.3.2 Endocytic accessory proteins in cargo recruitment

Data from the AP-2 knock down experiments described in section 3.2.2 suggest that at least under certain circumstances, AP-2 is not by itself sufficient to mediate the recruitment of all classes of endocytic cargo into CCPs and CCVs. Other cargo adaptors must exist that function independently of AP-2 or in cooperation with AP-2 in cargo recruitment. What properties should a protein have that would qualify it to function as an adaptor for cargo of CME? Any list of criteria would naturally be subjective, but minimally, the protein should: i) target to the plasma membrane, ii) localize to endocytic sites, iii) interact with components of the clathrin coat, and iv) interact with cargo proteins. In terms of cargo recognition, the adaptor could bind to one very specific cargo or it could bind to a broad class of cargo with a shared, characteristic feature. Moreover, such alternative adaptors could function completely independent of AP-2. Alternatively, they could bind to their specific cargo as well as AP-2, thus, increasing the variety of possible interactions and internalization motifs being served by AP-2.

One group of proteins that appear well suited to function as cargo adaptors in CME is the E/ANTH proteins. Several of them, including epsin, HIP1, and AP180 are localized to CCPs (Legendre-Guillimen et al. 2004). Moreover, they bind to PtdIns(4,5)P$_2$ and to clathrin and AP-2 (Fig. 3). Finally, many of these proteins contain additional binding activities that could allow them to recruit cargo into CCPs (Fig. 3).

Epsin. In its C-terminal region, epsin contains an ubiquitin-interacting motif (UIM), which are also found in the yeast epsin orthologues, Ent1p and Ent2p. UIMS are short peptide motifs that mediate interactions with ubiquitinated proteins (Polo et al. 2002). Interestingly, ubiquitination has emerged as a key signal regulating the endocytosis of a diverse group of proteins in both mammalian cells and yeast (Aguilar and Wendland 2003; Hicke and Dunn 2003). In yeast, Ent1p and Ent2p bind through their UIMs to ubiquitinated proteins at the plasma membrane and are likely to concentrate ubiquitinated cargo into CCVs (Aguilar et al. 2003). In mammalian systems, the EGF receptor is an example of a protein in which ubiquitination may be a key signal for CME (Hicke and Dunn 2003). Thus, epsin could bind to a variety of ubiquitinated proteins and could function as a general adaptor for this endocytic targeting signal (Fig. 3).

AP180. Disruption of the *C. elegans* AP180 orthologue, unc11, results in improper sorting of the synaptic vesicle protein synaptobrevin/VAMP. Normally, synaptobrevin is sorted into synaptic vesicles during their reformation by CCVs. In unc11 mutants, synaptobrevin is found diffusely over the presynaptic plasma membrane (Nonet et al. 1999). Thus, AP180, which is brain specific and interacts with clathrin and AP-2, could function as a cargo adaptor for certain classes of synaptic vesicle proteins (Fig. 3).

HIP1. In neurons, HIP1 is found in the post-synaptic compartment, where it binds to the GluR1 subunit of the AMPA type glutamate receptor (Metzler et al. 2003; Fig. 3). Interestingly, HIP1-knock out mice demonstrate a profound deficiency in CME of GluR1 following AMPA stimulation, whereas, constitutive endocytosis of transferrin receptor is not affected (Metzler et al. 2003). Thus, it is intriguing to speculate that HIP1 couples AMPA receptor complexes to CCPs and as such plays a critical role in the regulation of synaptic efficacy (Fig. 3). Whether HIP1 functions as an endocytic adaptor for a small number of specific endocytic cargos or has a more general cargo adaptor function remains to be determined.

Non-E/ANTH-bearing proteins. Proteins other than those containing E/ANTH domains have also have also been suggested to have cargo adaptor functions. Interestingly, the NPXY internalization motif of the LDL receptor, which is necessary for LDL receptor endocytosis via CCPs and CCVs, binds only weakly to the $\mu 2$-adaptin subunit of AP-2, but interacts strongly with the PTB (phosphotyrosine-binding) domain of disabled-2 (Dab2) (Morris and Cooper 2001; Mishra et al. 2002a; Fig. 3). The Dab2 gene has been primarily implicated in the regulation of growth control and the protein is downregulated in several human carcinomas (Morris and Cooper 2001). The PTB domain of Dab2 also binds to PtdIns(4,5)P$_2$, and additionally, the protein binds to clathrin and AP-2 and stimulates clathrin assembly (Morris and Cooper 2001; Mishra et al. 2002a; Fig. 3). Overexpression of the PTB domain impairs LDL receptor uptake with no influence on transferrin endocytosis (Mishra et al. 2002a). Moreover, Dab2 knock out mice display a deficiency in the uptake of the lipoprotein receptor megalin (Oleinikov et al. 2000; Morris et al. 2002). Thus, Dab2 has emerged as one of the clearest examples of an alternative cargo adaptor for CME.

A second PTB domain-containing protein that appears to function in LDL receptor endocytosis is ARH and intriguingly, mutations in its PTB domain cause autosomal recessive hypercholesterolemia (ARH) (He et al. 2002: Mishra et al. 2002b). The ARH disorder is characterized by defective CME of LDL receptor leading to very high plasma levels of LDL and premature atherosclerosis. As for Dab2, ARH binds through its PTB domain to the NPXY internalization motif in the LDL receptor and synchronously to PtdIns(4,5)P$_2$ (He et al. 2002; Mishra et al. 2002b; Fig. 3). ARH also binds to the terminal domain of the CHC and the $\beta 2$-ear of AP-2, and it co-localizes with clathrin and AP-2 at CCPs in the steady state (He et al. 2002: Mishra et al. 2002b; Fig. 3). Thus, the protein is well positioned to function as a cargo adaptor for CME of the LDL receptor.

Finally, CME of G-protein-coupled receptors (GPCRs) is a well-established example where AP-2 does not directly interact with cargo to mediate internalization. Instead, GPCR endocytosis is dependent on β-arrestin (Claing et al. 2002). Following activation of GPCRs, such as the β-adrenergic receptor, the liberated Gα subunits can activate GPCR receptor kinases including the β-adrenergic receptor kinase, leading to receptor phosphorylation and binding of β-arrestin (Claing et al. 2002). β-arrestin also binds to clathrin and AP-2 and the formation of a complex of β-adrenergic receptor with clathrin and AP-2 via β-arrestin is necessary for

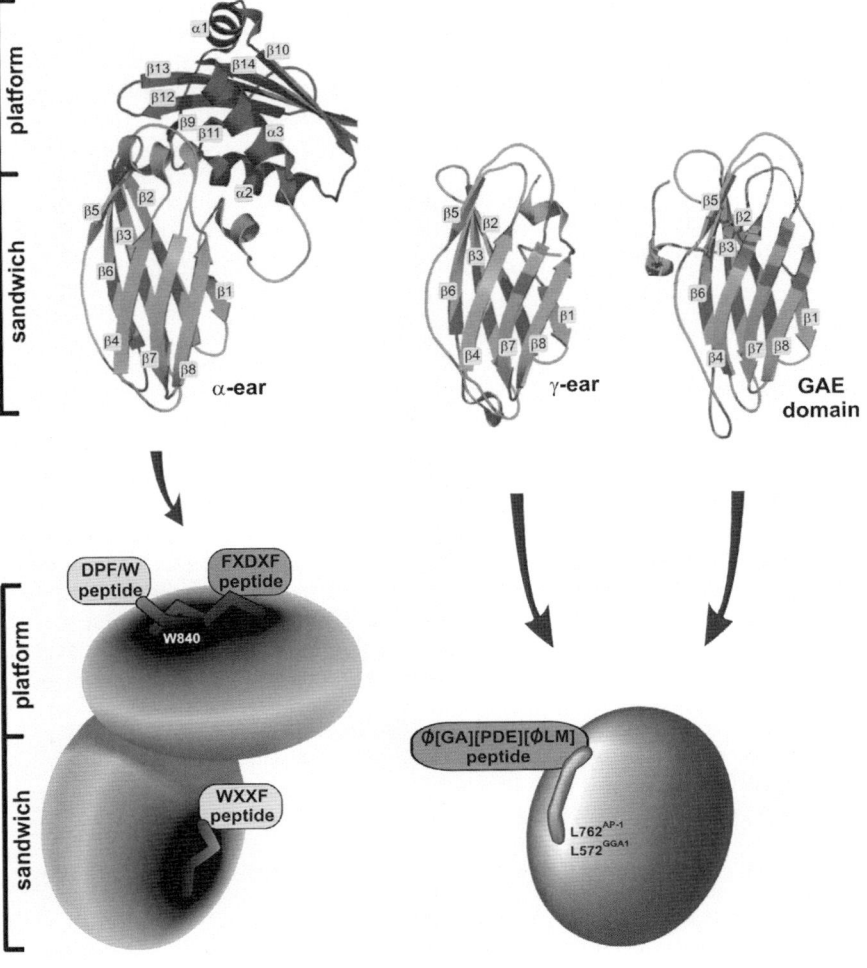

Fig. 4. Structures of the ear domains of the clathrin adaptors: ear-domain-binding motifs. The structure of the α-ear, γ-ear, and GGA1 GAE domain, taken from PDB entries 1QTS, 1IU1, and 1OM9, respectively, are illustrated. Schematic drawings of the structures are shown below. The binding site in the platform domain of the α-ear accommodates peptides matching the consensus α-ear binding motifs, DPF/W and FXDXF. Mutation of the tryptophan at position 840 (W840) of the α-ear disrupts AP-2-binding of proteins utilizing either of these motifs. The binding site in the sandwich domain of the α-ear interacts with WXXF-based peptides. Peptides that match the motif Ø[GA][PDE][ØLM] bind to both the γ-ear and the GAE domain, indicated as a single structure in the schematic drawing. Mutations of the leucine at position 762 (L762) of the γ-ear and the corresponding position, leucine 572 (L572) of the GAE domain, disrupt the interactions of TGN accessory proteins with AP-1 and GGAs, respectively.

receptor endocytosis (Ferguson et al. 1996; Goodman et al. 1996; Laporte et al. 2000). Thus, β-arrestin is well suited to function as an endocytic adaptor for the β-adrenergic receptor and other GPCRs (Fig. 3). The further characterization of the components of CCPs and CCVs will undoubtedly lead to the identification of additional proteins that can function as cargo adaptors in CME.

3.4 Interactions of endocytic accessory proteins with AP-2

3.4.1 Structure of the α-ear and identification of the DPF/W consensus α–ear-binding motif

The majority of the endocytic accessory proteins that contribute to CCV formation and function bind directly to AP-2 (Fig. 1 and 3). The interactions occur predominantly through the α-ear making this structure a critical component in the formation of protein networks that regulate CME (Fig. 1). Thus, to fully understand the molecular machinery for CME it is necessary to analyze the nature of α-ear interactions with endocytic accessory proteins. In fact, important insights into the mechanisms of CCV formation have come from studies aimed at the structural characterization of the α-ear and the identification of short peptide motifs that mediate α-ear binding.

The α-ear itself, located at the C-terminus of the α-adaptin subunit, is a two lobed globular structure that can be divided into two evenly sized subdomains, the more N-terminal (proximal) sandwich domain and the C-terminal (distal) platform domain (Fig. 4). The sandwich domain is folded into a two-sheet β sandwich formed from eight antiparallel β strands, which has a similar topology as proteins of the Ig superfamily (Owen et al. 1999; Traub et al. 1999; Fig. 4). The platform domain is folded into a five-strand antiparallel β sheet flanked by an α-helix on either side with an additional α-helix crossing on top of the sheet (Owen et al. 1999; Traub et al. 1999; Fig. 4). The two subdomains make extensive interactions with each other and as such there is no segmental mobility between them (Owen et al. 1999; Traub et al. 1999).

Analysis of the hydrophobic surface potential of the α-ear first suggested a potential protein interaction site in the platform domain located in a hydrophobic cavity around a tryptophan at position 840 (W840) (Owen et al. 1999; Traub et al. 1999; Fig. 4). Intriguingly, mutational analysis of W840 and residues within the hydrophobic pocket identified this site as the interface for binding to a wide range of endocytic accessory proteins including epsin, Eps15 and amphiphysin (Owen et al. 1999; Traub et al. 1999). This suggested that the proteins must share a common motif for binding to the α-ear. In fact, discrete regions for α-ear binding within these proteins had been previously described (Benmerah et al. 1996; Ramjaun et al. 1997; Slepnev et al. 1998; Chen et al. 1999). Interestingly, each of these regions contained one or multiple copies of the highly related tripeptides DPF or DPW. Using mutational analysis and peptides encoding DPF and DPW tripeptides, Owen et al. (1999) were able to demonstrate that the DPF/W motif was suf-

ficient for α-ear binding. In fact, other proteins that bound to the α-ear including AP180 and auxilin were also found to use this motif (Owen et al. 1999; Traub et al. 1999). Further details on the DPF/W motif and its interaction with the α-ear came with co-crystallization studies (Brett et al. 2002). DPW and DPF peptides show a very similar interaction with the platform-binding site as the hydrophobic pocket can accommodate both the phenyl side chain of the F and the indole ring of the W (Brett et al. 2002). The peptides bind in a nearly identical orientation and share the same integral contacts with the α-ear (Brett et al. 2002).

3.4.2 Identification of the FXDXF consensus α–ear-binding motif

Deletion analysis of amphiphysin was important to the original identification of the DPF/W α-ear-binding motif. However, further analysis of the protein revealed a minimal fragment that could bind AP-2 but contained no DPF/W-related sequences (Slepnev et al. 2000). Brett et al. (2002) co-crystallized the α-ear with a peptide encoding this fragment, revealing a distinct α-ear-binding motif, FXDXF, which uses two aromatic and one acidic residue to contact the platform domain. The binding site of the FXDXF motif overlaps with the DPF/W site and in fact the phenyl ring of the N-terminal F from the FXDXF peptide of amphiphysin is sequestered by the same hydrophobic pocket as that which sequesters the F/W in the DPF/W peptide (Brett et al. 2002). Consistent with this, mutation of W840 within the α-ear disrupts interactions with either motif (Brett et al. 2002). Thus, all endocytic accessory proteins that utilize these motifs must display some degree of competition for access to the α-ear (Fig. 4). This would contribute to a temporal and/or spatial regulation in the recruitment of these proteins to CCPs and CCVs.

One key aspect to the discovery of a new interaction motif for the α-ear of AP-2 was the ability to examine for the presence of the motif in other proteins, with the subsequent prediction that the protein should interact with AP-2. For example, following the discovery of the FXDXF motif, a sequence matching this motif was observed in the 170 kDa splice variant of synaptojanin, referred to as synaptojanin170 (McPherson et al. 1996; Ramjaun and McPherson 1996). The FXDXF motif in synaptojanin170 allows for AP-2 binding, indicating that the protein can be recruited to CCPs and CCVs through direct AP-2 interactions without the need for an additional endocytic accessory protein to mediate the binding (Brett et al. 2002; Fig. 1). Moreover, the presence of FXDXF motifs in HIP1 and Dab2 was critical to the characterization of these proteins as components of the endocytic machinery (Metzler et al. 2001; Mishra et al. 2002a, 2002b).

3.4.3 Identification of a novel WXXF-based α-ear-binding motif

In an effort to better define the molecular mechanisms responsible for clathrin-mediated membrane trafficking, we have recently performed a proteomics analysis of CCVs isolated from rat brain extracts (Blondeau et al. 2004). These studies led to the identification of a number of proteins associated with CCVs that had been previously described only as potential open-reading frames from cDNA databases

(Blondeau et al. 2004). Two such proteins were interesting in that they were highly homologous to each other but shared no homology or common domains with any previously characterized proteins. The proteins, which we named NECAP (adaptin ear binding coat associated protein) 1 and 2, were found to be highly enriched on the coats of isolated CCVs and to localize in part to CCPs at the cell surface (Ritter et al. 2003).

Interestingly, pull-down analysis revealed that a major interacting partner of the NECAPs is AP-2 with binding mediated through the α-ear. This result was surprising because the NECAPs do not contain any sequences that match the DPF/W or FXDXF consensus motifs. Through deletion analysis, we identified the sequence WVQF, found at the extreme C-terminus of both NECAP proteins, as necessary and sufficient for AP-2 binding (Ritter et al. 2003). Based on its distinct properties compared to the DPF/W and FXDXF motifs, we predicted that the WVQF sequence would bind to a different site on the α-ear. Evidence supporting this notion was provided by immunoprecipitation analysis using an antibody that recognizes an epitope in the platform domain of the α-ear and that disrupts the interaction of AP-2 with proteins using the platform binding site including amphiphysin and epsin (Ritter et al. 2003). Interestingly, this antibody strongly co-immunoprecipitates NECAP. Furthermore, a synthetic peptide from NECAP 1 that blocked NECAP interactions with the α-ear had no effect on AP-2 binding to proteins using the DPF/W and FXDXF motifs (Ritter et al. 2003). These results strongly suggest that the WVQF sequence is an example of a new peptide motif that utilizes a distinct interface on the α-ear. In fact, NMR studies reveal that WXXF-based motifs bind to a site in the sandwich domain of the α-ear (Ritter and McPherson, unpublished data) that was previously thought to bind to DPW motifs (Brett et al. 2002).

Using database searches, we were able to recognize a number of endocytic proteins that contain WVQF-like motifs. These include synaptojanin170, AAK1, auxilin 2, and stonin 2, a mammalian homologue of *Drosophila* stoned B, a presynaptic protein implicated in synaptic vesicle endocytosis (Fig. 1). Alignment of these sequences reveals a conserve core, WXXF. Interestingly, two recent studies have indicated that the WXXF-based motifs within these proteins are indeed active for AP-2 binding (Jha et al. 2003; Walther et al. 2004). Moreover, we have identified a WVXF-like sequences in a number of proteins that have not been previously demonstrated to interact with AP-2. It will be important to define the full range of amino acids that are allowable at each position within the WXXF motif in order to determine which of these proteins could potentially interact with AP-2 and to test those possibilities. It is already clear, however, that proteins that interact with AP-2 through the WXXF site do so without competing with the previously established α-ear-binding partners. This allows AP-2 to greatly increase the repertoire of endocytic accessory proteins with which it interacts.

4 Clathrin trafficking at the TGN

4.1 AP-1 and GGA adaptor complexes

Until this point, the major focus in this chapter has been on the mechanisms of clathrin-mediated budding occurring at the plasma membrane. A second major site of clathrin-mediated trafficking in cells is the TGN. Here, CCVs are generated that carry cargo from the secretory pathway to the endosomal and lysosomal systems. The molecular mechanisms that control clathrin-mediated budding and cargo selection at the TGN have not been as well defined as those at the plasma membrane. Interestingly, however, a series of recent studies have identified a consensus motif for interaction with the TGN clathrin adaptor AP-1. These studies have contributed to the identification of a series of proteins that appear to play regulatory roles in clathrin trafficking at the TGN, suggesting a molecular complexity for this event that may rival that seen at the plasma membrane.

The organization of the accessory protein network at the TGN is based around the clathrin adaptor AP-1, analogous to the role of AP-2 at the plasma membrane. AP-1 is a heterotetramer composed of two large subunits, γ- and $\beta 1$-adaptin, together with a medium sized $\mu 1$-adaptin and a small $\sigma 1$-adaptin (Fig. 5). γ- and $\beta 1$-adaptin, like α- and $\beta 2$-adaptin of the AP-2 complex, contain a globular ear domain at their C-termini connected to the main body of the adaptor complex through a flexible linker (Fig. 5).

CCV formation at the TGN also utilizes a second family of clathrin adaptors, Golgi-localized, gamma ear-containing, ADP ribosylation factor-binding protein (GGA) 1 through 3 (Robinson and Bonifacino 2001). The GGAs are multidomain monomeric proteins (Fig. 5). At their N-terminus, they have a modular VHS (Vps27p, Hrs, STAM) domain followed by a GAT (GGA and Tom) domain. Downstream of the GAT domain is a flexible linker region connected to a γ-ear like (GAE) domain, which is highly homologous to the γ-ear. Interestingly, recent data has demonstrated that the three GGA proteins have overlapping and cooperative functions in clathrin trafficking at the TGN (Gosh et al. 2003).

Like AP-2, the GGAs and AP-1 bind to clathrin, primarily through the linker regions in the GGAs and in the $\beta 1$-adaptin subunit of AP-1 (Fig. 5). Moreover, both adaptors bind to the TGN membrane and are thought to contribute to clathrin recruitment to the TGN (Robinson and Bonifacino 2001). For GGAs, TGN recruitment appears to be mediated primarily by binding of the GAT domain to the TGN localized small GTPase, ARF1 (Robinson and Bonifacino 2001; Fig. 5). AP-1 membrane recruitment is also dependent on ARF1 (Traub et al. 1993), likely through direct binding of ARF1 to the γ- and $\beta 1$-adaptin subunits (Austin et al. 2000; Fig. 5). Intriguingly, recent data has indicated that the inositol phospholipids, PtdIns(4)P also binds to AP-1 and has a critical role in membrane recruitment

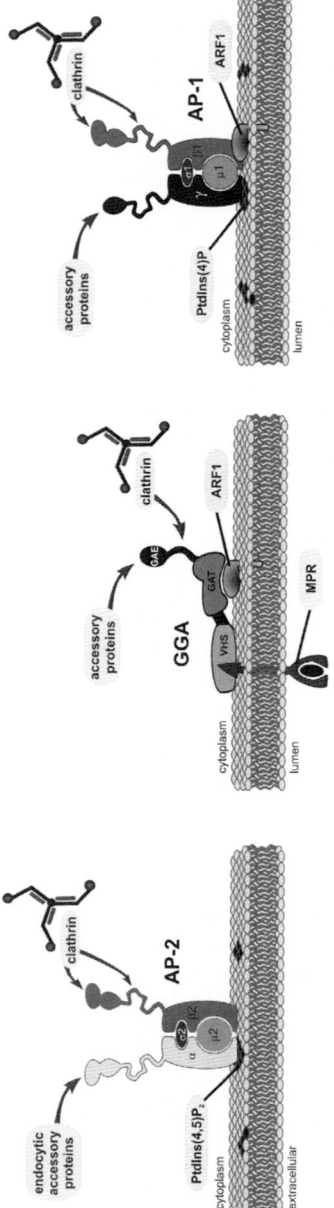

Fig. 5. Adaptor proteins in clathrin-mediated membrane budding. The structures and binding partners for clathrin adaptors used in clathrin-mediated membrane budding at the plasma membrane (AP-2), and the TGN (AP-1 and GGAs) are indicated.

of the adaptor (Wang et al. 2003). This is analogous to the role of PtdIns(4,5)P$_2$ in the recruitment of AP-2 to the plasma membrane (Fig. 5). Thus, inositol phospholipids may play an important role in determining the specific subcellular localization of the formation of clathrin bud sites on membranes.

In addition to their role in the recruitment of clathrin to the TGN, AP-1 and GGAs also function in cargo recruitment. The best characterized cargo for TGN-derived CCVs are the MPRs that bind to mannose-6-phosphate tagged lysosomal hydrolases and sort them into CCPs. MPR sorting appears to involve a specific sorting sequence in its cytoplasmic tail composed of acidic amino acid residues followed by two leucine residues, referred to as an acidic-cluster-dileucine motif (Puertollano et al. 2001; Zhu et al. 2001; Takatsu et al. 2001). The motif binds specifically to the VHS domains of the GGAs to target the MPRs to CCPs (Puertollano et al. 2001; Zhu et al. 2001; Takatsu et al. 2001; Fig. 5). It has been proposed that the GGAs may then transfer the receptor to AP-1 in a regulated manner (Doray et al. 2002). Thus, the GGAs and AP-1 appear to function in a cooperative manner to recruit clathrin to the TGN membrane and to subsequently recruit cargo into TGN-associated CCPs.

Finally, like AP-2, AP-1 and GGAs also have an important role in the recruitment of accessory proteins to sites of CCV formation. Some of the TGN accessory proteins, such as Eps15, may be overalpping with AP-2 binding proteins at the plasma membrane. However, several such as rabaptin-5, γ-synergin, p56, and enthoprotin appear to be specific to clathrin-mediated budding at the TGN.

4.2 Structural analysis of the γ-ear and GAE domain

The recruitment of accessory proteins for clathrin-mediated budding at the TGN appears to be based primarily on the γ-ear and the GAE domain. Thus, as for the α-ear, important information regarding the formation of accessory protein networks at the TGN can come from the structural analysis of these domains. The γ-ear is only 120 amino acids long (compared to 237 for the α-ear) and primary sequence comparison between the α- and γ-ears reveals only 12% sequence identity. However, X-ray crystallography studies have revealed a high degree of structural homology between the γ-ear and the sandwich domain of the α-ear (Kent et al. 2002; Nogi et al. 2002; Fig. 4). The GAE domain of the GGAs reveals an almost identical structure to the γ-ear (Lui et al. 2003; Fig. 4), which had been expected based on a high degree of sequence conservation.

Structure guided mutational analysis of the γ-ear was used to suggest a possible binding site for protein ligands. Kent and colleagues suggested a ligand-binding site composed of a shallow trough where β-sheets 4 and 5 meet (Kent et al. 2002). Nogi and colleagues proposed that ligands bind to an exposed surface composed of basic residues in β-sheets 4 and 7. In fact, NMR analysis of the γ-ear in a complex with a γ-ear-binding peptide from enthoprotin, which matches a consensus γ-ear/GAE domain-binding motif (see below), reveals that the ligand binding site is composed of residues from β-sheets 4, 5, 7, and 8, and covers an exposed face on

one side of the two-sheeted β-sandwich structure (Wasiak et al. 2003; Fig. 4). Similar results were seen for the binding site in the GAE domain of GGA1 and GGA3 (Collins et al. 2003; Miller et al. 2003; Fig. 4). Interestingly, a single mutation (leucine 762) within the ligand-binding site and the corresponding position, leucine 572 in the GAE domain, disrupts interactions with a wide variety of AP-1- and GGA-binding partners including enthoprotin, p56, γ-synergin, and Eps15 (Kent et al. 2002; Lui et al. 2003; Wasiak et al. 2003; Mills et al. 2003: Fig. 4). Thus, as for the α-ear, a variety of accessory proteins for clathrin-mediated budding at the TGN appear to use a similar interface on AP-1 and GGAs.

4.3 Enthoprotin (Clint/epsinR): identification of a novel γ-ear/GAE domain-binding consensus motif

Using a proteomics analysis of CCVs isolated from rat brain, we have recently identified a novel ENTH domain-containing protein that we named enthoprotin (Wasiak et al. 2002). Enthoprotin was independently identified and referred to as Clint (Kalthoff et al. 2002b) and epsinR (Hirst et al. 2003; Mills et al. 2003). Outside of its ENTH domain, enthoprotin has little homology with other known proteins but it does contain two clathrin-binding motifs and it binds directly to clathrin. Interestingly, enthoprotin also binds through its C-terminal region to the γ-ear and GAE domain and the protein is localized to the TGN and to membranes of the endosomal system (Wasiak et al. 2002; Kalthoff et al. 2002b; Hirst et al. 2003; Mills et al. 2003). Through its ENTH domain, enthoprotin has been demonstrated to bind to PtdIns(4)P (Hirst et al. 2003; Mills et al. 2003). Thus, an intriguing possibility is that enthoprotin binds to AP-1 with both enthoprotin and AP-1 binding simultaneously to PtdIns(4)P (Wang et al. 2003). This could increase the affinity of both AP-1 and enthoprotin for PtdIns(4)P positive TGN membranes, with the proteins working together as a stable recruitment platform for the formation of CCPs. As both AP-1 and enthoprotin also bind clathrin, they could cooperate in the recruitment of clathrin to the membrane. Finally, enthoprotin can contribute to the assembly of clathrin and could, thus, further aid in CCP formation (Wasiak et al. 2003; Legendre-Guillimen et al. 2004). Interestingly, overexpression of enthoprotin blocks the packaging of cathepsin D into CCVs and its trafficking from the TGN to endosomes (Mills et al. 2003). More recently, analysis of cells in which enthoprotin has been knocked down by RNAi has revealed a role for enthoprotin in the retrograde pathway from endosomes to the Golgi (Saint-Pol et al. 2004).

Duncan et al. (2003) recently performed a 2-hybrid screen using Gga2p and the γ-adaptin subunit of yeast AP-1 and isolated a protein, Ent3p, which appears to be the yeast homologue of enthoprotin. Like enthoprotin, Ent3p localizes in part to the Golgi and functions in Golgi/endosomal trafficking (Duncan et al. 2003). Interestingly, Ent3p has also been shown to bind via its ENTH domain to PtdIns(3,5)P2 and to mediate protein sorting into multivesicular bodies (Friant et al. 2003). In particular, Ent3p appears to cooperate with its homologue, Ent5p, and the FYVE domain-containing protein, Vps27p, in sorting ubiquitinated cargo from

the biosynthetic and endocytic pathway into invaginating vesicles of the multivesicular body (Eugster et al. 2004). Thus, enthoprotin/Ent3p appears to function broadly in clathrin-/AP-1-mediated trafficking.

Characterization of enthoprotin and Ent3p led to the identification of the short peptide sequences DFGDW and DDGFGDF, respectively, as critical sequences for γ-ear and GAE domain binding (Duncan et al. 2003; Mills et al. 2003; Wasiak et al. 2003). These sequences are similar to those found in a variety of AP-1- and GGA-binding proteins including γ-synergin (Page et al. 1999). Based on mutational analysis and alignments amongst proteins known to utilize this site, several consensus-binding motifs have been proposed including (D/E)-(G/A)$_{(0-1)}$-F-(G/A)-(D/E)-Ø (Duncan and Payne 2003). In two recent studies, Mattera et al. (2003) and Wasiak et al. (2003) have systematically defined the core of the motif through a combination of NMR, *in vitro* binding assays, and screens of phage-displayed recombinatorial peptide libraries. The resulting consensus core was found to be Ø[GA][PDE][ØLM] (Fig. 4). Intriguingly, this motif was then used in database searches to identify potentially new AP-1 and GGA-binding partners. One identified protein was a novel protein, which was named aftiphilin (Mattera et al. 2003). The interaction of aftiphilin with AP-1 and GGA was confirmed *in vitro* although its role in trafficking at the TGN and endosomes remains unclear. These results underline the value of consensus sequences as the identification of the protein machinery is crucial for the understanding of the mechanisms of accessory protein recruitment for clathrin-mediated transport at the TGN.

The function of many of the accessory proteins for clathrin-mediated trafficking at the TGN remains poorly defined. However, as evidenced by the identification of aftiphilin, it appears likely that the consensus γ-ear/GAE domain-binding motif will lead to new accessory proteins that regulate this important trafficking process. As for clathrin-mediated budding at the cell surface, it is likely that the function of these proteins will be intimately linked to that of the TGN clathrin adaptors.

5 Concluding remarks

As recently as 10 years ago, the formation of CCPs and CCVs at the cell surface was thought to be a relatively simple event. AP-2 recruited clathrin to the membrane and stimulated its formation into CCPs. AP-2 also bound to endocytic cargo causing it to concentrate in the nascent buds. The assembly of clathrin was sufficient to drive the membrane curvature that eventually led to a deeply invaginated CCP. Dynamin was then able to cause the liberation of the CCP from the membrane leading to a mature CCV.

It is now clear that the molecular machinery that controls this important cellular process is far more complicated involving a wide range of proteins with diverse functions. In retrospect, this seems obvious given the complexity of the event and the important physiological consequences of CME. New insights into the machin-

ery for clathrin-mediated budding at the TGN are likely to reveal a similar degree of molecular complexity.

Acknowledgments

The authors wish to thank The Montreal Neurological Institute, McGill University, and The Canadian Institutes for Health Research for support.

References

Aguilar RC, Wendland B (2003) Ubiquitin: not just for proteasomes anymore. Curr Opin Cell Biol 15:184-190

Aguilar RC, Watson HA, Wendland B (2003) The yeast Epsin Ent1 is recruited to membranes through multiple independent interactions. J Biol Chem 278:10737-10743

Ahle S, Ungewickell E (1986). Purification and properties of a new clathrin assembly protein. EMBO J 5:3143-3149

Austin C, Hinners I, Tooze SA (2000) Direct and GTP-dependent interaction of ADP-ribosylation factor 1 with clathrin adaptor protein AP-1 on immature secretory granules. J Biol Chem 275:21862-21869

Benmerah A, Begue B, Dautry-Varsat A, Cerf-Bensussan N (1996) The ear of alpha-adaptin interacts with the COOH-terminal domain of the Eps 15 protein. J Biol Chem 271:12111-12116

Blondeau F, Ritter B, Allaire PD, Wasiak S, Girard M, Hussain NK, Angers A, Legendre-Guillemin V, Roy L, Boismenu D, Kearney RE, Bell AW, Bergeron JJ, McPherson PS (2004) Tandem MS analysis of brain clathrin-coated vesicles reveals their critical involvement in synaptic vesicle recycling. Proc Natl Acad Sci USA 101:3833-3838

Brett TJ, Traub LM, Fremont DH (2002) Accessory protein recruitment motifs in clathrin-mediated endocytosis. Structure (Camb) 10:797-809

Brodsky FM, Chen CY, Knuehl C, Towler MC, Wakeham DE (2001) Biological basket weaving: formation and function of clathrin-coated vesicles. Annu Rev Cell Dev Biol 17:517-568

Bonifacino JS, Lippincott-Schwartz J (2003) Coat proteins: shaping membrane transport. Nat Rev Mol Cell Biol 4:409-414

Carroll RC, Beattie EC, von Zastrow M, Malenka RC (2001) Role of AMPA receptor endocytosis in synaptic plasticity. Nat Rev Neurosci 2:315-324

Chen H, Fre S, Slepnev VI, Capua MR, Takei K, Butler MH, Di Fiore PP, De Camilli P (1998) Epsin is an EH-domain-binding protein implicated in clathrin-mediated endocytosis. Nature 394:793-797

Claing A, Laporte SA, Caron MG, Lefkowitz RJ (2002) Endocytosis of G protein-coupled receptors: roles of G protein-coupled receptor kinases and beta-arrestin proteins. Prog Neurobiol 66:61-79

Collins BM, McCoy AJ, Kent HM, Evans PR, Owen DJ (2002) Molecular architecture and functional model of the endocytic AP2 complex. Cell 109:523–535

Collins BM, Praefcke GJ, Robinson MS, Owen DJ (2003) Structural basis for binding of accessory proteins by the appendage domain of GGAs. Nat Struct Biol 10:607-613

Conner SD, Schmid SL (2002) Identification of an adaptor-associated kinase, AAK1, as a regulator of clathrin-mediated endocytosis. J Cell Biol 156:921-929

Conner SD, Schmid SL (2003a) Regulated portals of entry into the cell. Nature 422:37-44

Conner SD, Schmid SL (2003b) Differential requirements for AP-2 in clathrin-mediated endocytosis. J Cell Biol 162:773-779

David C, McPherson PS, Mundigl O, de Camilli P (1996) A role of amphiphysin in synaptic vesicle endocytosis suggested by its binding to dynamin in nerve terminals. Proc Natl Acad Sci USA 93:331-335

Doray B, Ghosh P, Griffith J, Geuze HJ, Kornfeld S (2002) Cooperation of GGAs and AP-1 in packaging MPRs at the trans-Golgi network. Science 297:1700-1703

Drake MT, Traub LM (2001) Interaction of two structurally distinct sequence types with the clathrin terminal domain beta-propeller. J Biol Chem 276:28700-28709

Duncan MC, Payne GS (2003) ENTH/ANTH domains expand to the Golgi. Trends Cell Biol 13:211-215

Duncan MC, Costaguta G, Payne GS (2003)Yeast epsin-related proteins required for Golgi-endosome traffic define a gamma-adaptin ear-binding motif. Nat Cell Biol 5:77-81

Eugster A, Pecheur EI, Michel F, Winsor B, Letourneur F, Friant S (2004) Ent5p is required with Ent3p and Vps27p for ubiquitin-dependent protein sorting into the multivesicular body. Mol Biol Cell, in press

Ford MG, Pearse BM, Higgins MK, Vallis Y, Owen DJ, Gibson A, Hopkins CR, Evans PR, McMahon HT (2001) Simultaneous binding of PtdIns(4,5)P2 and clathrin by AP180 in the nucleation of clathrin lattices on membranes. Science 291:1051-1055

Ford MG, Mills IG, Peter BJ, Vallis Y, Praefcke GJ, Evans PR, McMahon HT (2002) Curvature of clathrin-coated pits driven by epsin. Nature 419:361-366

Ferguson SS, Downey WE 3rd, Colapietro AM, Barak LS, Menard L, Caron MG (1996) Role of beta-arrestin in mediating agonist-promoted G protein-coupled receptor internalization. Science 271:363-366

Friant S, Pecheur EI, Eugster A, Michel F, Lefkir Y, Nourrisson D, Letourneur F (2003) Ent3p Is a PtdIns(3,5)P2 effector required for protein sorting to the multivesicular body. Dev Cell 5:499-511.

Gaidarov I, Keen JH (1999) Phosphoinositide-AP-2 interactions required for targeting to plasma membrane clathrin-coated pits. J Cell Biol 146:755-764

Goodman OB Jr, Krupnick JG, Santini F, Gurevich VV, Penn RB, Gagnon AW, Keen JH, Benovic JL (1996) Beta-arrestin acts as a clathrin adaptor in endocytosis of the beta2-adrenergic receptor. Nature 383:447-450

Ghosh P, Griffith J, Geuze HJ, Kornfeld S (2003) Mammalian GGAs act together to sort mannose 6-phosphate receptors. J Cell Biol 163:755-766

Hao W, Luo Z, Zheng L, Prasad K, Lafer EM (1999) AP180 and AP-2 interact directly in a complex that cooperatively assembles clathrin. J Biol Chem 274:22785-22794

He G, Gupta S, Yi M, Michaely P, Hobbs HH, Cohen JC (2002) ARH is a modular adaptor protein that interacts with the LDL receptor, clathrin, and AP-2. J Biol Chem 277:44044-44049

Hicke L, Dunn R (2003) Regulation of membrane protein transport by ubiquitin and ubiquitin-binding proteins. Annu Rev Cell Dev Biol 19:141-172

Hinners I, Tooze SA (2003) Changing directions: clathrin-mediated transport between the Golgi and endosomes. J Cell Sci 116:763-771

Hinrichsen L, Harborth J, Andrees L, Weber K, Ungewickell EJ (2003) Effect of clathrin heavy chain- and alpha -adaptin specific small interfering RNAs on endocytic accessory proteins and receptor trafficking in HeLa cells. J Biol Chem 278:45160-45170

Hinshaw JE, Schmid SL (1995) Dynamin self-assembles into rings suggesting a mechanism for coated vesicle budding. Nature 374:190-192

Hirst J, Motley A, Harasaki K, Peak Chew SY, Robinson MS (2003) EpsinR: an ENTH domain-containing protein that interacts with AP-1. Mol Biol Cell 14:625-641

Huang F, Khvorova A, Marshall W, Sorkin A (2004) Analysis of Clathrin-mediated Endocytosis of epidermal growth factor receptor by RNA interference. J Biol Chem 279:16657-16661

Hussain NK, Yamabhai M, Ramjaun AR, Guy AM, Baranes D, O'Bryan JP, Der CJ, Kay BK, McPherson PS (1999) Splice variants of intersectin are components of the endocytic machinery in neurons and nonneuronal cells. J Biol Chem 274:15671-15677

Hyman J, Chen H, Di Fiore PP, De Camilli P, Brunger AT (2000) Epsin 1 undergoes nucleocytosolic shuttling and its eps15 interactor NH(2)-terminal homology (ENTH) domain, structurally similar to Armadillo and HEAT repeats, interacts with the transcription factor promyelocytic leukemia Zn(2)+ finger protein (PLZF). J Cell Biol 149:537-546

Itoh T, Koshiba S, Kigawa T, Kikuchi A, Yokoyama S, Takenawa T (2001) Role of the ENTH domain in phosphatidylinositol-4,5-bisphosphate binding and endocytosis. Science 291:1047-1051

Iversen TG, Skretting G, van Deurs B, Sandvig K (2003) Clathrin-coated pits with long, dynamin-wrapped necks upon expression of a clathrin antisense RNA. Proc Natl Acad Sci USA 100:5175-5180

Jha A, Agostinelli NR, Mishra SK, Keyel PA, Hawryluk MJ, Traub LM (2003) A novel AP-2 adaptor interaction motif initially identified in the long-splice isoform of synaptojanin 1, SJ170a. J Biol Chem [Epub ahead of print]

Kalthoff C, Alves J, Urbanke C, Knorr R, Ungewickell EJ (2002a) Unusual structural organization of the endocytic proteins AP180 and epsin 1. J Biol Chem 277:8209-8216

Kalthoff C, Groos S, Kohl R, Mahrhold S, Ungewickell EJ (2002b) Clint: A novel clathrin-binding ENTH-domain protein at the Golgi. Mol Biol Cell 13:4060-4073

Kay BK, Yamabhai M, Wendland B, Emr SD (1999) Identification of a novel domain shared by putative components of the endocytic and cytoskeletal machinery. Protein Sci 8:435-438

Kirchhausen T, Bonifacino JS, Riezman H (1997) Linking cargo to vesicle formation: receptor tail interactions with coat proteins. Curr Opin Cell Biol 9:488-495

Keen JH, Willingham MC, Pastan IH (1979) Clathrin-coated vesicles: isolation, dissociation and factor-dependent reassociation of clathrin baskets. Cell 16:303-312

Kent HM, McMahon HT, Evans PR, Benmerah A, Owen DJ (2002) Gamma-adaptin appendage domain: structure and binding site for Eps15 and gamma-synergin. Structure (Camb) 10:1139-1148

Krauss M, Kinuta M, Wenk MR, De Camilli P, Takei K, Haucke V (2003) ARF6 stimulates clathrin/AP-2 recruitment to synaptic membranes by activating phosphatidylinositol phosphate kinase type Igamma. J Cell Biol 162:113-124

Laporte SA, Oakley RH, Holt JA, Barak LS, Caron MG (2000) The interaction of beta-arrestin with the AP-2 adaptor is required for the clustering of beta 2-adrenergic receptor into clathrin-coated pits. J Biol Chem 275:23120-23126

Legendre-Guillemin V, Metzler M, Charbonneau M, Gan L, Chopra V, Philie J, Hayden MR, McPherson PS (2002) HIP1 and HIP12 display differential binding to F-actin, AP2, and clathrin. Identification of a novel interaction with clathrin light chain. J Biol Chem 277:19897-19904

Legendre-Guillemin V, Wasiak S, Hussain NK, Angers A, McPherson PS (2004) ENTH/ANTH proteins and clathrin-mediated membrane budding. J Cell Sci 117:9-18

Lui WW, Collins BM, Hirst J, Motley A, Millar C, Schu P, Owen DJ, Robinson MS (2003) Binding partners for the COOH-terminal appendage domains of the GGAs and gamma-adaptin. Mol Biol Cell 14:2385-2398

Mattera R, Ritter B, Sidhu SS, McPherson PS, Bonifacino JS (2003) Definition of the consensus motif recognized by gamma -adaptin ear domains. J Biol Chem [Epub ahead of print]

McPherson PS, Takei K, Schmid SL, De Camilli P (1994) p145, a major Grb2-binding protein in brain, is co-localized with dynamin in nerve terminals where it undergoes activity-dependent dephosphorylation. J Biol Chem 269:30132-30139

McPherson PS, Garcia EP, Slepnev VI, David C, Zhang X, Grabs D, Sossin WS, Bauerfeind R, Nemoto Y, De Camilli P (1996) A presynaptic inositol-5-phosphatase. Nature 379:353-357

McPherson PS, Kay BK, Hussain NK (2001) Signaling on the endocytic pathway. Traffic 2:375-384

Metzler M, Legendre-Guillemin V, Gan L, Chopra V, Kwok A, McPherson PS, Hayden MR (2001) HIP1 functions in clathrin-mediated endocytosis through binding to clathrin and adaptor protein 2. J Biol Chem 276:39271-39276

Metzler M, Li B, Gan L, Georgiou J, Gutekunst CA, Wang Y, Torre E, Devon RS, Oh R, Legendre-Guillemin V, Rich M, Alvarez C, Gertsenstein M, McPherson PS, Nagy A, Wang YT, Roder JC, Raymond LA, Hayden MR (2003) Disruption of the endocytic protein HIP1 results in neurological deficits and decreased AMPA receptor trafficking. EMBO J 22:3254-3266

Miller GJ, Mattera R, Bonifacino JS, Hurley JH (2003) Recognition of accessory protein motifs by the gamma-adaptin ear domain of GGA3. Nat Struct Biol 10:599-606

Mills IG, Praefcke GJ, Vallis Y, Peter BJ, Olesen LE, Gallop JL, Butler PJ, Evans PR, McMahon HT (2003) EpsinR: an AP1/clathrin interacting protein involved in vesicle trafficking. J Cell Biol 160:213-222

Mishra SK, Agostinelli NR, Brett TJ, Mizukami I, Ross TS, Traub LM (2001) Clathrin- and AP-2-binding sites in HIP1 uncover a general assembly role for endocytic accessory proteins. J Biol Chem 276:46230-46236

Mishra SK, Keyel PA, Hawryluk MJ, Agostinelli NR, Watkins SC, Traub LM (2002a) Disabled-2 exhibits the properties of a cargo-selective endocytic clathrin adaptor. EMBO J 21:4915-4926

Mishra SK, Watkins SC, Traub LM (2002b) The autosomal recessive hypercholesterolemia (ARH) protein interfaces directly with the clathrin-coat machinery. Proc Natl Acad Sci USA 99:16099-17104

Morgan JR, Prasad K, Hao W, Augustine GJ, Lafer EM (2000) A conserved clathrin assembly motif essential for synaptic vesicle endocytosis. J Neurosci 20:8667-8676

Morris SM, Cooper JA (2001) Disabled-2 colocalizes with the LDLR in clathrin-coated pits and interacts with AP-2. Traffic 2:111-123

Morris SM, Tallquist MD, Rock CO, Cooper JA (2002) Dual roles for the Dab2 adaptor protein in embryonic development and kidney transport. EMBO J 21:1555-1564

Moskowitz HS, Heuser J, McGraw TE, Ryan TA (2003) Targeted chemical disruption of clathrin function in living cells. Mol Biol Cell 14:4437-4447

Motley A, Bright NA, Seaman MN, Robinson MS (2003) Clathrin-mediated endocytosis in AP-2-depleted cells. J Cell Biol 162:909-918

Murthy VN, De Camilli P (2003) Cell biology of the presynaptic terminal. Annu Rev Neurosci 26:701-728

Nakashima, S, Morinaka K, Koyama S, Ikeda M, Kishida M, Okawa K, Iwamatsu A, Kishida S, Kikuchi A (1999) Small G protein Ral and its downstream molecules regulate endocytosis of EGF and insulin receptors. EMBO J 18:3629-3642

Nesterov A, Carter RE, Sorkina T, Gill GN, Sorkin A (1999) Inhibition of the receptor-binding function of clathrin adaptor protein AP-2 by dominant-negative mutant mu2 subunit and its effects on endocytosis. EMBO J 18:2489-2499

Nogi T, Shiba Y, Kawasaki M, Shiba T, Matsugaki N, Igarashi N, Suzuki M, Kato R, Takatsu H, Nakayama K, Wakatsuki S (2002) Structural basis for the accessory protein recruitment by the gamma-adaptin ear domain. Nat Struct Biol 9:527-531

Nonet ML, Holgado AM, Brewer F, Serpe CJ, Norbeck BA, Holleran J, Wei L, Hartwieg E, Jorgensen EM, Alfonso A (1999) UNC-11, a *Caenorhabditis elegans* AP180 homologue, regulates the size and protein composition of synaptic vesicles. Mol Biol Cell 10:2343-2360

Ohno H, Stewart J, Fournier MC, Bosshart H, Rhee I, Miyatake S, Saito T, Gallusser A, Kirchhausen T, Bonifacino JS (1995) Interaction of tyrosine-based sorting signals with clathrin-associated proteins. Science 269:1872-1875

Oleinikov AV, Zhao J, Makker SP (2000) Cytosolic adaptor protein Dab2 is an intracellular ligand of endocytic receptor gp600/megalin. Biochem J 347:613-621

Olusanya O, Andrews PD, Swedlow JR, Smythe E (2001) Phosphorylation of threonine 156 of the mu2 subunit of the AP2 complex is essential for endocytosis *in vitro* and *in vivo*. Curr Biol 11:896-900

Owen DJ, Vallis Y, Noble ME, Hunter JB, Dafforn TR, Evans PR, McMahon HT (1999) A structural explanation for the binding of multiple ligands by the alpha-adaptin appendage domain. Cell 97:805-815

Owen DJ, Vallis Y, Pearse BM, McMahon HT, Evans PR (2000) The structure and function of the beta 2-adaptin appendage domain. EMBO J 19:4216-4227

Page LJ, Sowerby PJ, Lui WW, Robinson MS (1999) Gamma-synergin: an EH domain-containing protein that interacts with gamma-adaptin. J Cell Biol 146:993-1004

Polo S, Sigismund S, Faretta M, Guidi M, Capua MR, Bossi G, Chen H, De Camilli P, Di Fiore PP (2002) A single motif responsible for ubiquitin recognition and monoubiquitination in endocytic proteins. Nature 416:451-455

Puertollano R, Aguilar RC, Gorshkova I, Crouch RJ, Bonifacino JS (2001) Sorting of mannose 6-phosphate receptors mediated by the GGAs. Science 292:1712-1716

Puertollano R, van der Wel NN, Greene LE, Eisenberg E, Peters PJ, Bonifacino JS (2003) Morphology and dynamics of clathrin/GGA1-coated carriers budding from the trans-Golgi network. Mol Biol Cell 14:1545-1557

Ramjaun AR, McPherson PS (1996) Tissue-specific alternative splicing generates two synaptojanin isoforms with differential membrane binding properties. J Biol Chem 271:24856-24861

Ramjaun AR, Micheva KD, Bouchelet I, McPherson PS (1997) Identification and characterization of a nerve terminal-enriched amphiphysin isoform. J Biol Chem 272:16700-16706

Ritter B, Philie J, Girard M, Tung EC, Blondeau F, McPherson PS (2003) Identification of a family of endocytic proteins that define a new alpha-adaptin ear-binding motif. EMBO Rep 4:1089-1095

Robinson MS, Bonifacino JS (2001) Adaptor-related proteins. Curr Opin Cell Biol 13:444-453

Robinson MS (2004) Adaptable adaptors for coated vesicles. Trends Cell Biol 14:167-174

Rosenthal JA, Chen H, Slepnev VI, Pellegrini L, Salcini AE, Di Fiore PP, De Camilli P (1999). The epsins define a family of proteins that interact with components of the clathrin coat and contain a new protein module. J Biol Chem 274:33959-33965

Saint-Pol A, Yelamos B, Amessou M, Mills IG, Dugast M, Tenza D, Schu P, Antony C, McMahon HT, Lamaze C, Johannes L (2004) Clathrin adaptor epsinR is required for retrograde sorting on early endosomal membranes. Dev Cell 6:525-538

Schlossman DM, Schmid SL, Braell WA, Rothman JE (1984) An enzyme that removes clathrin coats: purification of an uncoating ATPase. J Cell Biol 99:723-733

Schmid SL (1997) Clathrin-coated vesicle formation and protein sorting: an integrated process. Annu Rev Biochem 66:11-48

Sever S, Damke H, Schmid SL (2000) Garrotes, springs, ratchets, and whips: putting dynamin models to the test. Traffic 1:385-392

Slepnev VI, Ochoa GC, Butler MH, Grabs D, De Camilli P (1998) Role of phosphorylation in regulation of the assembly of endocytic coat complexes. Science 281:821-824

Slepnev VI, Ochoa GC, Butler MH, De Camilli P (2000) Tandem arrangement of the clathrin and AP-2 binding domains in amphiphysin 1 and disruption of clathrin coat function by amphiphysin fragments comprising these sites. J Biol Chem 275:17583-17589

Stahelin RV, Long F, Peter BJ, Murray D, De Camilli P, McMahon HT, Cho W (2003) Contrasting membrane interaction mechanisms of AP180 ANTH and Epsin ENTH domains. J Biol Chem 278:28993-28999

Takatsu H, Katoh Y, Shiba Y, Nakayama K (2001) Golgi-localizing, gamma-adaptin ear homology domain, ADP-ribosylation factor-binding (GGA) proteins interact with acidic dileucine sequences within the cytoplasmic domains of sorting receptors through their Vps27p/Hrs/STAM (VHS) domains. J Biol Chem 276:28541-28545

Takei K, McPherson PS, Schmid SL, De Camilli P (1995) Tubular membrane invaginations coated by dynamin rings are induced by GTP-gamma S in nerve terminals. Nature 374:186-190

Traub LM, Ostrom JA, Kornfeld S (1993) Biochemical dissection of AP-1 recruitment onto Golgi membranes. J Cell Biol 123:561-573

Traub LM, Downs MA, Westrich JL, Fremont DH (1999) Crystal structure of the alpha appendage of AP-2 reveals a recruitment platform for clathrin-coat assembly. Proc Natl Acad Sci USA 96:8907-8912

Traub LM (2003) Sorting it out: AP-2 and alternate clathrin adaptors in endocytic cargo selection. J Cell Biol 163:203-208

Ungewickell E, Ungewickell H, Holstein SE, Lindner R, Prasad K, Barouch W, Martin B, Greene LE, Eisenberg E (1995) Role of auxilin in uncoating clathrin-coated vesicles. Nature 378:632-635

Walther K, Diril MK, Jung N, Haucke V (2004) Functional dissection of the interactions of stonin 2 with the adaptor complex AP-2 and synaptotagmin. Proc Natl Acad Sci USA (In press)

Waelter S, Scherzinger E, Hasenbank R, Nordhoff E, Lurz R, Goehler H, Gauss C, Satha-sivam K, Bates GP, Lehrach H, Wanker EE (2001) The huntingtin interacting protein HIP1 is a clathrin and alpha-adaptin-binding protein involved in receptor-mediated endocytosis. Hum Mol Genet 10:1807-1817

Wang YJ, Wang J, Sun HQ, Martinez M, Sun YX, Macia E, Kirchhausen T, Albanesi JP, Roth MG, Yin HL (2003) Phosphatidylinositol 4 phosphate regulates targeting of clathrin adaptor AP-1 complexes to the Golgi. Cell 114:299-310

Warren RA, Green FA, Stenberg PE, Enns CA (1998) Distinct saturable pathways for the endocytosis of different tyrosine motifs. J Biol Chem 273:17056-17063

Wasiak S, Legendre-Guillemin V, Puertollano R, Blondeau F, Girard M, de Heuvel E, Bo-ismenu D, Bell AW, Bonifacino JS, McPherson PS (2002) Enthoprotin: a novel clathrin-associated protein identified through subcellular proteomics. J Cell Biol 158:855-862

Wasiak S, Denisov AY, Han Z, Leventis PA, de Heuvel E, Boulianne GL, Kay BK, Ge-hring K, McPherson PS (2003) Characterization of a gamma-adaptin ear-binding motif in enthoprotin. FEBS Lett 555:437-442

Wettey FR, Hawkins SF, Stewart A, Luzio JP, Howard JC, Jackson AP (2002) Controlled elimination of clathrin heavy-chain expression in DT40 lymphocytes. Science 297:1521-1515

Wu X, Zhao X, Baylor L, Kaushal S, Eisenberg E, Greene LE (2001) Clathrin exchange during clathrin-mediated endocytosis. J. Cell Biol 155:291-300

Zhu Y, Doray B, Poussu A, Lehto VP, Kornfeld S (2001) Binding of GGA2 to the ly-sosomal enzyme sorting motif of the mannose 6-phosphate receptor. Science 292:1716-1718

McPherson, Peter S.
Montreal Neurological Institute, McGill University, 3801 University, Montreal, QC, H3A 2B4 Canada
peter.mcpherson@mcgill.ca

Ritter, Brigitte
Montreal Neurological Institute, McGill University, 3801 University, Montreal, QC, H3A 2B4 Canada

Regulating membrane curvature

Joost C. M. Holthuis

Abstract

Membranes of living cells are subject to a variety of shape changes. Membrane deformation is essential for cell division, locomotion, organelle biogenesis, and creation of the high-curvature vesicles and tubules that mediate cargo transport along the endo- and exocytic pathways. Yet membrane bilayers are stabilized against deformations by a powerful hydrophobic effect. Moreover, the various membrane-bound organelles of a cell are composed of different mixtures of lipids, with each lipid species contributing differently to membrane elasticity and shape. So how do cells adopt the desired radii of curvature in their bilayers? A long-standing concept is that membrane curvature can be induced by assembly of a protein coat onto one side of the bilayer. Recent studies indicate that lipids too can serve as the molecular workers in accomplishing the structural changes needed to deform cellular membranes.

1 Introduction

Membrane deformation is a universal feature of living cells. It is crucial for cell division, organelle biogenesis, sustaining endo- and exocytosis, and for the release of enveloped viruses from infected cells. In many cases, (e.g. cell division, vesicle formation), but not always (e.g. formation of plasma membrane protrusions), membrane budding culminates in the physical separation, or fission, of the bud from the donor membrane. Fission requires strong membrane bending and the formation of a highly constricted neck where the opposing membranes eventually make contact and fuse. However, curling up a planar membrane, or flattening a curved one, is energetically unfavourable and does not occur spontaneously. So how does a cell induce the desired curvature in its bilayers?

It appears that cytoskeletal elements, apart from providing the network over which membrane traffic flows, are capable of driving membrane tubulation by pushing a developing bud into the extracellular environment or pulling it into the cytoplasm (Lippincott-Schwartz et al. 2000). Another classical paradigm for the acquisition of membrane curvature is the polymerisation of cytosolic proteins into a coat scaffold on the membrane surface (Kirchhausen 2000). Here, budding would be a consequence of the intrinsic curvature in the structure of the assembled coat. Consistent with this notion, *in vitro* studies with protein-free liposomes show that coat assembly is fully sufficient to deform lipid bilayers into buds and tubules,

Topics in Current Genetics, Vol. 10
S. Keränen, J. Jäntti (Eds.): Regulatory Mechanisms of Intracellular Membrane Transport
DOI 10.1007/b98566 / Published online: 25 May 2004
© Springer-Verlag Berlin Heidelberg 2004

and, in some cases, even to pinch off bilayer vesicles (Matsuoka et al. 1998; Takei et al. 1998; Bremser et al. 1999). Whether coat assembly or cytoskeletal mechanisms are also sufficient to pay the energy costs for membrane remodelling *in vivo* has remained an open issue. The matter is complicated by the fact that cellular membranes contain mixtures of 100 or more different lipids that are often asymmetrically distributed across the bilayer (Zachowski 1993; Holthuis et al. 2003), a situation that is difficult, if not impossible, to reproduce with model membranes. According to a recent estimate, the rigidity of the clathrin coat does not supersede the mechanical bending resistance of biological membranes (Nossal 2001). This would indicate that clathrin polymerisation by itself may stabilize budding vesicles, but unlikely provides a force sufficient to drive vesicle biogenesis *in vivo*.

Indeed, a growing body of evidence indicates that for remodelling cellular membranes, mechanisms applying external forces on the membrane (coat assembly, cytoskeletal mechanisms) often operate in conjunction with mechanisms exerting intrinsic ones, i.e. that cause a physical perturbation of the lipid bilayer. Several proteins have now been identified that function in concert with classical coat proteins and likely promote vesicle budding by inserting an amphipathic helix into one monolayer of the membrane, causing an intrinsic curvature by pushing the lipid head groups apart. Moreover, cells contain ATP-driven lipid translocases that facilitate or induce membrane curvature by controlling the lipid distribution across the bilayer. Enzymatically-induced changes in the geometry of the lipid constituents has emerged as yet another mechanism to modify membrane curvature directly. Finally, model membrane studies show that lipid phase separation and microdomain formation can cause membrane budding and fission events, raising the possibility that cells exploit lipid immiscibility to drive membrane vesiculation.

This review focuses on recent insights into the mechanisms by which cells control the radii of curvature in their bilayers. Particular attention is paid on how lipid composition affects membrane folding, how different membrane-deforming mechanisms may be combined to accomplish the dynamic shape transformations that are necessary to sustain membrane traffic in live cells, and how lipids and proteins cooperate to lower the energy costs of these transformations.

2 Topological considerations

Cellular membranes bud into two principal topological spaces, the cytoplasm and the extracytoplasmic space, with the latter comprising both the lumen of intracellular organelles and the extracellular environment (Fig. 1). Budding into the cytoplasm, or inward budding, is required for generating the transport vesicles and tubules that carry cargo between the different subcellular organelles (Lippincott-Schwartz et al. 2000), but also for creating the transverse tubular invaginations of the plasma membrane in striated muscle cells (Ishikawa 1968). Budding into the extracytoplasmic space, or outward budding, occurs during cell division, during apoptosis when microvesicles and blebs are shed from the plasma membrane

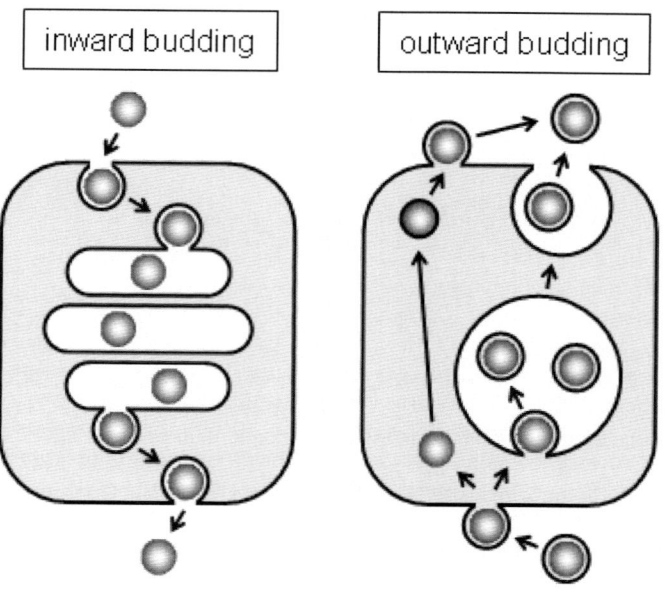

Fig. 1. Orientation of membrane budding events. Inward budding allows vesicular transport from the plasma membrane to intracellular compartments (endocytosis) and from intracellular compartments to the plasma membrane (exocytosis), with transport vesicles containing material taken up from the extracellular environment or the lumen of intracellular compartments, respectively. Outward budding allows the release of vesicles into the extracellular environment or into the lumen of intracellular compartments. These vesicles contain material taken up from the cytosol. Outward budding occurs during the shedding of milk-fat droplets or enveloped viruses form the cell surface, or during formation of the internal vesicles of multivesicular endosomes. Note that budding not necessarily culminates in fission and formation of a vesicle; budding also mediates formation of filopodia, microvilli and other cell surface projections as well as the biogenesis of T-tubules in muscle cells.

(Coleman et al. 2001), during the release of milk-fat globules from mammary gland epithelial cells (Deyrup-Olsen and Luchtel 1998), and during formation of filopodia, microvilli and other cell surface projections. Outward budding also concerns the assembly of enveloped viruses in infected cells (Pornillos et al. 2002) as well as the generation of the internal vesicles in multivesicular endosomes that, when released into the extracellular environment, are called exosomes (Katzmann et al. 2002). Exosomes contain MHC class II molecules and T cell co-stimulatory factors, and as such are potent modulators of the immune system (Thery et al. 2002). The variation in shape changes imposed on cellular membranes provides a strong indication that cells are equipped with more than one type of budding machinery.

3 Extrinsic forces affecting membrane curvature

3.1 Cytoskeleton

Cytoskeletal elements seem to play a prominent role in the regulation of membrane curvature. A direct participation of the microtubule network in the formation and movement of tubular transport intermediates as well as in the tubular dynamics of the endoplasmic reticulum (ER) has been well established (Dabora and Sheetz 1988; Vale and Hotani 1988; Lippincott-Schwartz et al. 2000). The interactions of membranes with microtubules are mediated by several classes of proteins, notably the motor proteins of the dynein and kinesin families (Hirokawa 1998). While moving along a preformed microtubule track, motor proteins would exert forces on the membrane to which they are attached, resulting in the formation of membrane tubes and tubular networks (Fig. 2 A). Consistent with this view, depletion of kinesin heavy chain or depolymerization of microtubules causes a retraction of the ER toward the cell centre and abolishes formation of new tubes (Lee et al. 1989; Feiguin et al. 1994). Importantly, recent work has demonstrated that synthetic giant liposomes, kinesin-coated beads, immobilized microtubules and ATP provide a minimal system for generating tubular structures that resemble tubes formed from cellular membranes (Roux et al. 2002; Koster et al. 2003). The forces required for pulling a tube are higher than can be generated by individual motor proteins, indicating that multiple motor proteins must work together. Membrane tube formation is sensitive to lipid composition; when the bending rigidity of the liposome is increased by the addition of cholesterol, the threshold concentration of motor proteins for tubulation is doubled (Koster et al. 2003).

Polymerisation forces generated by the cytoskeleton itself provide an alternative to motor proteins in affecting membrane curvature (Fig. 2 B). Animal cells change shape and move by polymerising actin at the leading edge of lamelipodia and filipodia (Lauffenburger and Horwitz 1996; Mitchison and Cramer 1996). Model systems using synthetic liposomes encapsulating actin or tubulin indicate that self-assembly of these proteins in principle can generate sufficient mechanical force to deform a lipid bilayer. Hence, actin polymerisation causes spherical liposomes to transform into stable disk or dumbbell-shapes (Miyata and Hotani 1992; Palmer et al. 2003), whereas, tubulin polymerisation results in the development of a bipolar shape (Hotani and Miyamoto 1990). Maintenance of the shape changes induced by polymerised tubulin requires microtubule-associated proteins (Kaneko et al. 1998).

The force required for pulling or pushing a tube from a giant vesicle is determined not only by the bending rigidity, but also the tension of the membrane. When more and longer tubes are being pulled, the membrane tension will rise in part because of the fixed area-to-volume ratio of the giant liposome. This tension can be released by the introduction of pores in the bilayer (Koster et al. 2003). Hence, regulation of lumenal volume may also contribute to tubule shape, and an obvious way of doing this would be with an ion pump. Whether cells exploit ion pumps in regulating membrane curvature is unclear. However, note that membrane tubulation can occur independently of the cytoskeleton. In *Xenopus* egg

a

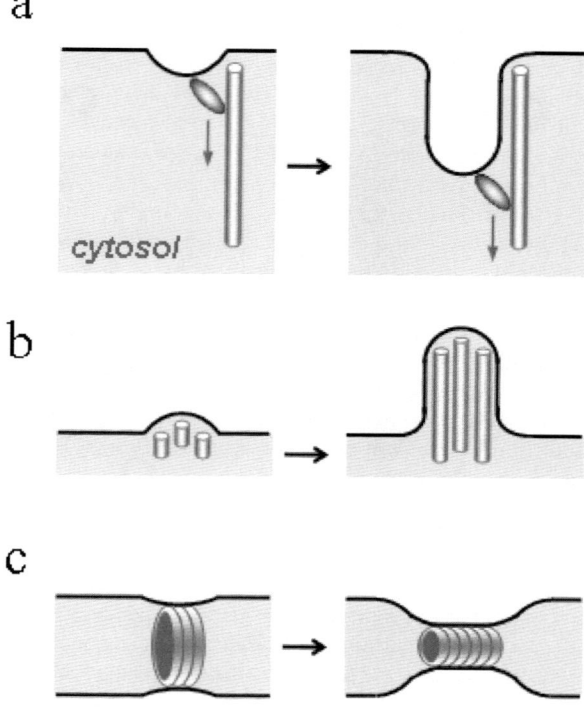

b

c

Fig. 2. Cytoskeleton-induced membrane curvature. (a) Membrane tubule pulled by a cytoskeletal motor protein. (b) Membrane tubule pushed by the polymerisation of actin or tubulin. (c) Membrane shaft constricted by a contractile ring composed of actin and myotubulin.

extracts, ER networks form *de novo* in the absence of microtubules or an actin scaffold (Dreier and Rapoport 2000). The exact mechanism by which this happens remains to be established. Moreover, cytosolic proteins with potent tubulogenic activities have been identified that do not form part of the cytoskeleton (see sections 4.1 and 4.2).

Finally, contractive forces generated by the cytoskeleton may also influence membrane curvature. A well-known example is the contractile ring of overlapping actin and bipolar myosin II filaments that forms during cytokinesis and exerts a force on the plasma membrane to generate a cleavage furrow at the site where the two daughter cells will be separated (Fig. 2 C). A redistribution of phosphatidyl ethanolamine (PE) from the inner to the outer leaflet of the plasma membrane appears to be involved in regulating the disassembly of the contractive ring and is essential for a proper progression of cytokinesis (Emoto et al. 1996; Emoto and Umeda 2000). Since PE has an effective cone shape (section 4.2), its redistribution to the outer leaflet may serve to facilitate a further inward bending of the plasma membrane.

3.2 Adhesion to curved particles

The acquisition of membrane curvature can also result from enwrapping a cyto-plasmic or exoplasmic particle (Fig. 3 A). Here, the driving force would derive from adhesion of the membrane to an already curved particle. Examples include cells eating beads (Desjardins and Griffiths 2003), the outward budding of milk-fat droplets from the plasma membrane (Deyrup-Olsen and Luchtel 1998) and the biogenesis of dense-core secretory granules at the trans-Golgi network (TGN; Burgoyne and Morgan 2003). In the latter case, membrane deformation is thought to result from a pH and calcium-dependent aggregation of secretory proteins in the TGN lumen. Chromogranin A (CgA) appears to be a key mediator of this process, since CgA depletion in neuroendocrine cells reduces granule number while CgA expression in fibroblasts induces dense-core granule formation (Kim et al. 2001). Hence, the aggregative properties of CgA, coupled to its propensity to interact with membranes (Huttner et al. 1991) would allow wrapping of the TGN membrane around a forming CgA aggregate. Although CgA alone seems capable of triggering secretory granule formation, its restricted expression in neuroendocrine tissues implies that other proteins may fulfil this role in other cell types. Indeed, heterologous expression of pro-von Willebrand factor in neuroendocrine or epithelial cell lines causes the induction of rod shaped structures resembling Weibel-Palade bodies, the organelles in which von Willebrand factor (vWF) is normally stored within endothelial cells (Wagner et al. 1991; Voorberg et al. 1993). Formation of these bodies requires the ability of vWF proteins to condensate into multimers. How nascent granule buds are subsequently pinched off to form secretory granules is unclear. Cholesterol is essential for this process as cholesterol depletion blocks granule biogenesis at a late stage with dense-core buds observable at the TGN (Wang et al. 2000). The role of cholesterol in vesicle fission may be direct, by facilitating the strong membrane curvature at the bud neck, or indirect, as a key component of lipid microdomains in the recruitment of proteins that drive vesicle fission (e.g. dynamin).

3.3 Coat assembly

The idea that cytosolic coat proteins bind membranes and then deform them to initiate vesicle budding is a long-standing one (Fig. 3 B). This concept, first developed for the clathrin coat, and then extended to other vesicle coats like COPI and COPII, is supported by studies revealing an intrinsic curvature in the structure of assembled coat complexes (Musacchio et al. 1999; Bi et al. 2002). Indeed, coat assembly, membrane budding and even fission can occur on protein-free liposomes, indicating that interfaces between lipids and cytosolic coat proteins are sufficient for pinching off bilayer vesicles (Matsuoka et al. 1998; Spang et al. 1998; Takei et al. 1998; Bremser et al. 1999). However, recent progress in the field has led to a partial revision of this view.

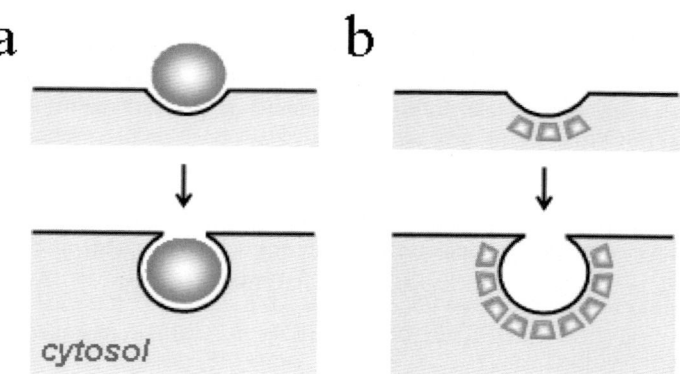

Fig. 3. Membrane curvature induced by adhesion to a body with intrinsic curvature. (a) Membrane budding driven by adhesion to a curved particle, for example during phagocytosis of bacteria or the biogenesis of secretory granules. (b) Membrane budding driven by coat proteins polymerising with an intrinsic curvature, for example during COPI-, COPII-, or clathrin-coated vesicle formation.

First, real-time imaging of green fluorescent protein (GFP)-cargo proteins in mammalian cells revealed the existence of clathrin- or COPI-containing membrane carriers that are larger and more pleiomorphic than the conventional, 60-100 nm-diameter coated vesicles produced *in vitro*. These carriers include vesicles up to 1 µm in diameter, tubules several µm in length and vesicular-tubular structures of various sizes and shapes (Lippincott-Schwartz et al. 2000). Second, electron microscopy studies demonstrated that coat complexes adopt a larger variety of shapes than previously anticipated. For example, early endosomes are partially decorated with flat 'bilayered' clathrin coats. Rather than giving rise to clathrin-coated vesicles, these structures seem to function in retaining lysosomally targeted cargo proteins when recycling proteins are being removed by vesicles or tubules that bud from early endosomes (Sachse et al. 2002). Third, fluorescence recovery after photobleaching (FRAP) experiments on GFP-tagged clathrin and COPI indicate that protein coats are much more dynamic than previously envisioned. Instead of providing a rigid assembly that is put on and taken off only once in a single round of coated-vesicle formation and fusion, protein coats undergo a continuous exchange between membranes and the cytosol by a process that can be uncoupled from vesicle formation (Presley et al. 2002). Fourth, in order to drive membrane curvature efficiently, the rigidity of an assembled coat polymer has to supersede the mechanical bending resistance of the membrane. This notion has been challenged for clathrin, because of estimations that the rigidity of the clathrin triskelia is at best similar to the mechanical bending resistance of the plasma membrane (Nossal 2001). Finally, at least two clathrin-binding proteins, epsin and amphiphysin, induce membrane curvature in the absence of clathrin (Farsad et al. 2001; Ford et al. 2002). Remarkably, the curvature-stimulating properties of these pro-

teins do not depend on oligomerization, but rather on a unique and intimate inter-action with the membrane bilayer (section 4.1).

Coat proteins are multifunctional, and likely stabilize nascent transport vesicles and tubules, select and concentrate their contents, prevent their fusion prior to completion of the budding process, and carry out post-budding reactions through the recruitment of accessory proteins that mediate interactions with the cytoskele-ton. However, the precise function of coat assembly in driving the formation of the highly curved membrane carriers in cells is incompletely understood. There is growing evidence that mechanisms in addition to coat-protein lattice formation play a role in the biogenesis of coated transport intermediates.

4 Intrinsic forces affecting membrane curvature

4.1 Transbilayer area asymmetry

4.1.1 Physical considerations

Due to the high-energy price associated with exposing their carbon chains to wa-ter, the lipid molecules in a bilayer are tightly packed. In this condensed state, lipid bilayers display solid- or liquid-like material behaviour with the common feature of limited surface compressibility, i.e. great resistance to change in surface density (Tanford 1978; Rawicz et al. 2000). Yet budding a vesicle requires mem-brane bending that must be associated with an imbalance in surface area between the two membrane leaflets. At the level of the head groups, the outer leaflet of a 60-nm diameter synaptic vesicle accommodates 1.5 times the number of lipids of the inner leaflet. If lipids were free to cross the membrane, a change in the transbi-layer lipid distribution would be easily accomplished and vesicle budding would happen as a manifestation of the thermal fluctuations exerted by the membrane. But in a lipid bilayer, it generally costs a lot of energy for a phospholipid to trav-erse the membrane. In model membranes, half times of spontaneous flip-flop of phosphatidylcholine range from several hours to days at physiological tempera-tures (Kornberg and McConnell 1971; Wimley and Thompson 1991), a time scale incompatible with the rate of vesicle budding in living cells.

Hence, vesicle budding requires transbilayer area asymmetry, and one way of achieving this is by the insertion of additional lipid in one of the two membrane halves. An artificial method for selectively increasing the lipid content of one leaf-let involves the addition of lyso-phosphatidylcholine (lyso-PC) to a giant liposome or erythrocyte. Lyso-PC molecules will penetrate and selectively expand the outer leaflet since the spontaneous flip-flop rate of lyso-PC is extremely slow. Because the two leaflets of a lipid bilayer are coupled by a powerful hydrophobic effect, a progressive modification of the ratio between the inner and outer areas will force the membrane to bend (Sheetz and Singer 1974). In fact, insertion of a small amount of lyso-PC (equivalent to 0.1-1% of total lipid) already creates a mismatch in lateral pressure between the two leaflets sufficient to drive formation of a

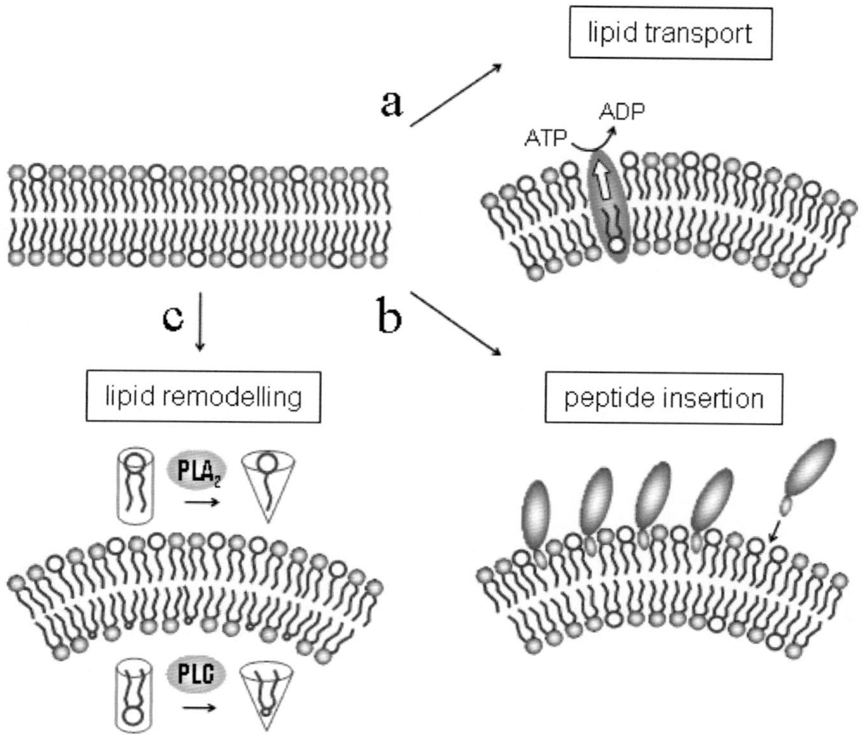

Fig. 4. Membrane curvature induced by a physical perturbation of the lipid-bilayer. (a) Transbilayer area asymmetry generated by a lipid translocase. (b) Transbilayer area asymmetry generated by proteins capable of physically penetrating into one face of the bilayer. (c) Transbilayer curvature asymmetry generated by lipid remodelling enzymes. PLA$_2$: phospholipase A$_2$; PLC: phospholipase C or sphingomyelinase.

microvesicle at the surface of a 50 μm-diameter giant liposome (Farge and Devaux 1992; Devaux 2000). Giant liposomes containing a small fraction (1 mol%) of phosphatidic acid or phosphatidylglycerol undergo similar shape changes when a transmembrane pH gradient is applied (Mathivet et al. 1996). In a living cell, net translocation of phospholipids can be achieved by lipid translocases that catalyse a transbilayer movement of phospholipids at the expense of ATP hydrolysis. Hence, lipid translocases may induce transbilayer area asymmetry and thus regulate membrane curvature *in vivo* (Fig. 4 A). As discussed below, a strong correlation between the activity of lipid translocases and the rate of endo- and exocytosis is consistent with this view.

4.1.2 Inward-directed lipid translocases

Eukaryotic plasma membranes generally display an asymmetric lipid arrangement across the bilayer. Whereas phosphatidylcholine (PC), sphingomyelin (SM), and glycosphingolipids are largely localized to the outer leaflet, the aminophospholipids phosphatidylserine (PS) and phosphatidylethanolamine (PE) reside predominantly in the inner leaflet (Zachowski 1993). Maintenance of this lipid asymmetry is a dynamic process involving two types of ATP-driven lipid transporters, namely an inward-directed transporter specific for PS and PE, known as the aminophospholipid translocase, and a less specific transporter responsible for a compensatory outward movement of lipids (Daleke 2003). By controlling the cytosolic levels of ATP, which is normally in the range of 2-3 mM, it is possible to modify the steady state distribution of the aminophospholipids and in turn to trigger shape changes in erythrocytes by shifting a fraction of the lipids from the inner to the outer leaflet and vice versa. When ATP levels are raised to 5-6 mM, erythrocytes form endocytic vesicles (Birchmeier et al. 1979). This process can be quantified by monitoring the acetylcholine esterase activity exposed on the cell surface, the reduction of which is an indication of the amount of plasma membrane internalised. Strikingly, membrane invagination and endovesiculation are enhanced by external addition and subsequent translocation of PS or PE, but inhibited when PC or SM are added to the outer leaflet (Seigneuret and Devaux 1984; Muller et al. 1994). On the basis of these findings, it has been postulated that active lipid transport catalysed by lipid translocases can act as a driving force in vesicle biogenesis (Devaux 1991; Devaux 2000). As discussed below, recent studies lend further support to this idea.

Erythrocytes do not represent the model of choice for studying endocytosis since they lack the ability to do so under physiological conditions. In fact, the aminophospholipid translocase activity in erythrocytes is rather weak in comparison to cells with natural endocytic activity; initial PS translocation rates in nucleated cells are nearly two orders of magnitude higher than in erythrocytes (Cribier et al. 1993; Pomorski et al. 1996). Addition of PS to the external surface of erythroleukemia K562 cells accelerates the endocytic uptake of fluoresceinated membrane proteins, whereas addition of lyso-PS, which is not transported by the aminophospholipid translocase, inhibits endocytosis (Farge et al. 1999). Complementary evidence for a role of lipid translocases in membrane vesiculation came from recent work in yeast. Disruption of the *DNF1* and *DNF2* genes abolishes inward phospholipid transport across the yeast plasma membrane and causes an accumulation of PE in the outer leaflet, especially when a related gene, *DRS2,* is also removed (Pomorski et al. 2003). Concomitantly, *dnf1dnf2* mutant cells, but in particular *dnf1dnf2drs2* cells exhibit a defect in the internalisation step of bulk phase and receptor-mediated endocytosis (Pomorski et al. 2003). Together with two additional yeast proteins (Neo1p and Dnf3p), Dnf1p, Dnf2p and Drs2p belong to a subfamily of P-type ATPases that also includes ATPase II, a putative aminophospholipid translocase purified from bovine chromaffin granules (Tang et al. 1996). Whereas Dnf1p and Dnf2p are primarily associated with the yeast plasma membrane, Drs2p resides in a late compartment of the yeast Golgi complex (Chen et al. 1999; Hua et al. 2002; Pomorski et al. 2003). Deletion of the *DRS2* gene is lethal

when combined with mutations in the genes for clathrin heavy chain or ADP-ribosylation factor 1, and perturbs formation of a specific class of clathrin-coated vesicles carrying invertase and acid phosphatase to the plasma membrane (Chen et al. 1999; Gall et al. 2002). Neo1p is localised to an early Golgi compartment and required for efficient COPI-dependent protein transport from the Golgi to the ER (Hua and Graham 2003).

All five members of the yeast Drs2/Neo1 P-type ATPase family have now been implicated in membrane trafficking at different steps along the endocytic and secretory pathways. Since these proteins most likely function as inward-directed lipid pumps (Tang et al. 1996; Gomes et al. 2000; Pomorski et al. 2003), it appears that a dynamic regulation of the transbilayer lipid arrangement plays a fundamental role in the formation of coated transport vesicles. Exactly how P-type ATPase-dependent lipid pumping participates in vesicle biogenesis remains to be established. A high concentration of aminophospholipids in the cytoplasmic leaflet may be necessary for an efficient recruitment of peripheral proteins such as ARFs, clathrin, amphiphysin, and endophilins (e.g. Takei et al. 1998). Yet this scenario is somewhat difficult to reconcile with the observation that cells lacking multiple Drs2/Neo1 family members down regulate the aminophospholipid content of their membranes, a response that would be counter productive for the rate of vesicle budding (Pomorski et al. 2003). Moreover, stimulation of inward lipid transport appears sufficient to trigger endocytic vesicle formation in erythrocytes, a cell type lacking the machinery for generating coated vesicles. This would imply that P-type ATPase-dependent lipid pumping primarily serves to facilitate inward vesicle budding by expanding the cytosolic leaflet of the membrane at the expense of the exoplasmic leaflet. According to this scenario, coat assembly would help localize the process, rather than providing the primary driving force. Aminophospholipid asymmetry would then merely be a consequence of the fact that, if there had been no lipid selectivity, the translocase would be trapped in a continuous pumping of the entire bilayer.

4.1.3 Outward-directed lipid translocases

Whereas Neo1/Drs2-related P-type ATPases appear to function as inward-directed lipid pumps, prime candidates for outward-directed lipid pumps are members of the ABC family of transport ATPases (Smit et al. 1993; Ruetz and Gros 1994; van Helvoort et al. 1996; Decottignies et al. 1998). Overexpression of ABC transporters with outward lipid translocase activity slows down endocytosis (Kean et al. 1997; Decottignies et al. 1998). Conversely, loss of ABCA1, a putative phospholipid transporter affected in Tangier disease, stimulates endocytosis (Zha et al. 2001). By catalysing an efflux rather than an influx of lipids across the bilayer, ABC transporters would promote an outward bending of cellular membranes. ABCA1 is required for the engulfment of apoptotic cells, a process characterized by an elaborate protrusion of the plasma membrane (Marguet et al. 1999).

Note that the action of a unidirectional lipid transporter would lead to a global, rather than a local difference in lateral pressure between the two membrane leaflets. However, a global difference in lateral pressure maintained by lipid pumps

could drive local membrane bending, with protein coats, membrane inserting peptides (below), or lipid microdomains (section 4.3) acting as the nucleation sites.

4.1.4 Bidirectional lipid flippases

Contrary to the situation in the plasma membrane and Golgi complex, flip-flop of phospholipids in the ER is a fast process with half times in the order of seconds or minutes (Seigneuret and Devaux 1984; Buton et al. 1996, 2002; Marx et al. 2000). This flip-flop is believed to be protein-mediated, involving one or more lipid flippases (Bretscher 1973; Bishop and Bell 1985; Menon et al. 2000). ER flippases operate independently of metabolic energy and facilitate a non-vectorial movement of most, if not all phospholipid classes in both directions. Consequently, they would promote a symmetric phospholipid distribution across the bilayer. Whether the rapid flip-flop in the ER is mediated by a single dedicated flippase, a group of flippases, or accomplished by the mere presence of integral membrane proteins in a particular lipid environment remains to be established. Peptides mimicking the transmembrane α-helices of proteins are capable of inducing phospholipid flip-flop in model membranes (Kol et al. 2003). This helix-mediated flip-flop is strongly inhibited by cholesterol, an abundant component of the plasma membrane. Cholesterol may exert this inhibitory effect by increasing the packing density of the acyl chains through which the polar head group has to travel. It has been postulated that the gradual increase in cholesterol levels along the secretory pathway serves as a regulatory device to switch from constitutive flip-flop in the ER to a more tightly controlled translocation of lipids in the Golgi and plasma membrane (Kol et al. 2002). An important implication of the fact that phospholipids can freely cross the ER bilayer in both directions is that a transbilayer lipid arrangement permissive for vesicle budding should be easily accomplished. Hence, unlike the situation at the plasma membrane where flip-flop is constrained, assembly of a coat on the surface of the ER may well be sufficient to drive membrane vesiculation.

Note that the plasma membrane of animal cells contains a Ca^{2+}-dependent scramblase that, when induced, mediates a complete intermixing of lipids between the bilayer leaflets, irrespective of head group specificity (Williamson et al. 1992). This activity is turned on during apoptosis, and perhaps facilitates the outward blebbing of the plasma membrane observed in apoptotic cells.

4.1.5 Amphipathic peptides

Recent work has led to the identification of proteins capable of mediating transbilayer area asymmetry through burying an amphipathic helix into one face of the membrane (Fig. 4 B), hence bypassing the requirement for a lipid rearrangement across the bilayer. A prominent example is epsin 1, an accessory protein that collaborates with clathrin and AP2 adopter complexes in the budding of clathrin-coated vesicles from the plasma membrane (Chen et al. 1998). Epsin 1 contains an epsin amino terminal homology (ENTH) domain that comprises a phosphoinositide-binding site. Upon binding phosphatidylinositol 4,5-bisphosphate (PIP$_2$), the

ENTH domain forms an amphipathic α-helix, H_0, in which the hydrophobic residues are facing the exterior (Ford et al. 2002). When added to PIP_2-containing giant liposomes, the ENTH domain causes a strong degree of membrane curvature, leading to the formation of 20 nm-diameter tubules. Mutation of hydrophobic residues in H_0 abolishes the tubulogenic behaviour of ENTH, suggesting that a physical penetration of H_0 in the outer membrane leaflet is responsible for inducing curvature (Ford et al. 2002). A striking example of the membrane deforming potential of an amphipathic helix can be found in the ability of a *de novo* designed, 18-mer amphipathic peptide to transform spherical liposomes of varying lipid composition into nanotubes up to several μm in length (Lee et al. 2001). By virtue of its ENTH domain and clathrin/adaptor-binding motifs, epsin 1 would provide both a driving force and a molecular link between membrane invagination, clathrin polymerisation and AP2 complex recruitment. By influencing local phosphoinositide metabolism, transduction events at the plasma membrane may create sites where epsin 1 would bind. This would allow cells to specify locations of membrane vesiculation.

The proposed working mechanism of epsin 1 may apply to other proteins implicated in clathrin-coated vesicle formation. EpsinR binds clathrin, the γ-appendage of the AP1 adapter complex, and contains an ENTH domain with a phosphatidylinositol-4 phosphate-binding site. Several lines of evidence indicate that epsinR is functionally equivalent to epsin 1, but in clathrin-coated vesicle budding from Golgi/endosomes rather than from the plasma membrane (Mills et al. 2003). Amphiphysin and endophilin, two major interactors of the GTPase dynamin, are capable of reshaping liposomes into narrow membrane tubules. Strikingly, the amino termini of these tubulogenic proteins contain a so-called N-BAR domain with an amino acid stretch predicted to form an amphipathic helix necessary for lipid bilayer tubulation (Farsad et al. 2001). Co-incubation of endophilin, clathrin and liposomes results in the formation of tubules capped with clathrin-coated buds in spite of the fact that endophilin has no known clathrin-binding properties. This suggests that proteins capable of driving membrane curvature on their own might facilitate clathrin-mediated bud formation by altering bilayer structure to favour this process (Farsad and De Camilli 2003). Membrane deforming proteins involved in cellular processes other than clathrin-coated vesicle formation have also been reported. A member of the amphiphysin protein family, amphiphysin 2, localises to the T-tubule system in striated muscle and induces tubular plasma membrane invaginations when expressed in non-muscle cells (Lee et al. 2002). Induction of the nanotubular network from the parasitophorous vacuole of *Toxoplasma gondi* requires a parasite secretory protein, Gra2, that has two amphipathic α-helical regions the integrity of which is essential for correct formation of the network (Mercier et al. 2002).

4.1.6 Lipid remodelling enzymes

Finally, transbilayer area asymmetry can also be achieved through chemical modifications affecting the molecular area or biophysical properties of the lipids in one leaflet. ATP depletion turns erythrocytes into echinocytes whose spiky appearance

has been ascribed to the breakdown of phosphatidylinositol-4,5-biphosphate and subsequent shrinkage of the inner membrane leaflet by 0.6% (Ferrell and Huestis 1984). Microinjection of sphingomyelinase into sphingomyelin-containing giant liposomes results in the appearance of microvesicles on the liposomal surface (Holopainen et al. 2000) and treatment of macrophages with externally added sphingomyelinase triggers ATP-independent endocytosis through the formation of large endocytic vesicles without discernable coats (Zha et al. 1998). Sphingomyelin constitutes nearly 10% of all phospholipids in mammalian cells and occurs primarily in the outer leaflet of the plasma membrane (Holthuis et al. 2001). While removal of its head group by sphingomyelinase would reduce crowding in the outer leaflet (Fig. 4 C), the breakdown product ceramide could flip across the bilayer due to its hydrophobicity (Pagano 1990) and hence induce a relative increase in lateral pressure in the inner leaflet. Both of these processes would promote inward curvature. Sphingomyelinase-induced endovesiculation provides a mechanism potentially used by pathogenic bacteria like *Staphylococcus aureus* and *Neisseria gonorrhoeae* to invade human cells (Walev et al. 1996; Grassme et al. 1997).

4.2 Transbilayer curvature asymmetry

4.2.1 Lipid polymorphism and membrane shape

If there are no forces acting on its surface, a lipid monolayer adopts a curvature called the spontaneous or intrinsic curvature. Conventionally, the curvature of a monolayer is defined as positive if the monolayer bends in the direction of the polar heads, and negative for the opposite direction of bending. Unlike spontaneous bilayer curvature, which depends also on lipid number (see above), spontaneous monolayer curvature is primarily determined by the effective shapes of the lipids (Cullis and de Kruijff 1979; Israelachvili et al. 1980). Hence, cone-shaped lipids whose head groups are larger than the cross-sectional areas of their acyl chains (e.g. lyso-PC) will promote a positive spontaneous curvature. Inverted cone- or wedge-shaped lipids whose head groups are smaller than the cross-sectional areas of their acyl chains (e.g. unsaturated PE) will promote a negative spontaneous curvature. Cylindrically-shaped lipids with equally sized cross-sectional areas of the head group and acyl chains (e.g. di-monounsaturated PC, sphingomyelin) will assemble into monolayers with a close to zero spontaneous curvature. The curvature preference of lipids is not exclusively determined by their molecular structure, but also influenced by additional factors such as temperature, pH and ion concentrations.

Lipids promoting a pronounced positive or negative spontaneous curvature are also called non-bilayer lipids, although an equimolar mixture of a cone- and wedge-shaped lipid, for example lyso-PC and di-unsaturated PE, will form a stable bilayer (Madden and Cullis 1982). Cellular membranes consist of mixtures of bilayer and non-bilayer (primarily inverted cone-shaped) lipids. The latter are forced to remain in the lipid-bilayer, a situation generally referred to as 'frustration'. The ratio of bilayer to non-bilayer lipids is strictly regulated, and likely

helps determine the propensity of cellular membranes to undergo fusion (Burger 2000). A planar membrane may be composed of either two monolayers having negative (or positive) spontaneous curvatures, or two monolayers having zero spontaneous curvatures. Because of symmetry, the net bilayer spontaneous curvature will be zero for both membranes, even though the membrane composed of lipids with non-cylindrical effective shapes will be stressed. But a difference in spontaneous monolayer curvature between the two membrane halves, or transbilayer curvature asymmetry, will result in membrane bending. This principle is perhaps best illustrated by the shape changes that occur in erythrocytes when natural, mono-unsaturated PC present in the outer leaflet is exchanged for synthetic di-saturated or di-unsaturated PC using a PC-specific transfer protein. While hardly affecting the surface area of the outer leaflet, the PC-exchange reaction causes dramatic shape changes with di-saturated PC triggering an outward membrane bending and echinocyte formation, and di-unsaturated PC producing the opposite effect (Kuypers et al. 1984; Christiansson et al. 1985).

By virtue of their ability to alter the lipid composition in the membrane, lipid transfer proteins, lipid translocases and lipid remodellases may all influence the spontaneous curvature of the membrane leaflets, and hence regulate membrane bending *in vivo* (Burger 2000).

4.2.2 Lipid hydrolases and lipid acyltransferases

The idea that local lipid metabolism and the generation of transbilayer curvature asymmetry are used *in vivo* to regulate membrane bending is supported by the recent discovery that phospholipid hydrolases and acyltransferases co-ordinately regulate physiological shape changes of the Golgi complex. Thus, inhibition of a cytoplasmic Ca^{2+}-independent phospholipase A_2 (PLA$_2$) prevents the formation of Golgi membrane tubules that form both constitutively and in response to brefeldin A (BFA) treatment (de Figueiredo et al. 1998, 1999). PLA$_2$ antagonists also inhibit retrograde protein transport from the Golgi to the ER (de Figueiredo et al. 2000) and the membrane tubule-mediated reassembly of a nocodazole-fragmented Golgi complex (Drecktrah et al. 2003). Peptide-induced stimulation of a cytoplasmic PLA$_2$ has the opposite effect, that of inducing Golgi membrane tubulation (Polizotto et al. 1999). Hence, by catalysing the conversion of cylindrical (or cone-shaped) phospholipids into inverted cone-shaped lyso-phospholipids in the cytosolic membrane leaflet, PLA$_2$ would promote an outward bending of the bilayer that at its most extreme would trigger membrane tubulation (Fig. 4 C). Consistent with this model, inhibition of a Golgi-associated lyso-phospholipid acyltransferase (LPAT), which reacylates lyso-phospholipids back to phospholipids, increases the levels of lyso-phosphatidylcholine in Golgi membranes and stimulates membrane tubulation and retrograde trafficking (Drecktrah et al. 2003). Importantly, preincubation of cells with PLA$_2$ antagonists prevents LPAT antagonists from inducing Golgi tubulation, providing further evidence that this process is directly controlled by the phospholipid/lyso-phospholipid ratio in the cytosolic leaflet of the Golgi. A similar mechanism seems to regulate tubulation of endosomes

and the recycling of transferrin receptors from endocytic compartments to the cell surface (de Figueiredo et al. 2001).

Other work indicates that LPATs are part of the fission machineries involved in Golgi tubule fragmentation and endocytic vesicle formation. Thus, a lyso-phosphatidic acid (lyso-PA) specific LPAT, CtBP/BARS, induces fission and vesicle formation from Golgi membrane tubules (Weigert et al. 1999). Likewise, inhibition of the intrinsic lyso-PA specific LPAT activity of endophilin reduces its ability to trigger endocytic vesicle formation (Schmidt et al. 1999), even though its tubulogenic properties do not require LPAT activity (Farsad et al. 2001). For both proteins, it has been postulated that condensation of membrane-embedded lyso-PA and cytosolic fatty acyl-CoA to phosphatidic acid (PA) would promote a negative spontaneous curvature in the cytosolic leaflet, in particular when unsaturated fatty acyl-CoA such as arachidonoyl-CoA is used. This, in turn, would drive inward bending of the membrane at the neck of a budding vesicle, thus, aiding in its fission (Burger 2000; Kooijman et al. 2003). In the Golgi, CtBP/BARS-derived PA is likely metabolised rapidly into diacylglycerol (DAG), a lipid with an even larger negative spontaneous curvature. Due to its hydrophobicity, DAG would readily equilibrate between the two membrane halves. A recent theoretical analysis suggests that the lipid transformation sequence lyso-PA→PA→DAG and subsequent partitioning of DAG may drive the progressive constriction of Golgi membrane tubules and explain the CtBP/BARS-induced 'pearling' of these tubules (Shemesh et al. 2003). While PA and DAG may serve as the direct mechanical workers in deforming the membrane, alternative scenarios cannot be excluded. For example, PA or DAG may recruit and activate downstream effectors involved in membrane bending or fission. Protein kinase D binds DAG and is required for secretory vesicle formation at the TGN (Baron and Malhotra 2002). The GTPase dynamin regulates membrane constriction and fission during receptor-mediated endocytosis (Hinshaw 2000). Dynamin is a membrane active protein whose ability to penetrate in between lipid head groups is greatly stimulated by PA (Burger et al. 2000). Hence, the PA generated by dynamin-bound endophilin may induce deep penetration of dynamin into the neck region of a budding vesicle or tubule, and hence promote fission indirectly. Note that cells may exploit the above principles simultaneously to accomplish membrane fission.

4.2.3 Lysobisphosphatidic acid in multivescicular endosome formation

Lysobisphosphatidic acid (LBPA) is an abundant component of internal vesicles in multivesicular endosomes, accounting for nearly 15 mole percent of total organelle phospholipids (Kobayashi et al. 1998). Recent work revealed that this unconventional cone-shaped lipid possesses the capacity to drive vesicle budding and fission within acidified liposomes. A cytosolic LBPA-binding protein, called Alix, controls this invagination process *in vitro* and is involved in the biogenesis of multivesicular endosomes *in vivo* (Matsuo et al. 2004). Precisely how LBPA and Alix participate in multivesicular endosome formation remains to be established, but the intrinsic capacity of LBPA to stimulate internal vesicle formation

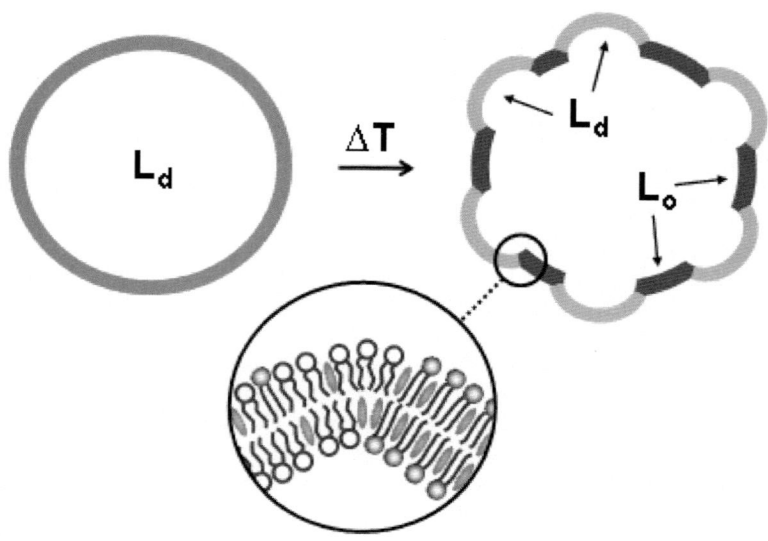

Fig. 5. Lipid domain-induced membrane curvature. Giant liposomes prepared from a ternary mixture of glycerolipids, sphingolipids and cholesterol are subjected to a temperature shift, ΔT, causing a lipid phase separation. As a result, two different liquid phases, or domains, are formed: a liquid ordered (L_o) domain enriched in sphingolipids and sterols, and a liquid disordered (L_d) domain containing the bulk of glycerolipids. A line tension builds up, which is proportional to the length of the boundaries between the two domains. This tension can be partially relieved by increasing the membrane curvature in one of the two domains. Due to a higher packing density of the lipids, the bending resistance of L_o domains is higher than that of L_d domains. Budding from L_d domains would, therefore, provide the most efficient way of reducing the line tension.

within acidic liposomes provides an important clue. At neutral pH, LBPA has a negatively charged head group that would impair its spontaneous movement across the bilayer. Lowering the pH would cause a reduction in charge and increase the flip-flop rate of LBPA. Acidification of the lumen of LBPA-containingliposomes would then result in a preferential accumulation of this cone-shaped lipid in the outer leaflet, thus promoting a negative spontaneous curvature of the outer leaflet at the expense of the inner leaflet. This, in turn, may drive membrane invagination and formation of internal vesicles. Consistent with an active role for LBPA in multivesicular endosome biogenesis, some patients afflicted with antiphospholipid syndrome produce auto-antibodies against LBPA that, when applied to cells, accumulate in late endosomes and disrupt their multivesicular morphology (Kobayashi et al. 1998).

4.3 Lipid domain-induced curvature

Studies on model membranes established that lipids have a strong self-organizing capacity; lipid immiscibility can trigger phase separation and create domains with unique lipid compositions and physical properties (Brown 1998). Thus, mixtures of sphingolipids, unsaturated glycerolipids, and cholesterol can segregate spontaneously into two fluid domains where the sphingolipids and part of the cholesterol coalesce into a liquid ordered (L_o) domain and break away from the unsaturated glycerolipids in a liquid disordered (L_d) phase. This phase behaviour has a clear physico-chemical basis. Due to their long and saturated fatty acyl chains, sphingolipids contact their neighbours along a greater and flatter surface than unsaturated glycerolipids, resulting in a dramatic increase in the van der Waals attraction between neighbouring sphingolipid molecules (McIntosh et al. 1992). Intriguingly, high-resolution fluorescence imaging of giant liposomes composed of a ternary mixture of sphingomyelin, dioleoyl-PC and cholesterol has demonstrated that domain formation can drive vesicle budding and fission (Fig. 5; Baumgart et al. 2003). Here, the driving force is generated by the line tension, an energy associated with the edge of the domain that is proportional to the length of the edge. Domain-induced budding is a spontaneous process, governed by a competition between a decrease in line tension and an increase in bending energy (Julicher and Lipowsky 1993).

As there is evidence for the existence of phase-separated domains in cellular membranes (Simons and Ikonen 1997; Brown and London 1998; Anderson and Jacobson 2002), domain-induced budding may also occur *in vivo*. This concept could be particularly relevant for the Golgi complex, which is a major site of lipid sorting. Sphingolipids are synthesised in the Golgi, cholesterol in the ER (van Meer 1998; Huitema et al. 2004). In spite of extensive membrane trafficking at the ER-Golgi interface, both lipids accumulate in the plasma membrane. Cholesterol has a high affinity for sphingolipids, and anterograde sorting of newly synthesized sphingolipids in the Golgi has been put forward as a mechanism to deplete cholesterol from the ER and to promote its concentration in the plasma membrane (van Meer 1998). Anterograde sphingolipid sorting could be achieved if sphingolipids were prevented from leaving the Golgi cisternae in which they are made; cisternal maturation would then ensure their unidirectional transport toward the plasma membrane (Holthuis et al. 2001; Pelham 2001). By triggering a phase separation in Golgi membranes, newly synthesized sphingolipids may stimulate budding of retrograde COPI vesicles or tubules from which they themselves are excluded; note that this budding would preferentially occur from domains enriched in unsaturated glycerolipids, since it is negatively influenced by fatty acyl chain saturation and cholesterol (Spang et al. 1998; Brugger et al. 2000; Koster et al. 2003). A domain-induced budding mechanism has the particular appeal of being inherently self-organizing. Consequently, it may represent one of the cell's most primordial mechanisms for membrane vesiculation.

5 Concluding remarks

Cells clearly evolved more than one type of strategy to overcome the energy barrier of membrane deformation. So far, attention has been focused primarily on the oligomerization-driven mechanical deformation of lipid bilayers by 'classical' coat proteins, even though several of the intracellular budding events might not require a supramolecular coat complex. Indeed, two of the five endocytic pathways described in mammalian cells involve buds and vesicles that lack a discernable coat (Pelkmans and Helenius 2003). The same holds true for the internal vesicles of multivesicular endosomes, the secretory granules formed in neuroendocrine cells and the milk-fat droplets shed from mammary epithelial cells. At the same time, emerging data emphasize that lipids serve a more dynamic and active role in membrane deformation than previously envisioned. In fact, model membrane studies reveal that budding and fission can occur in the absence of added protein. In live cells, lipid remodelling and unidirectional lipid transport across the bilayer each seem to affect shape changes that support membrane tube formation and vesiculation. Current challenges include the unambiguous identification of the lipid remodellases and translocases involved. Reconstitution of these activities in giant liposomes will be important to further elucidate the functional link between lipid remodelling, lipid translocation and membrane folding.

One may wonder why cells invented so many different kinds of membrane-deforming mechanisms. Where and when do these mechanisms operate? Do cellular budding machineries apply multiple types of membrane bending simultaneously? If so, how are the actions of the proteins and lipids involved coordinated? At present, these issues are incompletely understood. The physical constraints associated with membrane folding are dependent on lipid composition. Our comprehension of this process in cells is hampered by a relative ignorance of the steady-state lipid compositions of the various organelles. The fact that lipids can freely cross the ER bilayer in both directions implies that this organelle would readily adopt a transbilayer lipid arrangement permissive for vesicle formation. This is in contrast to the situation at the plasma membrane where the lipid flip-flop rate is constrained due to high levels of sterols and sphingolipids. Hence, it is unlikely that the same membrane-deforming mechanism can be used for all organelles. This may help explain the multiplicity of budding machineries that are currently uncovered in cells.

Acknowledgements

I am grateful to Gerrit van Meer and Hein Sprong for critically reading the manuscript. Support was from a grant of the Dutch NWO-CW VIDI programme and from SENTER, the Dutch Ministry of Economic Affairs.

References

Anderson RG, Jacobson K (2002) A role for lipid shells in targeting proteins to caveolae, rafts, and other lipid domains. Science 296:1821-1825

Baron CL, Malhotra V (2002) Role of diacylglycerol in PKD recruitment to the TGN and protein transport to the plasma membrane. Science 295:325-328

Baumgart T, Hess ST, Webb WW (2003) Imaging coexisting fluid domains in biomembrane models coupling curvature and line tension. Nature 425:821-824

Bi X, Corpina RA, Goldberg J (2002) Structure of the Sec23/24-Sar1 pre-budding complex of the COPII vesicle coat. Nature 419:271-277

Birchmeier W, Lanz JH, Winterhalter KH, Conrad MJ (1979) ATP-induced endocytosis in human erythrocyte ghosts. Characterization of the process and isolation of the endocytosed vesicles. J Biol Chem 254:9298-9304

Bishop WR, Bell RM (1985) Assembly of the endoplasmic reticulum phospholipid bilayer: the phosphatidylcholine transporter. Cell 42:51-60

Bremser M, Nickel W, Schweikert M, Ravazzola M, Amherdt M, Hughes CA, Sollner TH, Rothman JE, Wieland FT (1999) Coupling of coat assembly and vesicle budding to packaging of putative cargo receptors. Cell 96:495-506

Bretscher MS (1973) Membrane structure: some general principles. Science 181:622-629

Brown DA, London E (1998) Functions of lipid rafts in biological membranes. Annu Rev Cell Dev Biol 14:111-136

Brown RE (1998) Sphingolipid organization in biomembranes: what physical studies of model membranes reveal. J Cell Sci 111 Pt 1:1-9

Brugger B, Sandhoff R, Wegehingel S, Gorgas K, Malsam J, Helms JB, Lehmann WD, Nickel W, Wieland FT (2000) Evidence for segregation of sphingomyelin and cholesterol during formation of COPI-coated vesicles. J Cell Biol 151:507-518

Burger KN (2000) Greasing membrane fusion and fission machineries. Traffic 1:605-613

Burger KN, Demel RA, Schmid SL, de Kruijff B (2000) Dynamin is membrane-active: lipid insertion is induced by phosphoinositides and phosphatidic acid. Biochemistry 39:12485-12493

Burgoyne RD, Morgan A (2003) Secretory granule exocytosis. Physiol Rev 83:581-632

Buton X, Herve P, Kubelt J, Tannert A, Burger KN, Fellmann P, Muller P, Herrmann A, Seigneuret M, Devaux PF (2002) Transbilayer movement of monohexosylsphingolipids in endoplasmic reticulum and Golgi membranes. Biochemistry 41:13106-13115

Buton X, Morrot G, Fellmann P, Seigneuret M (1996) Ultrafast glycerophospholipid-selective transbilayer motion mediated by a protein in the endoplasmic reticulum membrane. J Biol Chem 271:6651-6657

Chen H, Fre S, Slepnev VI, Capua MR, Takei K, Butler MH, Di Fiore PP, De Camilli P (1998) Epsin is an EH-domain-binding protein implicated in clathrin-mediated endocytosis. Nature 394:793-797

Chen CY, Ingram MF, Rosal PH, Graham TR (1999) Role for Drs2p, a P-type ATPase and potential aminophospholipid translocase, in yeast late Golgi function. J Cell Biol 147:1223-1236

Christiansson A, Kuypers FA, Roelofsen B, Op den Kamp JA, van Deenen LL (1985) Lipid molecular shape affects erythrocyte morphology: a study involving replacement of native phosphatidylcholine with different species followed by treatment of cells with sphingomyelinase C or phospholipase A2. J Cell Biol 101:1455-1462

Coleman ML, Sahai EA, Yeo M, Bosch M, Dewar A, Olson MF (2001) Membrane bleb-bing during apoptosis results from caspase-mediated activation of ROCK I. Nat Cell Biol 3:339-345

Cribier S, Sainte-Marie J, Devaux PF (1993) Quantitative comparison between aminophos-pholipid translocase activity in human erythrocytes and in K562 cells. Biochim Bio-phys Acta 1148:85-90

Cullis PR, de Kruijff B (1979) Lipid polymorphism and the functional roles of lipids in bio-logical membranes. Biochim Biophys Acta 559:399-420

Dabora SL, Sheetz MP (1988) The microtubule-dependent formation of a tubulovesicular network with characteristics of the ER from cultured cell extracts. Cell 54:27-35

Daleke DL (2003) Regulation of transbilayer plasma membrane phospholipid asymmetry. J Lipid Res 44:233-242

de Figueiredo P, Doody A, Polizotto RS, Drecktrah D, Wood S, Banta M, Strang MS, Brown WJ (2001) Inhibition of transferrin recycling and endosome tubulation by phospholipase A2 antagonists. J Biol Chem 276:47361-47370

de Figueiredo P, Drecktrah D, Katzenellenbogen JA, Strang M, Brown WJ (1998) Evidence that phospholipase A2 activity is required for Golgi complex and trans Golgi network membrane tubulation. Proc Natl Acad Sci USA 95:8642-8647

de Figueiredo P, Drecktrah D, Polizotto RS, Cole NB, Lippincott-Schwartz J, Brown WJ (2000) Phospholipase A2 antagonists inhibit constitutive retrograde membrane traffic to the endoplasmic reticulum. Traffic 1:504-511

de Figueiredo P, Polizotto RS, Drecktrah D, Brown WJ (1999) Membrane tubule-mediated reassembly and maintenance of the Golgi complex is disrupted by phospholipase A2 antagonists. Mol Biol Cell 10:1763-1782

Decottignies A, Grant AM, Nichols JW, de Wet H, McIntosh DB, Goffeau A (1998) AT-Pase and multidrug transport activities of the overexpressed yeast ABC protein Yor1p. J Biol Chem 273:12612-12622

Desjardins M, Griffiths G (2003) Phagocytosis: latex leads the way. Curr Opin Cell Biol 15:498-503

Devaux PF (1991) Static and dynamic lipid asymmetry in cell membranes. Biochemistry 30:1163-1173

Devaux PF (2000) Is lipid translocation involved during endo- and exocytosis? Biochimie 82:497-509

Deyrup-Olsen I, Luchtel DL (1998) Secretion of mucous granules and other membrane-bound structures: a look beyond exocytosis. Int Rev Cytol 183:95-141

Drecktrah D, Chambers K, Racoosin EL, Cluett EB, Gucwa A, Jackson B, Brown WJ (2003) Inhibition of a Golgi complex lysophospholipid acyltransferase induces mem-brane tubule formation and retrograde trafficking. Mol Biol Cell 14:3459-3469

Dreier L, Rapoport TA (2000) In vitro formation of the endoplasmic reticulum occurs inde-pendently of microtubules by a controlled fusion reaction. J Cell Biol 148:883-898

Emoto K, Kobayashi T, Yamaji A, Aizawa H, Yahara I, Inoue K, Umeda M (1996) Redis-tribution of phosphatidylethanolamine at the cleavage furrow of dividing cells during cytokinesis. Proc Natl Acad Sci USA 93:12867-12872

Emoto K, Umeda M (2000) An essential role for a membrane lipid in cytokinesis. Regula-tion of contractile ring disassembly by redistribution of phosphatidylethanolamine. J Cell Biol 149:1215-1224

Farge E, Devaux PF (1992) Shape changes of giant liposomes induced by an asymmetric transmembrane distribution of phospholipids. Biophys J 61:347-357

Farge E, Ojcius DM, Subtil A, Dautry-Varsat A (1999) Enhancement of endocytosis due to aminophospholipid transport across the plasma membrane of living cells. Am J Physiol 276:C725-C733

Farsad K, De Camilli P (2003) Mechanisms of membrane deformation. Curr Opin Cell Biol 15:372-381

Farsad K, Ringstad N, Takei K, Floyd SR, Rose K, De Camilli P (2001) Generation of high curvature membranes mediated by direct endophilin bilayer interactions. J Cell Biol 155:193-200

Feiguin F, Ferreira A, Kosik KS, Caceres A (1994) Kinesin-mediated organelle translocation revealed by specific cellular manipulations. J Cell Biol 127:1021-1039

Ferrell JE Jr, Huestis WH (1984) Phosphoinositide metabolism and the morphology of human erythrocytes. J Cell Biol 98:1992-1998

Ford MG, Mills IG, Peter BJ, Vallis Y, Praefcke GJ, Evans PR, McMahon HT (2002) Curvature of clathrin-coated pits driven by epsin. Nature 419:361-366

Gall WE, Geething NC, Hua Z, Ingram MF, Liu K, Chen SI, Graham TR (2002) Drs2p-dependent formation of exocytic clathrin-coated vesicles *in vivo*. Curr Biol 12:1623-1627

Gomes E, Jakobsen MK, Axelsen KB, Geisler M, Palmgren MG (2000) Chilling tolerance in *Arabidopsis* involves ALA1, a member of a new family of putative aminophospholipid translocases. Plant Cell 12:2441-2454

Grassme H, Gulbins E, Brenner B, Ferlinz K, Sandhoff K, Harzer K, Lang F, Meyer TF (1997) Acidic sphingomyelinase mediates entry of *N. gonorrhoeae* into nonphagocytic cells. Cell 91:605-615

Hinshaw JE (2000) Dynamin and its role in membrane fission. Annu Rev Cell Dev Biol 16:483-519

Hirokawa N (1998) Kinesin and dynein superfamily proteins and the mechanism of organelle transport. Science 279:519-526

Holopainen JM, Angelova MI, Kinnunen PK (2000) Vectorial budding of vesicles by asymmetrical enzymatic formation of ceramide in giant liposomes. Biophys J 78:830-838

Holthuis JC, Pomorski T, Raggers RJ, Sprong H, Van Meer G (2001) The organizing potential of sphingolipids in intracellular membrane transport. Physiol Rev 81:1689-1723

Holthuis JC, van Meer G, Huitema K (2003) Lipid microdomains, lipid translocation and the organization of intracellular membrane transport . Mol Membr Biol 20:231-241

Hotani H, Miyamoto H (1990) Dynamic features of microtubules as visualized by darkfield microscopy. Adv Biophys 26:135-156

Hua Z, Fatheddin P, Graham TR (2002) An essential subfamily of Drs2p-related P-type ATPases is required for protein trafficking between Golgi complex and endosomal/vacuolar system. Mol Biol Cell 13:3162-3177

Hua Z, Graham TR (2003) Requirement for neo1p in retrograde transport from the Golgi complex to the endoplasmic reticulum. Mol Biol Cell 14:4971-4983

Huitema K, Van Den Dikkenberg J, Brouwers JF, Holthuis JC (2004) Identification of a family of animal sphingomyelin synthases. EMBO J 23:33-44

Huttner WB, Gerdes HH, Rosa P (1991) The granin (chromogranin/secretogranin) family. Trends Biochem Sci 16:27-30

Ishikawa H (1968) Formation of elaborate networks of T-system tubules in cultured skeletal muscle with special reference to the T-system formation. J Cell Biol 38:51-66

Israelachvili JN, Marcelja S, Horn RG (1980) Physical principles of membrane organization. Q Rev Biophys 13:121-200

Julicher F, Lipowsky R (1993) Domain-induced budding of vesicles. Phys Rev Lett 70:2964-2967

Kaneko T, Itoh TJ, Hotani H (1998) Morphological transformation of liposomes caused by assembly of encapsulated tubulin and determination of shape by microtubule-associated proteins (MAPs). J Mol Biol 284:1671-1681

Katzmann DJ, Odorizzi G, Emr SD (2002) Receptor downregulation and multivesicular-body sorting. Nat Rev Mol Cell Biol 3:893-905

Kean LS, Grant AM, Angeletti C, Mahé Y, Kuchler K, Fuller RS, Nichols JW (1997) Plasma membrane translocation of fluorescent-labeled phosphatidylethanolamine is controlled by transcription regulators, *PDR1* and *PDR3*. J Cell Biol 138:255-270

Kim T, Tao-Cheng JH, Eiden LE, Loh YP (2001) Chromogranin A, an "on/off" switch controlling dense-core secretory granule biogenesis. Cell 106:499-509

Kirchhausen T (2000) Three ways to make a vesicle. Nat Rev Mol Cell Biol 1:187-198

Kobayashi T, Stang E, Fang KS, de Moerloose P, Parton RG, Gruenberg J (1998) A lipid associated with the antiphospholipid syndrome regulates endosome structure and function. Nature 392:193-197

Kol MA, de Kruijff B, de Kroon AI (2002) Phospholipid flip-flop in biogenic membranes: what is needed to connect opposite sides. Semin Cell Dev Biol 13:163-170

Kol MA, van Laak AN, Rijkers DT, Killian JA, de Kroon AI, de Kruijff B (2003) Phospholipid flop induced by transmembrane peptides in model membranes is modulated by lipid composition. Biochemistry 42:231-237

Kooijman EE, Chupin V, de Kruijff B, Burger KN (2003) Modulation of membrane curvature by phosphatidic acid and lysophosphatidic acid. Traffic 4:162-174

Kornberg RD, McConnell HM (1971) Inside-outside transitions of phospholipids in vesicle membranes. Biochemistry 10:1111-1120

Koster G, VanDuijn M, Hofs B, Dogterom M (2003) Membrane tube formation from giant vesicles by dynamic association of motor proteins. Proc Natl Acad Sci USA 100:15583-15588

Kuypers FA, Roelofsen B, Berendsen W, Op den Kamp JA, van Deenen LL (1984) Shape changes in human erythrocytes induced by replacement of the native phosphatidylcholine with species containing various fatty acids. J Cell Biol 99:2260-2267

Lauffenburger DA, Horwitz AF (1996) Cell migration: a physically integrated molecular process. Cell 84:359-369

Lee C, Ferguson M, Chen LB (1989) Construction of the endoplasmic reticulum. J Cell Biol 109:2045-2055

Lee E, Marcucci M, Daniell L, Pypaert M, Weisz OA, Ochoa GC, Farsad K, Wenk MR, De Camilli P (2002) Amphiphysin 2 (Bin1) and T-tubule biogenesis in muscle. Science 297:1193-1196

Lee S, Furuya T, Kiyota T, Takami N, Murata K, Niidome Y, Bredesen DE, Ellerby HM, Sugihara G (2001) *De novo*-designed peptide transforms Golgi-specific lipids into Golgi-like nanotubules. J Biol Chem 276:41224-41228

Lippincott-Schwartz J, Roberts TH, Hirschberg K (2000) Secretory protein trafficking and organelle dynamics in living cells. Annu Rev Cell Dev Biol 16:557-589

Madden TD, Cullis PR (1982) Stabilization of bilayer structure for unsaturated phosphatidylethanolamines by detergents. Biochim Biophys Acta 684:149-153

Marguet D, Luciani MF, Moynault A, Williamson P, Chimini G (1999) Engulfment of apoptotic cells involves the redistribution of membrane phosphatidylserine on phagocyte and prey. Nat Cell Biol 1:454-456

Marx U, Lassmann G, Holzhutter HG, Wustner D, Muller P, Hohlig A, Kubelt J, Herrmann A (2000) Rapid flip-flop of phospholipids in endoplasmic reticulum membranes studied by a stopped-flow approach. Biophys J 78:2628-2640

Mathivet L, Cribier S, Devaux PF (1996) Shape change and physical properties of giant phospholipid vesicles prepared in the presence of an AC electric field. Biophys J 70:1112-1121

Matsuo H, Chevallier J, Mayran N, Le Blanc I, Ferguson C, Faure J, Blanc NS, Matile S, Dubochet J, Sadoul R, Parton RG, Vilbois F, Gruenberg J (2004) Role of LBPA and Alix in multivesicular liposome formation and endosome organization. Science 303:531-534

Matsuoka K, Orci L, Amherdt M, Bednarek SY, Hamamoto S, Schekman R, Yeung T (1998) COPII-coated vesicle formation reconstituted with purified coat proteins and chemically defined liposomes. Cell 93:263-275

McIntosh TJ, Simon SA, Needham D, Huang CH (1992) Structure and cohesive properties of sphingomyelin/cholesterol bilayers. Biochemistry 31:2012-2020

Menon AK, Watkins WE, Hrafnsdottir S (2000) Specific proteins are required to translocate phosphatidylcholine bidirectionally across the endoplasmic reticulum. Curr Biol 10:241-252

Mercier C, Dubremetz JF, Rauscher B, Lecordier L, Sibley LD, Cesbron-Delauw MF (2002) Biogenesis of nanotubular network in Toxoplasma parasitophorous vacuole induced by parasite proteins. Mol Biol Cell 13:2397-2409

Mills IG, Praefcke GJ, Vallis Y, Peter BJ, Olesen LE, Gallop JL, Butler PJ, Evans PR, McMahon HT (2003) EpsinR: an AP1/clathrin interacting protein involved in vesicle trafficking. J Cell Biol 160:213-222

Mitchison TJ, Cramer LP (1996) Actin-based cell motility and cell locomotion. Cell 84:371-379

Miyata H, Hotani H (1992) Morphological changes in liposomes caused by polymerization of encapsulated actin and spontaneous formation of actin bundles. Proc Natl Acad Sci USA 89:11547-11551

Muller P, Pomorski T, Herrmann A (1994) Incorporation of phospholipid analogues into the plasma membrane affects ATP-induced vesiculation of human erythrocyte ghosts. Biochem Biophys Res Commun 199:881-887

Musacchio A, Smith CJ, Roseman AM, Harrison SC, Kirchhausen T, Pearse BM (1999) Functional organization of clathrin in coats: combining electron cryomicroscopy and X-ray crystallography. Mol Cell 3:761-770

Nossal R (2001) Energetics of clathrin basket assembly. Traffic 2:138-147

Pagano RE (1990) Lipid traffic in eukaryotic cells: mechanisms for intracellular transport and organelle-specific enrichment of lipids. Curr Opin Cell Biol 2:652-663

Palmer AF, Wingert P, Nickels J (2003) Atomic force microscopy and light scattering of small unilamellar actin-containing liposomes. Biophys J 85:1233-1247

Pelham HR (2001) Traffic through the Golgi apparatus. J Cell Biol 155:1099-1101

Pelkmans L, Helenius A (2003) Insider information: what viruses tell us about endocytosis. Curr Opin Cell Biol 15:414-422

Polizotto RS, de Figueiredo P, Brown WJ (1999) Stimulation of Golgi membrane tubulation and retrograde trafficking to the ER by phospholipase A(2) activating protein (PLAP) peptide. J Cell Biochem 74:670-683

Pomorski T, Lombardi R, Riezman H, Devaux PF, van Meer G, Holthuis JC (2003) Drs2p-related P-type ATPases Dnf1p and Dnf2p are required for phospholipid translocation across the yeast plasma membrane and serve a role in endocytosis. Mol Biol Cell 14:1240-1254

Pomorski T, Muller P, Zimmermann B, Burger K, Devaux PF, Herrmann A (1996) Trans-bilayer movement of fluorescent and spin-labeled phospholipids in the plasma membrane of human fibroblasts: a quantitative approach. J Cell Sci 109 Pt 3:687-698

Pornillos O, Garrus JE, Sundquist WI (2002) Mechanisms of enveloped RNA virus budding. Trends Cell Biol 12:569-579

Presley JF, Ward TH, Pfeifer AC, Siggia ED, Phair RD, Lippincott-Schwartz J (2002) Dissection of COPI and Arf1 dynamics in vivo and role in Golgi membrane transport. Nature 417:187-193

Rawicz W, Olbrich KC, McIntosh T, Needham D, Evans E (2000) Effect of chain length and unsaturation on elasticity of lipid bilayers. Biophys J 79:328-339

Roux A, Cappello G, Cartaud J, Prost J, Goud B, Bassereau P (2002) A minimal system allowing tubulation with molecular motors pulling on giant liposomes. Proc Natl Acad Sci USA 99:5394-5399

Ruetz S, Gros P (1994) Phosphatidylcholine translocase: a physiological role for the mdr2 gene. Cell 77:1071-1081

Sachse M, Urbe S, Oorschot V, Strous GJ, Klumperman J (2002) Bilayered clathrin coats on endosomal vacuoles are involved in protein sorting toward lysosomes. Mol Biol Cell 13:1313-1328

Schmidt A, Wolde M, Thiele C, Fest W, Kratzin H, Podtelejnikov AV, Witke W, Huttner WB, Soling HD (1999) Endophilin I mediates synaptic vesicle formation by transfer of arachidonate to lysophosphatidic acid. Nature 401:133-141

Seigneuret M, Devaux PF (1984) ATP-dependent asymmetric distribution of spin-labeled phospholipids in the erythrocyte membrane: relation to shape changes. Proc Natl Acad Sci USA 81:3751-3755

Sheetz MP, Singer SJ (1974) Biological membranes as bilayer couples. A molecular mechanism of drug-erythrocyte interactions. Proc Natl Acad Sci USA 71:4457-4461

Shemesh T, Luini A, Malhotra V, Burger KN, Kozlov MM (2003) Prefission constriction of Golgi tubular carriers driven by local lipid metabolism: a theoretical model. Biophys J 85:3813-3827

Simons K, Ikonen E (1997) Functional rafts in cell membranes. Nature 387:569-572

Smit JJ, Schinkel AH, Oude Elferink RP, Groen AK, Wagenaar E, van Deemter L, Mol CA, Ottenhoff R, van der Lugt NM, van Roon MA, et al. (1993) Homozygous disruption of the murine mdr2 P-glycoprotein gene leads to a complete absence of phospholipid from bile and to liver disease. Cell 75:451-462

Spang A, Matsuoka K, Hamamoto S, Schekman R, Orci L (1998) Coatomer, Arf1p, and nucleotide are required to bud coat protein complex I-coated vesicles from large synthetic liposomes. Proc Natl Acad Sci USA 95:11199-11204

Takei K, Haucke V, Slepnev V, Farsad K, Salazar M, Chen H, De Camilli P (1998) Generation of coated intermediates of clathrin-mediated endocytosis on protein-free liposomes. Cell 94:131-141

Tanford C (1978) The hydrophobic effect and the organization of living matter. Science 200:1012-1018

Tang X, Halleck MS, Schlegel RA, Williamson P (1996) A subfamily of P-type ATPases with aminophospholipid transporting activity. Science 272:1495-1497

Thery C, Zitvogel L, Amigorena S (2002) Exosomes: composition, biogenesis and function. Nat Rev Immunol 2:569-579

Vale RD, Hotani H (1988) Formation of membrane networks *in vitro* by kinesin-driven microtubule movement. J Cell Biol 107:2233-2241

van Helvoort A, Smith AJ, Sprong H, Fritzsche I, Schinkel AH, Borst P, van Meer G (1996) MDR1 P-glycoprotein is a lipid translocase of broad specificity, while MDR3 P-glycoprotein specifically translocates phosphatidylcholine. Cell 87:507-517

van Meer G (1998) Lipids of the Golgi membrane. Trends Cell Biol 8:29-33

Voorberg J, Fontijn R, Calafat J, Janssen H, van Mourik JA, Pannekoek H (1993) Biogenesis of von Willebrand factor-containing organelles in heterologous transfected CV-1 cells. EMBO J 12:749-758

Wagner DD, Saffaripour S, Bonfanti R, Sadler JE, Cramer EM, Chapman B, Mayadas TN (1991) Induction of specific storage organelles by von Willebrand factor propolypeptide. Cell 64:403-413

Walev I, Weller U, Strauch S, Foster T, Bhakdi S (1996) Selective killing of human monocytes and cytokine release provoked by sphingomyelinase (beta-toxin) of *Staphylococcus aureus*. Infect Immun 64:2974-2979

Wang Y, Thiele C, Huttner WB (2000) Cholesterol is required for the formation of regulated and constitutive secretory vesicles from the trans-Golgi network. Traffic 1:952-962

Weigert R, Silletta MG, Spano S, Turacchio G, Cericola C, Colanzi A, Senatore S, Mancini R, Polishchuk EV, Salmona M, Facchiano F, Burger KN, Mironov A, Luini A, Corda D (1999) CtBP/BARS induces fission of Golgi membranes by acylating lysophosphatidic acid. Nature 402:429-433

Williamson P, Kulick A, Zachowski A, Schlegel RA, Devaux PF (1992) Ca^{2+} induces transbilayer redistribution of all major phospholipids in human erythrocytes. Biochemistry 31:6355-6360

Wimley WC, Thompson TE (1991) Transbilayer and interbilayer phospholipid exchange in dimyristoylphosphatidylcholine/dimyristoylphosphatidylethanolamine large unilamellar vesicles. Biochemistry 30:1702-1709

Zachowski A (1993) Phospholipids in animal eukaryotic membranes: transverse asymmetry and movement. Biochem J 294 Pt 1:1-14

Zha X, Genest J Jr, McPherson R (2001) Endocytosis is enhanced in Tangier fibroblasts: possible role of ATP-binding cassette protein A1 in endosomal vesicular transport. J Biol Chem 276:39476-39483

Zha X, Pierini LM, Leopold PL, Skiba PJ, Tabas I, Maxfield FR (1998) Sphingomyelinase treatment induces ATP-independent endocytosis. J Cell Biol 140:39-47

Holthuis, Joost C. M.
Department of Membrane Enzymology, Faculty of Chemistry, Padualaan 8, 3584 CH Utrecht, The Netherlands
j.c.holthuis@chem.uu.nl

Functions of Rab GTPases in organelle biogenesis

Christopher G. Burd and Ruth N. Collins

Abstract

Rab GTPases regulate diverse aspects of the biogenesis of secretory and endocytic organelles. The conformational changes that accompany GTP binding and hydrolysis by Rab are harnessed to recruit and activate specific effector proteins that regulate vesicle tethering and fusion, cargo sorting, and cytoskeleton-dependent organelle transport. We review recent progress in understanding how Rab GTPases are regulated and how they regulate membrane trafficking.

1 General considerations

In the late 1980's, a convergence of observations implicating GTP hydrolysis by small G proteins in regulating membrane trafficking culminated with the discovery by Salminen and Novick that the yeast Ras-related GTPase Sec4 regulates fusion of secretory vesicles with the plasma membrane (Salminen and Novick 1987). This watershed discovery focused research into the functions of Rab GTPases as key regulators of the secretory and endocytic pathways. All molecularly characterized eukaryotic genomes encode Rab GTPases: the yeast *Saccharomyces cerevisiae* contains 11 (often called "Ypt" instead of "Rab"), *Drosophila melanogaster* 29, *Caenorhabditis elegans* 29, *Arabidopsis thaliana* 57, and human at least 60 Rabs (Pereira-Leal and Seabra 2000; Vernoud et al. 2003). Rabs are localized to nearly every membrane-bounded organelle and their general functions can be said to be in regulating vesicle-mediated trafficking between them. In this review, we shall discuss what has been learned regarding the functions of Rab GTPases by focusing on the following major questions: What are the signals and factors that activate Rabs? What are the functions of effectors that recognize activated Rabs? What is the role of GTP hydrolysis by Rab? How are Rabs targeted to distinct sites within the cell? Remarkably, the answers to these questions are different for every Rab, so the diversity in their regulation and functions makes the study of individual Rabs all the more interesting and important.

Topics in Current Genetics, Vol. 10
S. Keränen, J. Jäntti (Eds.): Regulatory Mechanisms of Intracellular Membrane Transport
DOI 10.1007/b97781 / Published online: 23 March 2004
© Springer-Verlag Berlin Heidelberg 2004

2 The GTPase cycle

The Rab GTPase cycle consists of GTP-for-GDP exchange, which activates Rab signaling and GTP hydrolysis, which terminates Rab signaling (Fig. 1). This cycle is often described as a molecular switch, an analogy that is based on the observation that downstream signaling by Ras occurs when GTP occupies the active site, but is inactive when GDP occupies the active site. While this paradigm is generally true for Rab GTPases, an important distinction to be made is that fusion of biological membranes is not strictly dependent upon GTP hydrolysis by Rab, at least for the cases that have been rigorously tested (Geppert et al. 1994; Rybin et al. 1996). Moreover, overexpression of appropriate SNAREs or a mutation in the Sec1/SM family protein Sly1 can bypass the essential requirement for the yeast Ypt1 Rab in ER-to-Golgi transport (Dascher et al. 1991; Lian et al. 1994). The physiological function of Rab appears to be to cycle between the GTP-bound and the GDP-bound conformations, which allows it to recruit multiple effectors to particular sites on cellular membranes and to facilitate their assembly into larger complexes through cooperative interactions. In this context, the rate of GTP hydrolysis by Rab constitutes a molecular timer that facilitates and monitors the formation of macromolecular complexes of proteins and lipids required for fusion (Aridor and Balch 1996; Rybin et al. 1996; Zerial and McBride 2001).

An additional layer of complexity to the Rab GTPase cycle is imparted from exchange of Rab between organelle membranes and the cytosol where it is bound to the Rab-specific chaperone, Rab GDP dissociation inhibitor (GDI) (Fig. 1). Membrane anchoring of Rab is mediated by covalently attached long chain isoprenoid lipids and the role of GDI is to maintain a soluble reservoir of prenylated Rab in the inactive GDP-bound state. Membrane-associated, inactive Rab-GDP is solubilized by extraction from the membrane by GDI, and the GDI-Rab complex dissociates upon delivery of Rab to a membrane (Ullrich et al. 1993). Thus, Rabs are only transiently associated with organelle membranes, indicating that organelle targeting occurs continuously.

GTP binding by Rab is facilitated by guanine nucleotide exchange factors (GEFs), and GTP hydrolysis is facilitated by GTPase activating proteins (GAPs). The identification of the cognate GEF for every Rab is crucial for understanding the upstream events that activate Rab signaling, yet only a handful of Rab GEFs have been identified so far and they are unrelated in terms of primary sequence. Null mutations in the genes encoding Rab GEFs generally phenocopy null mutations in the corresponding Rab genes, indicating that GTP loading of Rab is essential for Rab signaling. The mechanisms by which Rab GEFs activate Rab are not yet known, but they probably induce a conformational change in Rab that stabilizes the nucleotide-empty conformation, thus favoring GTP loading from the cytosol (Goody and Hofmann-Goody 2002).

In the GTP-bound state, Rabs recruit molecules termed effectors that are specific for one or a subset of related Rabs and these molecules carry out the downstream functions associated with each Rab. There is recent evidence suggesting that the GDP form of some Rabs have functions that are distinct from the

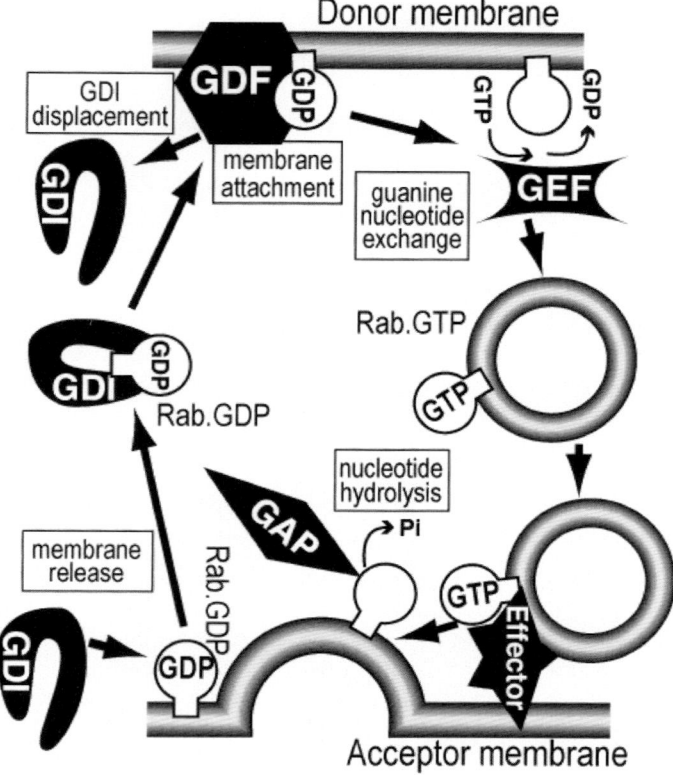

Fig. 1. The Rab GTPase cycle. A Rab GTPase is activated by GTP-for-GDP nucleotide exchange on the organelle membrane that is facilitated by a guanine nucleotide exchange factor (GEF). In the activated state, Rab-GTP recruits specific effectors. Hydrolysis of GTP is stimulated by a GTPase activating protein (GAP) and returns Rab to the inactive GDP-bound form. Rab-GDP is a ligand for GDI, which extracts Tab from the membrane and recycles it back to the appropriate organelle by loading it onto the organelle membrane. Attachment of Rab onto the organelle membrane is facilitated by the action of specific GDI displacement factors (GDF).

GTP-bound effector functions. More than 50 proteins have been identified that are associated with Rab5-GTP (directly or indirectly), suggesting that each Rab probably has many different effectors (Zerial and McBride 2001). These molecules have typically been identified by Rab-GTP affinity chromatography and by yeast two-hybrid interaction trap screening. Analysis of the primary sequences of Rab effectors have not revealed common motifs, so it is likely that recognition of different Rabs occurs through diverse mechanisms.

In contrast to the diversity of Rab GEFs and effectors, nearly all Rab GAPs possess a TBC (Tre17 Bub2 Cdc16) domain (Richardson and Zon 1995) that contains the catalytic activity (Bernards 2003). Structure and function studies of a

TBC domain derived from yeast Gyp1 GAP indicates that GAP activity is mediated by an Arginine finger-type mechanism where the side chain of the Arginine residue directly participates in GTP hydrolysis (Albert et al. 1999; Rak et al. 2000). Analysis of yeast TBC proteins has revealed that none of them are encoded by essential genes and that *in vivo* they have GAP activity towards multiple Rabs, suggesting that *in vitro* their functions may be regulated by localization and other factors (Albert and Gallwitz 1999). Recently, two GAPs for the yeast Sec4 GTPases were found to link signaling by the Rho GTPase Cdc42 to polarized secretion (Gao et al. 2003).

2.1 Structural features of Rab

There is a high degree of amino acid identity between all Ras-related GTPases and this is due to their conserved functions in binding and hydrolyzing GTP. Within the past several years the structures of a handful of Rab GTPases have been elucidated and their three dimensional structures have been found to be generally the same as Ras and other Ras-related GTPases, as expected due to the highly conserved GTP-binding fold (Dumas et al. 1999; Ostermeier and Brunger 1999; Esters et al. 2000; Stroupe and Brunger 2000; Merithew et al. 2001; Constantinescu et al. 2002; Rak et al. 2003; Zhu et al. 2003). Given the high degree of amino acid identity and structural similarity in the GTPase domains of all Ras-related GTPases, it can be difficult to classify a particular Ras-related GTPase as a Rab, however, five short sequences called RabF motifs that are clustered about the switch1 and switch2 regions that undergo nucleotide-dependent conformational changes have been described that are diagnostic of Rab (Pereira-Leal and Seabra 2000). Because the structural characteristics of these regions are general features of all Rabs, indeed of all Ras-related GTPases, the location of the RabF motifs has two implications. First, there are subtle differences in the conformational changes that accompany nucleotide exchange and hydrolysis that are unique to the Rab family (Merithew et al. 2001), and second, distinct recognition of a particular Rab by its effectors must involve multiple surfaces of the enzyme. Indeed, recognition of Rab3A by Rabphilin is by two distinct binding surfaces, the switch1/switch2 region and a second region whose sequence is diagnostic of the Rab3 sub-family (Ostermeier and Brunger 1999).

Sub-families of GTPases can be clearly distinguished within the entire Rab family that correspond to functionally related Rabs that regulate secretion or endocytosis (Fig. 2). Analysis of Rab protein sequences can, therefore, be used to predict the general functions of uncharacterized Rabs. With regard to the primary sequences of Rabs, the most divergent regions when compared to one another are the amino-terminal extensions (generally 5-20 amino acids in length) and the carboxy-terminal 20-40 amino acids that comprise the so-called hypervariable (HV) region (Pereira-Leal and Seabra 2000). Each of these regions do not appear to be structured in free Rabs, but each region has been implicated in Rab localization, raising the possibility that they are recognized by other proteins or interact with membranes (Chavrier et al. 1991; Brennwald and Novick 1993; Dunn et al. 1993).

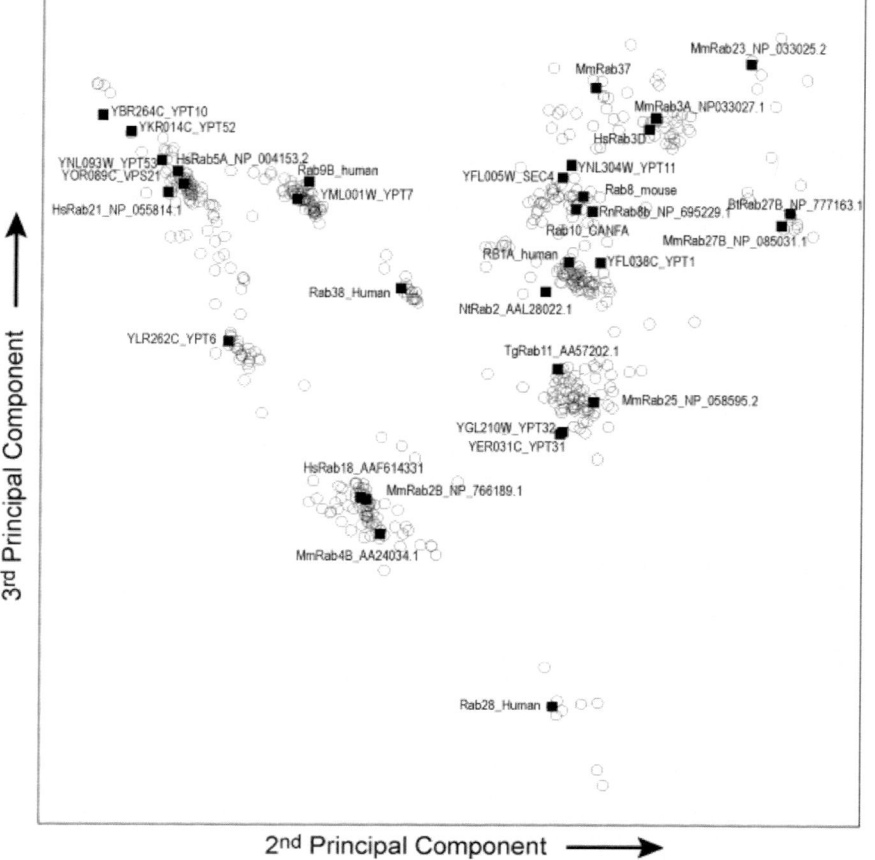

Fig. 2. Two-dimensional version of an evolutionary tree showing the sequence distances between 560 complete and unique Rab GTPase sequences identified in the database as of December 2003. Genomic sequences were manually inspected in conjunction with alignment and EST data order to identify correct gene start and splice sites, in several cases these differed from automatic computer identification. A ClustalW alignment was performed to calculate degree of identity and divergence matrices, which were used to generate a two-dimensional principal components plot, with the x-axis representing the second and the y-axis the third principal component of the relative sequence distances. Endocytic and exocytic Rab subfamilies can be clearly distinguished as clusters that project away from the center. A subset of Rab GTPases (squares) are labeled with accession numbers, these include all the Rab proteins of *S. cerevisiae* and cover most of the known Rab GTPase sequence space (circles).

The structure of yeast Ypt1 in a complex with the yeast Rab escort protein (REP, discussed below) shows that its HV region wraps loosely around the surface of REP, making only minimal molecular contacts with the surface of REP (Rak et al.

2003). Given the importance of HV regions in Rab localization, it is anticipated that organelle-specific factors recognize the HV regions of different Rabs and facilitate Rab localization.

2.2 Prenylation and localization of Rab GTPases

Rab proteins are localized to the cytoplasmic leaflet of different organelles and to the cytosol. General membrane association is mediated by two isoprenoid lipids that are covalently and irreversibly attached to two cysteine residues located at or very near the carboxy terminus of Rab in one of the following sequences: CXC, CC, CCX, CCXX, CCXXX (Maurer-Stroh et al. 2003). Rab isoprenylation is catalyzed by Rab GGTase II (geranylgeranyl transferase II, RGGT), consisting of two core subunits and a third subunit REP (Rab Escort Protein), which does not participate in the catalytic reaction, but serves as a chaperone to deliver the prenyltransferase to its Rab protein substrate (Seabra et al. 1992). Incompletely prenylated (i.e. monoprenylated) Rabs are non-functional and are mislocalized to the endoplasmic reticulum (ER), indicating that the double prenylation motif that is unique to Rab proteins also plays a specific functional role in Rab localization (Calero et al. 2003; Gomes et al. 2003). Interestingly, the yeast integral membrane protein Yip1 has been shown to bind doubly-prenylated, but not mono-prenylated Rabs and is required for localization of Ypt1p to the Golgi (Calero et al. 2003).

Newly prenylated Rab is delivered to the correct organelle membrane in a complex with REP, but little is known regarding specific targeting mechanisms (Alexandrov et al. 1994). Because Rab proteins continuously exchange between organelle membranes and the cytosol, and because Rabs are localized to transport intermediates, Rab localization is dynamic and likely to reflect the contributions of many different variables. Localization studies of chimeric Rabs have shown that Rab proteins themselves contain all the information for their intracellular localization, the specificity determinants being unique to each family member. Although multiple features of Rabs contribute to their localization, the predominant organelle targeting signals appear to reside in the COOH-terminal (HV) regions that end with the prenylated cysteine residues. In support of this, chimeric Rabs containing the HV regions of Rab5 or Rab7 were targeted to early and late endosomes, respectively (Chavrier et al. 1991), and the HV region of yeast Rab Ypt1 partly targeted Sec4p to post-Golgi vesicles when it replaced the Sec4 HV region (Brennwald and Novick 1993; Dunn et al. 1993).

Rabs that accompany transport vesicles must be recycled after each round of vectorial membrane transport between a donor and acceptor compartment and the membrane attachment/detachment cycles of Rabs are coupled to this process (Fig. 1). Through its preferential interaction with Rab proteins in their GDP-bound conformation, Rab-GDI provides directionality to the process of Rab protein retrieval by collecting the "spent" GDP-bound Rab protein after each round of transport and transferring it through the cytosol back onto the donor compartment where it can be activated for another round of transport (Barbero et al. 2002). In support of this, several studies have shown that the cytosolic GDI-Rab dimer is sufficient to

deliver Rab GTPases to their correct membrane-bound compartments but very little is known regarding mechanisms (Dirac-Svejstrup et al. 1994; Soldati et al. 1995). Membrane loading does not require nucleotide exchange, but does require membrane-associated proteins that facilitate Rab membrane loading, perhaps by binding and/or displacing GDI from the GDI-Rab complex, freeing the geranyl-geranyl moieties to insert into the membrane (Dirac-Svejstrup et al. 1997; Luan et al. 1999; Chou and Jahn 2000; Gilbert and Burd 2001). Integral membrane proteins that interact with prenylated Rabs, but not non-prenylated Rabs, have been identified and an activity with the property of a GDI displacement factor (GDF) for endosomal Rab proteins has been identified (Martincic et al. 1997; Hutt et al. 2000; Calero et al. 2001, 2002; Calero and Collins 2002; Sivars et al. 2003). Rab proteins can be activated by their exchange factors only after displacement from GDI, indicating that membrane loading and activation are two distinct events (Collins et al. 1997; Fukui et al. 1997).

3 Regulation of inter-organelle trafficking and organelle transport

For most Rabs, surprisingly little is known regarding the specific mechanisms by which Rabs control vesicle-mediated protein sorting. In general terms, Rabs regulate the formation of functionally specialized domains within organelle membranes by recruiting proteins and lipids to specific sites. Individual Rabs function in different aspects of trafficking, including the sorting of cargo proteins into budding vesicles, the regulation of membrane fusion, and the transport of organelles (Fig. 3).

3.1 Cargo sorting and vesicle budding

Proteins that are to be packaged into transport vesicles for export to other organelles need to be distinguished from resident proteins and concentrated into a budding patch of membrane. These tasks are carried out primarily by vesicle coat proteins that recognize sorting signals in the cytoplasmic domains of cargos and organize them into a vesicle bud. Members of the Sar and Arf families of GTPases are well known regulators of these events (Duden 2003), however, there is also evidence, described below, that several Rabs regulate cargo sorting via binding to cargo molecules or adapter proteins that bind Rab and cargo.

In polarized epithelial cells, Rab3b regulates transcytosis of polymeric immunoglobulin (pIg) from the basolateral membrane to the apical membrane where it is secreted into the lumen of the gut. Transcytosis is mediated by the polymeric immunoglobulin receptor (pIgR) and van Ijzendoorn et al. (2002) found that Rab3b judges the occupancy state of the pIgR and regulates it's trafficking itinerary accordingly: Rab3b-GTP bound the cytoplasmic domain of pIgR when it was unoccupied and targeted these vesicles back to the basal membrane, but when

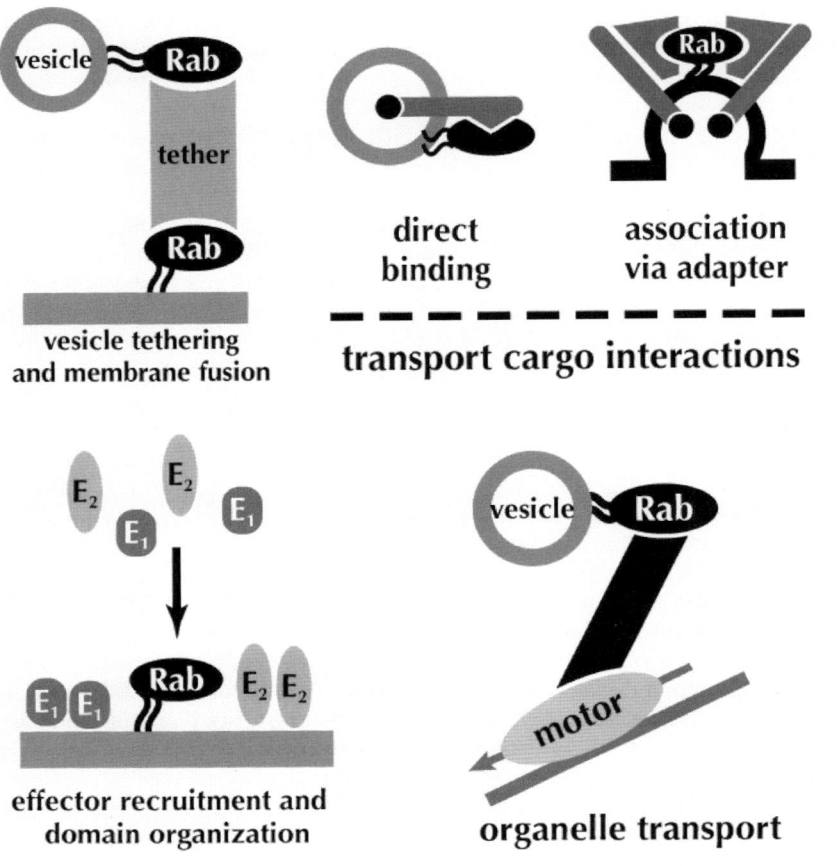

Fig. 3. Diversity of Rab functions. The roles of Rab GTPases in organelle biogenesis include: regulation of vesicle tethering and fusion factors, sorting of cargo proteins into transport vesicles by directly binding the cargo or by binding an adapter protein that binds the Rab and the cargo, recruitment of multiple effectors (E1, E2, etc) to membranes, and by linking cytoskeletal motors to organelles by binding adapters that bind the Rab and the motor.

dIgA occupied the receptor, Rab3b was displaced and the receptor-dIgA was transcytosed to the apical membrane. The cytoplasmic tail of the Angiotensin II Type1A receptor has been reported to bind Rab5a and this interaction promoted transport of the receptor-ligand complex along the endocytic/lysosomal pathway (Seachrist et al. 2002). Recognition of cargo molecules by Rabs could be a mechanism for targeting the cognate vesicles to particular organelles through interactions with other Rab-binding factors, such as tethering proteins (discussed below).

A more typical role for Rabs in cargo sorting is through binding of an adapter protein that binds the cargo protein and the Rab. Packaging of mannose-6-

phosphate receptors (M6PR) into Golgi-targeted recycling vesicles that bud from late endosomes is facilitated by an oligomeric protein named TIP47 that binds M6PRs and Rab9-GTP, suggesting that TIP47 links cargo selection and vesicle targeting (Carroll et al. 2001; Sincock et al. 2003). For the epidermal growth factor receptor (EGFR), the SH2 domain-containing protein RIN1 binds to tyrosine phosphorylated EGFR and activates Rab5 via its Rab5 GEF domain, which stimulates the rate of endocytosis of activated EGFR (Tall et al. 2001; Barbieri et al. 2003). Rab5 has also been implicated in sequestration of transferrin receptor in clathrin-coated pits (McLauchlan et al. 1998), and yeast Ypt1 has been implicated in sorting of GPI-anchored proteins in ER but the underlying molecular mechanisms for either of these sorting events is not yet known (Morsomme and Riezman 2002).

3.2 Regulation of membrane tethering and fusion

Fusion of transport vesicles with the membranes of appropriate target organelles occurs through a complex step-wise series of regulated biochemical reactions that end in SNARE-mediated fusion of the vesicle and organelle membranes. In biochemically reconstituted membrane fusion assays, one of the earliest events is the capture of a vesicle by proteins associated with the target organelle, a process referred to as vesicle "tethering." Many tethering reactions are critically dependent upon Rab GTPases and it has been suggested that this may be a widespread and general function of Rabs (Waters and Pfeffer 1999). Although the concept of tethering is relatively new and the biochemical nature of tethering is poorly understood, it seems that the significance of tethering is that it could promote the formation of appropriate trans-SNARE complexes by facilitating the approach of a vesicle with the appropriate organelle membrane, thus providing an initial layer of specificity that precedes the formation of trans-SNARE complexes (Whyte and Munro 2002).

Tethering of ER-derived COPII vesicles to Golgi membranes is regulated by Rab1 via its effectors, including p115, a long rod-shaped coiled-coil protein that localizes to the cis-Golgi at steady state (Barroso et al. 1995). In an *in vitro* assay that reconstituted budding of COPII vesicles from the ER, p115 was found to be recruited to budding COPII vesicles by Rab1-GTP, suggesting that this interaction may program fusion of these vesicles with the Golgi because Rab1 and p115 also bind the Golgi proteins GM130 and GRASP65 (Allan et al. 2000; Moyer et al. 2001; Yoshimura et al. 2001). In yeast, Golgi-localized Ypt1 regulates tethering of vesicles to the Golgi through multiple effectors, including the p115-related protein, Uso1, although it is not known if Ypt1 and Uso1 directly bind each other (Cao et al. 1998). Other Rabs, such as Rab2, may also regulate ER-to-Golgi transport, through orchestrating interactions with putative tethering factors (Short et al. 2001). Interestingly, Rab2 has been reported to stimulate recruitment of COPI and atypical protein kinase Cι/λ to pre-Golgi compartments, suggesting that it coordinates the functions of a wide variety of factors required for retrograde transport and Golgi structure (Tisdale and Balch 1996; Tisdale and Jackson 1998; Tisdale

2003). A second Golgi tethering factor for ER-derived vesicles called the TRAPP complex has been identified in yeast. It is a multi-subunit protein complex that binds to COPII vesicles independently of Ypt1 but it exhibits Ypt1 GEF activity, so it is positioned to signal the arrival of a vesicle at the Golgi and to initiate Ypt1-dependent downstream events, including Uso1-mediated tethering (Wang et al. 2000; Sacher et al. 2001).

The overall structure of the Golgi arises in part from the balanced flow of cargo into and out of Golgi compartments, so it is logical to expect that the arrival of cargo at an early compartment might be coupled to export of cargo. Interestingly, Ypt1-GTP affinity-purified material was found to exhibit GEF activity for Ypt32, a Rab required for the formation of secretory vesicles from the trans Golgi (Benli et al. 1996; Jedd et al. 1997; Wang and Ferro-Novick 2002). A Ypt32-GTP effector, Sec2, is a GEF for Sec4 required for loading of Sec4 onto secretory vesicles (Ortiz et al. 2002). Sec4 regulates polarized targeting of vesicles with the plasma membrane in part through binding its effector Sec15, a component of the exocyst complex that is localized to sites of polarized growth (Guo et al. 1999). These observations support the idea that a cascade of Rab GTPases, Ypt1-to-Ypt32-to-Sec4, functions to sequentially activate downstream-acting Rabs in the secretory pathway, ultimately leading to the Sec4-regulated tethering and fusion of secretory vesicles with the plasma membrane (Ortiz et al. 2002).

Rab-regulated tethering plays an important role in membrane fusion within the endosomal/lysosomal systems. After internalization of membrane by endocytosis, primary endocytic vesicles fuse with each other and with early endosomes and these events are regulated by members of the Rab5 sub-family (Rab5a-d) (Bucci et al. 1992; Stenmark et al. 1994). Rab5-GTP mediates these effects in part through the recruitment of one of its effectors, Early Endosome Autoantigen (EEA1), a tethering factor that also binds Syntaxin13 (McBride et al. 1999). A similar complex was first described in yeast where Vac1p, an EEA1-related protein, binds the yeast Rab5 ortholog Vps21-GTP, the tSNARE Pep12p, and the Sec1/SM family protein Vps45p (Burd et al. 1997; Peterson et al. 1999). EEA1 and Vac1p also bind phosphatidylinositol 3-phosphate (PtdIns(3)P), a lipid that is enriched in endosomal membranes, via their FYVE domains and together these results suggest that signaling by Rab5 family proteins coordinates the formation of conserved protein and lipid complexes that directly regulate fusion of endosomes (Burd and Emr 1998; Gaullier et al. 1998; Patki et al. 1998). Importantly, phosphatidylinositol 3-kinase (PI3K) has been identified as a component of Rab5-GTP effector complexes (Christoforidis et al. 1999; Murray et al. 2002) and this could enhance the recruitment of Rab5 effectors that bind PtdIns(3)P, such as EEA1 and Rabenosyn-5 (Nielsen et al. 2000). PI3K has also been implicated as an effector of Rab7 (Stein et al. 2003).

A great deal has been learned from *in vitro* studies of homotypic fusion of yeast lysosome-like vacuoles where tethering can be directly visualized as the clustering of fluorescently labeled vacuoles (Mayer and Wickner 1997). Vacuole tethering is regulated by the Ypt7 Rab and *in vitro* tethering was found to be sensitive to addition of recombinant Rab GDI, which removes Ypt7 from vacuole membranes, but not sensitive to addition of the GAP domain of Gyp1, which accelerates GTP hy-

drolysis by Ypt7 and should inactivate Ypt7 (Wang et al. 2003). These observations suggest that Ypt7-GDP was functional for tethering and it is some of the first evidence for an active function of the GDP-bound conformation of a Rab. Ypt7 was found to be enriched at the junctions of tethered vacuoles ("vertex rings"), where it remained throughout the fusion reaction, and nucleotide exchange on Ypt7, mediated through a complex of proteins called Class C Vps complex/HOPS, was required for the ordered recruitment and enrichment of other fusion factors including SNAREs within the vertex ring (Wurmser et al. 2000; Wang et al. 2003).

Overexpression of some Rabs, such as Rab5, leads to increased vesicle fusion, while overexpression of others, such as Rab3 family proteins, appear to limit vesicle fusion. Rab3 isoforms (Rab3a-d) are components of the calcium-regulated secretion machinery in a variety of cell types, and the most compelling evidence for negative regulation of fusion by a Rab comes from an analysis of synaptic vesicle exocytosis in hippocampal cells derived from mice lacking Rab3a. Upon excitation, hippocampal cells derived from wild type mice exocytose one vesicle, while in cells lacking Rab3a multiple rounds of fusion occur, suggesting that Rab3a may coordinate the fusion machinery to ensure the fidelity (i.e. number of vesicles, location of fusion) of vesicle fusion (Geppert et al. 1994). Perhaps Rab3a exerts this effect through a sort of "anti-tethering" mechanism that is activated after fusion of one vesicle and prevents access of more vesicles to the fusion machinery.

3.3 Rabs and membrane microdomains

It has been proposed that within an organelle membrane Rabs are not symmetrically distributed, but rather that they are enriched in particular regions of the membrane that constitute molecular domains with specialized functions (Zerial and McBride 2001; Pfeffer 2003). For a few Rabs in the endosomal system there is clear evidence for asymmetrically localized Rabs that is based on high resolution fluorescence imaging of transfected GFP-tagged Rabs. In trafficking of recycling cargos such as the transferrin receptor (TfR), Rab5 regulates fusion of primary endocytic vesicles with early endosomes and Rab4 and Rab11 cooperate to regulate recycling from early endosomes back to the plasma membrane. Endosomes have distinct regions containing Rab5, Rab4, and Rab11, but these Rabs also co-localize in common domains that might reflect transitional domains where cargo sorting takes place (Sonnichsen et al. 2000; De Renzis et al. 2002). The protein Rabenosyn-5 is proposed to be a molecular link between the Rab5 and Rab4 domains because it binds both Rabs and its overexpression leads to an expansion of the common Rab5-Rab4 domain (De Renzis et al. 2002). These observations are provocative because the localization of Rab5 and Rab4 parallel that of the trafficking pathway for recycling molecules. Similarly, late endosomes contain overlapping and distinct regions enriched in Rab7 and Rab9 or both Rabs (Barbero et al. 2002). How widespread and significant the existence of unique and overlapping Rab domains is not yet known. So far, only the endosomal system has been investigated in this respect and an important consideration regarding these studies

is that they have all employed GFP-tagged, overexpressed Rabs and it is not known how they compete with endogenous Rabs for GEFS and effectors, which could dramatically influence their functionality and microlocalization.

3.4 Organelle transport

As unique constituents of different organelles, Rabs are ideal candidates to be recognized by organelle-specific factors that cooperate with the cytoskeleton to transport and localize organelles within the cell. The actin- and microtubule-based cytoskeletal networks are used in conjunction with myosins (actin), dynein/dynactin (microtubules), and kinesins (microtubules) to move organelles, and several Rab GTPases have been identified that are components of organelle transport machineries.

Melanocytes, the major pigmented cells of mammalian skin, contain lysosome-related organelles called melanosomes that are maintained at the periphery of the cell by MyosinVa and the actin cytoskeleton (Marks and Seabra 2001). Rab27 is localized to melanosomes and it recruits MyosinVa via a bridging protein, melanophilin, that binds Rab27-GTP (Strom et al. 2002; Wu et al. 2002). Each of these components of the melanosome actin-based transport system were discovered via genetic analysis of mouse pigmentation mutants, *dilute (Myo5a^d)*, *ashen (Rab27a^{ash})*, *leaden (Mlph^{ln})*, a phenotype that arises from aberrant localization of melanosomes (Mercer et al. 1991; Wilson et al. 2000; Matesic et al. 2001). In addition, analysis of the mouse coat color mutant *chocolate* implicated Rab 38, which is localized to melanosomes, in sorting of proteins to melanosomes (Loftus et al. 2002). Rab27 and related isoforms are also expressed in the pigmented cells of the eye, pancreatic beta cells, lymphocytes, and other cell types, where they may be used in a similar manner for coordinating short-range actin-based movement of organelles (Fukuda and Kuroda 2002; Desnos et al. 2003; Waselle et al. 2003).

The dynein-dynactin motor complex is responsible in part for microtubule-dependent transport of many different secretory and endocytic compartments. Three different Rab GTPases, Rab6 on Golgi membranes and Rab5 and Rab7 on late endosomes/lysosomes, have been found to be key organelle components that are used to capture organelles by the dynein-dynactin complex. Microtubule-dependent movement of late endosomes and lysosomes appears to be regulated by the Rab7 effector, RILP, whose overexpression leads to recruitment of dynein-dynactin to lysosomes and clustering of lysosomes at the microtubule organizing center (Cantalupo et al. 2001; Jordens et al. 2001). Rab5 also appears to regulate minus end-directed microtubule movement of early endosomes (Nielsen et al. 1999). For Rab6-dependent minus-end directed movement of Golgi compartments, bicaudal-D proteins that are Rab6 effectors link the organelle to the motor (Hoogenraad et al. 2001; Matanis et al. 2002; Short et al. 2002). Another Rab6-binding protein, Rabkinesin-6, is related to kinesin and may directly move Golgi membrane towards the plus ends of microtubules (Echard et al. 1998).

Human Rab32 and yeast Ypt11 have been localized to mitochondria and implicated in cytoskeleton-dependent transport of mitochondria (Alto et al. 2002; Itoh et al. 2002). In yeast, Myo2 associates with Ypt11 and this interaction is important for transport of mitochondria into growing bud (Itoh et al. 2002). In human cells overexpressing a GTP-binding-defective Ypt32 mutant, mitochondria accumulated at the microtubule organizing center suggesting that this mutant Rab interferes with microtubule plus end-directed transport of mitochondria (Alto et al. 2002). Moreover, mitochondria in these cells were elongated raising the possibility that rab32 might function to regulate mitochondrial membrane fission (Alto et al. 2002).

The motile behavior of many organelles may be explained in part by the Rab GTP cycle. Most organelles move in a stop and go fashion and abruptly change direction. This is due to the properties of the motor molecules, however, for organelles whose movement is regulated by a Rab, the GTP cycle could also contribute to this behavior. Since activated Rab-GTP is the form of Rab that couples the organelle to the motor protein, hydrolysis of GTP would sever the link to the motor, resulting in a halt to organelle movement and allowing access to a competing motor.

3.5 The roles of Rab GTPases in signaling pathways

With a general appreciation of the functions and the mechanisms of Rab GTPase regulation in hand, investigation of Rab GTPases is turning to broader questions of how Rabs integrate the functions of multiple organelles into the physiology of the cell. A wide variety of growth factors, nutrients, and cell-cell contacts initiate signal transduction cascades that elicit diverse changes in cell physiology and Rab-regulated membrane trafficking events participate in these changes in various ways. Receptor signaling often leads to an increase in the rate of endocytosis, a change in receptor trafficking itineraries, and/or transport to lysosomes for degradation. Rab7 regulates delivery of material to late endosomes and lysosmes for degradation and it was recently shown to be required for lysosome-mediated degradation of nutrient transporters in growth factor-deprived cells, which is required for apoptosis in response to growth factor withdrawal (Edinger et al. 2003).

Activation of the EGF receptor leads to tyrosine auto-phosphorylation of the intracellular domain of the receptor, creating binding sites for SH2-domain containing adapter proteins to bind the receptor. One such protein that binds the EGFR is called RIN1, a protein first identified on the basis of it's ability to attenuate Ras-mediated downstream signaling (Han and Colicelli 1995; Barbieri et al. 2003). In addition to an SH2 domain, RIN1 also contains a domain that functions as a GEF for Rab5A and RIN1 potentiates internalization of activated EGFR through activation of Rab5 (Tall et al. 2001). RIN1 has also been shown to bind c-Abl and 14-3-3 proteins, so it provides an excellent example of an adapter protein that orchestrates signal transduction and membrane trafficking.

As Rab GTPases are important general regulators of protein trafficking, mutant forms of Rab proteins that exert dominant-negative effects on receptor trafficking

are emerging as useful tools to investigate the roles of receptor signaling. For example, an emerging theme of research into the developmental biology of *Drosophila* is the use of endocytic mutants to localize the sites of signaling and to investigate the role of intracellular transport in signaling (Piddini and Vincent 2003). As our knowledge of Rab GTPase function and the ability to manipulate Rab signaling continues to expand, the frontier of Rab GTPase research will expand into many new and exciting areas of cell biology.

4 Rab proteins in disease and development

Mutations in the genes encoding several Rab GTPases and Rab-associated factors cause human disease (Table 1). Griscelli syndrome (GS) is a rare autosomal recessive disorder that associates hypopigmentation, characterized by a silver-gray sheen of the hair and the presence of large clusters of pigment in the hair shaft, and the occurrence of either a primary neurological impairment or a severe immune disorder. Genetic analysis of GS has uncovered several components of the Rab27-regulated exocytic machinery. Two different genetic forms, GS1 and GS2, respectively, account for the mutually exclusive neurological and immunological phenotypes. Mutations in the gene encoding the molecular motor protein Myosin Va (MyoVa) cause GS1 and the dilute mutant in mice (Mercer et al. 1991; Pastural et al. 1997), whereas, mutations in the gene encoding the small GTPase Rab27a are responsible for GS2 and the ashen mouse model (Wilson et al. 2000). In addition, a third form of GS (GS3), whose phenotype is restricted to the characteristic hypopigmentation of GS, results from mutation in the gene that encodes melanophilin (Mlph), the ortholog of the gene mutated in *leaden* mice (Menasche et al. 2003). This spectrum of GS conditions pinpoints the distinct molecular pathways used by melanocytes, neurons, and immune cells in secretory granule exocytosis, which remain to be elucidated. Two genetic diseases involve the Rab geranylgeranylation machinery: choroideremia, an X-linked retinal degeneration resulting from loss-of-function mutations in REP-1 (Seabra et al. 1993), and *gunmetal*, a mouse model of Hermansky-Pudlak syndrome resulting from mutations in the alpha-subunit of RGGT (Detter et al. 2000).

5 Perspective

Past research on Rab GTPases has provided a conceptual framework for understanding their functions in the regulation of vesicle-mediated trafficking, organelle biogenesis, and organelle motility. The common functional features of all Rabs include the role of the GTP cycle in dynamic targeting of Rabs and their effectors to cellular membranes, the requirement for GDI in Rab recycling, and the general functions of Rabs in vesicle and organelle trafficking (Pfeffer 2001). Despite these commonalities, one of the most important themes to emerge is the diversity of

Table 1. Rab GTPases implicated in disease

Disease pheno-type/abnormality	Rab protein	Rab accessory molecule	Reference
Alteration of circadian period and homeostatic response to sleep loss	Rab3a		(Kapfhamer et al. 2002)
Charcot-Marie-Tooth type 2B neuropathy	Rab7		(Verhoeven et al. 2003)
amyotrophic lateral sclerosis	Rab5	ALS2 (Rab5 GEF)	(Otomo et al. 2003)
Griscelli syndrome	Rab27a	myosin Va	(Pastural et al. 1997; Menasche et al. 2000)
CTL release of lytic granules	Rab27a		(Stinchcombe et al. 2001)
Pigmentation, ashen and chocolate mice mutants	Rab27a, Rab38	Mlph (Rab27 effector)	(Wilson et al. 2000; Loftus et al. 2002; Provance et al. 2002)
Cardiomyopathy	Rab1		(Wu et al. 2001)
Cancer	Rab11	TBC-domain protein	(Goldenring et al. 1999; Pei et al. 2002; Zhou et al. 2002)
Drosophila embryo development	Rab-GDI, Rab11		(Ricard et al. 2001; Dollar et al. 2002)
sonic hedgehog-mediated neural patterning	Rab23		(Eggenschwiler et al. 2001)
infectious disease, H. pylori	Rab7		(Papini et al. 1997)
X-linked mental retardation		GDI	(D'Adamo et al. 2002)
Choroideremia		REP-1	(Seabra et al. 1993)
Hermansky-Pudlak syndrome, gunmetal mouse mutants		RabGGT - subunit	(Detter et al. 2000)

specific functions of individual Rabs. For example, more than 50 Rab5-associated proteins (effectors and regulators) have been identified and they directly link Rab5 regulation to protein sorting, fusion of endosomal membranes, lipid kinase activity, and mitogenic signal transduction. Thus, Rab5 can influence a wide variety of cellular events. For most Rabs, however, remarkably little is known about the auxiliary factors that regulate them or the effectors that mediate downstream signaling. Future studies of uncharacterized Rabs will need to address the common set of core questions regarding identification of Rab GEFs and effectors and for this, proteomic-based approaches should soon provide a windfall of information that will invigorate Rab research. This information may also help to answer the long-standing question concerning the mechanisms responsible for targeting Rabs to their resident organelles. Steady state Rab localization arises from the contribu-

tions of many different factors, so a complete appreciation of Rab interaction networks should provide important insights into this question.

As more is learned about the upstream signals that lead to Rab activation and the downstream effectors of all Rabs, it should become apparent how Rab signaling networks are integrated with other cellular pathways. In this regard, proteins containing a TBC Rab GAP domain are particularly interesting because many of them also contain other modular domains associated with signal transduction (Bernards 2003), suggesting that they may link Rab signaling to signaling cascades that regulate cell growth and differentiation. By analogy to the ARF GAP proteins that also contain modular signaling domains (Nie et al. 2003), Rab GAPs may be key adapters that mediate crosstalk between signal transduction and trafficking pathways.

Most, if not all Rab GTPases in the human genome have now been identified, and the analysis of orthologous Rabs in model organisms has provided a useful framework with which to predict the cellular functions of a small subset of human Rabs. The large number of Rabs in human and plant cells portends cell type-specific functions that are presently unknown. Future Rab research will need to address these functions by examining the roles of Rabs in their native cell type. For Rabs that are conserved between mouse and human, the use of mouse knockout genetics should be especially useful for identifying loss of function phenotypes that have been the cornerstone of genetic analysis of Rab function. These studies should reveal how Rab regulation of cellular physiology contributes to the formation and function of organs in the whole organism. The future of Rab research is, therefore, at a threshold that promises to be even more exciting than the past.

Acknowledgements

Work in the authors' laboratories is supported by grants from The National Institutes of Health (GM61221 to C.G.B.), the National Science Foundation (MCB-0079045 to R.C.), and The American Heart Association (0030316T to R.C.).

References

Albert S, Gallwitz D (1999) Two new members of a family of Ypt/Rab GTPase activating proteins. Promiscuity of substrate recognition. J Biol Chem 274:33186-33189

Albert S, Will E, Gallwitz D (1999) Identification of the catalytic domains and their functionally critical arginine residues of two yeast GTPase-activating proteins specific for Ypt/Rab transport GTPases. EMBO J 18:5216-5225

Alexandrov K, Horiuchi H, Steele-Mortimer O, Seabra MC, Zerial M (1994) Rab escort protein-1 is a multifunctional protein that accompanies newly prenylated rab proteins to their target membranes. EMBO J 13:5262-5273

Allan BB, Moyer BD, Balch WE (2000) Rab1 recruitment of p115 into a cis-SNARE complex: programming budding COPII vesicles for fusion. Science 289:444-448

Alto NM, Soderling J, Scott JD (2002) Rab32 is an A-kinase anchoring protein and participates in mitochondrial dynamics. J Cell Biol 158:659-668

Aridor M, Balch WE (1996) Timing is everything. Nature 383:220-221

Barbero P, Bittova L, Pfeffer SR (2002) Visualization of Rab9-mediated vesicle transport from endosomes to the trans-Golgi in living cells. J Cell Biol 156:511-518

Barbieri MA, Kong C, Chen PI, Horazdovsky BF, Stahl PD (2003) The SRC homology 2 domain of Rin1 mediates its binding to the epidermal growth factor receptor and regulates receptor endocytosis. J Biol Chem 278:32027-32036

Barroso M, Nelson DS, Sztul E (1995) Transcytosis-associated protein (TAP)/p115 is a general fusion factor required for binding of vesicles to acceptor membranes. Proc Natl Acad Sci USA 92:527-531

Benli M, Doring F, Robinson DG, Yang X, Gallwitz D (1996) Two GTPase isoforms, Ypt31p and Ypt32p, are essential for Golgi function in yeast. EMBO J 15:6460-6475

Bernards A (2003) GAPs galore! A survey of putative Ras superfamily GTPase activating proteins in man and *Drosophila*. Biochim Biophys Acta 1603:47-82

Brennwald P, Novick P (1993) Interactions of three domains distinguishing the ras-related GTP-binding proteins Ypt1 and Sec4. Nature 362:560-563

Bucci C, Parton RG, Mather IH, Stunnenberg H, Simons K, Hoflack B, Zerial M (1992) The small GTPase rab5 functions as a regulatory factor in the early endocytic pathway. Cell 70:715-728

Burd CG, Emr SD (1998) Phosphatidylinositol(3)-phosphate signaling mediated by specific binding to RING FYVE domains. Mol Cell 2:157-162

Burd CG, Peterson M, Cowles CR, Emr SD (1997) A novel Sec18p/NSF-dependent complex required for Golgi-to-endosome transport in yeast. Mol Biol Cell 8:1089-1104

Calero M, Chen CZ, Zhu W, Winand N, Havas KA, Gilbert PM, Burd CG, Collins RN (2003) Dual prenylation is required for Rab protein localization and function. Mol Biol Cell 14:1852-1867

Calero M, Collins RN (2002) *Saccharomyces cerevisiae* Pra1p/Yip3p interacts with Yip1p and Rab proteins. Biochem Biophys Res Commun 290:676-681

Calero M, Whittaker GR, Collins RN (2001) Yop1p, the yeast homolog of the polyposis locus protein 1, interacts with Yip1p and negatively regulates cell growth. J Biol Chem 276:12100-12112

Calero M, Winand NJ, Collins RN (2002) Identification of the novel proteins Yip4p and Yip5p as Rab GTPase interacting factors. FEBS Lett 515:89-98

Cantalupo G, Alifano P, Roberti V, Bruni CB, Bucci C (2001) Rab-interacting lysosomal protein (RILP): the Rab7 effector required for transport to lysosomes. EMBO J 20:683-693

Cao X, Ballew N, Barlowe C (1998) Initial docking of ER-derived vesicles requires Uso1p and Ypt1p but is independent of SNARE proteins. EMBO J 17:2156-2165

Carroll KS, Hanna J, Simon I, Krise J, Barbero P, Pfeffer SR (2001) Role of Rab9 GTPase in facilitating receptor recruitment by TIP47. Science 292:1373-1376

Chavrier P, Gorvel JP, Stelzer E, Simons K, Gruenberg J, Zerial M (1991) Hypervariable C-terminal domain of rab proteins acts as a targeting signal. Nature 353:769-772

Chou JH, Jahn R (2000) Binding of Rab3A to synaptic vesicles. J Biol Chem 275:9433-9440

Christoforidis S, Miaczynska M, Ashman K, Wilm M, Zhao L, Yip SC, Waterfield MD, Backer JM, Zerial M (1999) Phosphatidylinositol-3-OH kinases are Rab5 effectors. Nat Cell Biol 1:249-252

Collins RN, Brennwald P, Garrett M, Lauring A, Novick P (1997) Interactions of nucleotide release factor Dss4p with Sec4p in the post-Golgi secretory pathway of yeast. J Biol Chem 272:18281-18289

Constantinescu AT, Rak A, Alexandrov K, Esters H, Goody RS, Scheidig AJ (2002) Rab-subfamily-specific regions of Ypt7p are structurally different from other RabGTPases. Structure (Camb) 10:569-579

D'Adamo P, Welzl H, Papadimitriou S, Raffaele di Barletta M, Tiveron C, Tatangelo L, Pozzi L, Chapman PF, Knevett SG, Ramsay MF, Valtorta F, Leoni C, Menegon A, Wolfer DP, Lipp HP, Toniolo D (2002) Deletion of the mental retardation gene Gdi1 impairs associative memory and alters social behavior in mice. Hum Mol Genet 11:2567-2580

Dascher C, Ossig R, Gallwitz D, Schmitt HD (1991) Identification and structure of four yeast genes (*SLY*) that are able to suppress the functional loss of *YPT1*, a member of the RAS superfamily. Mol Cell Biol 11:872-885

De Renzis S, Sonnichsen B, Zerial M (2002) Divalent Rab effectors regulate the sub-compartmental organization and sorting of early endosomes. Nat Cell Biol 4:124-133

Desnos C, Schonn JS, Huet S, Tran VS, El-Amraoui A, Raposo G, Fanget I, Chapuis C, Menasche G, de Saint Basile G, Petit C, Cribier S, Henry JP, Darchen F (2003) Rab27A and its effector MyRIP link secretory granules to F-actin and control their motion towards release sites. J Cell Biol 163:559-570

Detter JC, Zhang Q, Mules EH, Novak EK, Mishra VS, Li W, McMurtrie EB, Tchernev VT, Wallace MR, Seabra MC, Swank RT, Kingsmore SF (2000) Rab geranylgeranyl transferase alpha mutation in the gunmetal mouse reduces Rab prenylation and platelet synthesis. Proc Natl Acad Sci USA 97:4144-4149

Dirac-Svejstrup AB, Soldati T, Shapiro AD, Pfeffer SR (1994) Rab-GDI presents functional rab9 to the intracellular transport machinery and contributes selectivity to rab9 membrane recruitment. J Biol Chem 269:15427-15430

Dirac-Svejstrup AB, Sumizawa T, Pfeffer SR (1997) Identification of a GDI displacement factor that releases endosomal rab GTPases from rab-GDI. EMBO J 16:465-472

Dollar G, Struckhoff E, Michaud J, Cohen RS (2002) Rab11 polarization of the *Drosophila* oocyte: a novel link between membrane trafficking, microtubule organization, and oskar mRNA localization and translation. Development 129:517-526

Duden R (2003) ER-to-Golgi transport: COP I and COP II function (Review). Mol Membr Biol 20:197-207

Dumas JJ, Zhu Z, Connolly JL, Lambright DG (1999) Structural basis of activation and GTP hydrolysis in Rab proteins. Structure Fold Des 7:413-423

Dunn B, Stearns T, Botstein D (1993) Specificity domains distinguish the ras-related GTPases Ypt1 and Sec4. Nature 362:563-565

Echard A, Jollivet F, Martinez O, Lacapere JJ, Rousselet A, Janoueix-Lerosey I, Goud B (1998) Interaction of a Golgi-associated kinesin-like protein with Rab6. Science 279:580-585

Edinger AL, Cinalli RM, Thompson CB (2003) Rab7 prevents growth factor-independent survival by inhibiting cell-autonomous nutrient transporter expression. Dev Cell 5:571-582

Eggenschwiler JT, Espinoza E, Anderson KV (2001) Rab23 is an essential negative regulator of the mouse Sonic hedgehog signalling pathway. Nature 412:194-198

Esters H, Alexandrov K, Constantinescu AT, Goody RS, Scheidig AJ (2000) High-resolution crystal structure of *S. cerevisiae* Ypt51(DeltaC15)-GppNHp, a small GTP-binding protein involved in regulation of endocytosis. J Mol Biol 298:111-121

Fukuda M, Kuroda TS (2002) Slac2-c (synaptotagmin-like protein homologue lacking C2 domains-c), a novel linker protein that interacts with Rab27, myosin Va/VIIa, and actin. J Biol Chem 277:43096-43103

Fukui K, Sasaki T, Imazumi K, Matsuura Y, Nakanishi H, Takai Y (1997) Isolation and characterization of a GTPase activating protein specific for the Rab3 subfamily of small G proteins. J Biol Chem 272:4655-4658

Gao XD, Albert S, Tcheperegine SE, Burd CG, Gallwitz D, Bi E (2003) The GAP activity of Msb3p and Msb4p for the Rab GTPase Sec4p is required for efficient exocytosis and actin organization. J Cell Biol 162:635-646

Gaullier JM, Simonsen A, D'Arrigo A, Bremnes B, Stenmark H, Aasland R (1998) FYVE fingers bind PtdIns(3)P. Nature 394:432-433

Geppert M, Bolshakov VY, Siegelbaum SA, Takei K, De Camilli P, Hammer RE, Sudhof TC (1994) The role of rab3A in neurotransmitter release. Nature 369:493-497

Gilbert PM, Burd CG (2001) GDP dissociation inhibitor domain II required for Rab GTPase recycling. J Biol Chem 276:8014-8020

Goldenring JR, Ray GS, Lee JR (1999) Rab11 in dysplasia of Barrett's epithelia. Yale J Biol Med 72:113-120

Gomes AQ, Ali BR, Ramalho JS, Godfrey RF, Barral DC, Hume AN, Seabra MC (2003) Membrane targeting of Rab GTPases is influenced by the prenylation motif. Mol Biol Cell 14:1882-1899

Goody RS, Hofmann-Goody W (2002) Exchange factors, effectors, GAPs and motor proteins: common thermodynamic and kinetic principles for different functions. Eur Biophys J 31:268-274

Guo W, Roth D, Walch-Solimena C, Novick P (1999) The exocyst is an effector for Sec4p, targeting secretory vesicles to sites of exocytosis. EMBO J 18:1071-1080

Han L, Colicelli J (1995) A human protein selected for interference with Ras function interacts directly with Ras and competes with Raf1. Mol Cell Biol 15:1318-1323

Hoogenraad CC, Akhmanova A, Howell SA, Dortland BR, De Zeeuw CI, Willemsen R, Visser P, Grosveld F, Galjart N (2001) Mammalian Golgi-associated Bicaudal-D2 functions in the dynein-dynactin pathway by interacting with these complexes. EMBO J 20:4041-4054

Hutt DM, Da-Silva LF, Chang LH, Prosser DC, Ngsee JK (2000) PRA1 inhibits the extraction of membrane-bound rab GTPase by GDI1. J Biol Chem 275:18511-18519

Itoh T, Watabe A, Toh EA, Matsui Y (2002) Complex formation with Ypt11p, a rab-type small GTPase, is essential to facilitate the function of Myo2p, a class V myosin, in mitochondrial distribution in *Saccharomyces cerevisiae*. Mol Cell Biol 22:7744-7757

Jedd G, Mulholland J, Segev N (1997) Two new Ypt GTPases are required for exit from the yeast trans-Golgi compartment. J Cell Biol 137:563-580

Jordens I, Fernandez-Borja M, Marsman M, Dusseljee S, Janssen L, Calafat J, Janssen H, Wubbolts R, Neefjes J (2001) The Rab7 effector protein RILP controls lysosomal transport by inducing the recruitment of dynein-dynactin motors. Curr Biol 11:1680-1685

Kapfhamer D, Valladares O, Sun Y, Nolan PM, Rux JJ, Arnold SE, Veasey SC, Bucan M (2002) Mutations in Rab3a alter circadian period and homeostatic response to sleep loss in the mouse. Nat Genet 32:290-295

Lian JP, Stone S, Jiang Y, Lyons P, Ferro-Novick S (1994) Ypt1p implicated in v-SNARE activation. Nature 372:698-701

Loftus SK, Larson DM, Baxter LL, Antonellis A, Chen Y, Wu X, Jiang Y, Bittner M, Hammer JA, 3rd, Pavan WJ (2002) Mutation of melanosome protein RAB38 in chocolate mice. Proc Natl Acad Sci USA 99:4471-4476

Luan P, Balch WE, Emr SD, Burd CG (1999) Molecular dissection of guanine nucleotide dissociation inhibitor function *in vivo*. Rab-independent binding to membranes and role of Rab recycling factors. J Biol Chem 274:14806-14817

Marks MS, Seabra MC (2001) The melanosome: membrane dynamics in black and white. Nat Rev Mol Cell Biol 2:738-748

Martincic I, Peralta ME, Ngsee JK (1997) Isolation and characterization of a dual prenylated Rab and VAMP2 receptor. J Biol Chem 272:26991-26998

Matanis T, Akhmanova A, Wulf P, Del Nery E, Weide T, Stepanova T, Galjart N, Grosveld F, Goud B, De Zeeuw CI, Barnekow A, Hoogenraad CC (2002) Bicaudal-D regulates COPI-independent Golgi-ER transport by recruiting the dynein-dynactin motor complex. Nat Cell Biol 4:986-992

Matesic LE, Yip R, Reuss AE, Swing DA, O'Sullivan TN, Fletcher CF, Copeland NG, Jenkins NA (2001) Mutations in Mlph, encoding a member of the Rab effector family, cause the melanosome transport defects observed in leaden mice. Proc Natl Acad Sci USA 98:10238-10243

Maurer-Stroh S, Washietl S, Frank E (2003) Protein prenyltransferases: anchor size, pseudogenes and parasites. Biol Chem 384:977-989

Mayer A, Wickner W (1997) Docking of yeast vacuoles is catalyzed by the Ras-like GTPase Ypt7p after symmetric priming by Sec18p (NSF). J Cell Biol 136:307-317

McBride HM, Rybin V, Murphy C, Giner A, Teasdale R, Zerial M (1999) Oligomeric complexes link Rab5 effectors with NSF and drive membrane fusion via interactions between EEA1 and syntaxin 13. Cell 98:377-386

McLauchlan H, Newell J, Morrice N, Osborne A, West M, Smythe E (1998) A novel role for Rab5-GDI in ligand sequestration into clathrin-coated pits. Curr Biol 8:34-45

Menasche G, Ho CH, Sanal O, Feldmann J, Tezcan I, Ersoy F, Houdusse A, Fischer A, de Saint Basile G (2003) Griscelli syndrome restricted to hypopigmentation results from a melanophilin defect (GS3) or a MYO5A F-exon deletion (GS1). J Clin Invest 112:450-456

Menasche G, Pastural E, Feldmann J, Certain S, Ersoy F, Dupuis S, Wulffraat N, Bianchi D, Fischer A, Le Deist F, de Saint Basile G (2000) Mutations in RAB27A cause Griscelli syndrome associated with haemophagocytic syndrome. Nat Genet 25:173-176

Mercer JA, Seperack PK, Strobel MC, Copeland NG, Jenkins NA (1991) Novel myosin heavy chain encoded by murine dilute coat colour locus. Nature 349:709-713

Merithew E, Hatherly S, Dumas JJ, Lawe DC, Heller-Harrison R, Lambright DG (2001) Structural plasticity of an invariant hydrophobic triad in the switch regions of Rab GTPases is a determinant of effector recognition. J Biol Chem 276:13982-13988

Morsomme P, Riezman H (2002) The Rab GTPase Ypt1p and tethering factors couple protein sorting at the ER to vesicle targeting to the Golgi apparatus. Dev Cell 2:307-317

Moyer BD, Allan BB, Balch WE (2001) Rab1 interaction with a GM130 effector complex regulates COPII vesicle cis--Golgi tethering. Traffic 2:268-276

Murray JT, Panaretou C, Stenmark H, Miaczynska M, Backer JM (2002) Role of Rab5 in the recruitment of hVps34/p150 to the early endosome. Traffic 3:416-427

Nie Z, Hirsch DS, Randazzo PA (2003) Arf and its many interactors. Curr Opin Cell Biol 15:396-404

Nielsen E, Christoforidis S, Uttenweiler-Joseph S, Miaczynska M, Dewitte F, Wilm M, Hoflack B, Zerial M (2000) Rabenosyn-5, a novel Rab5 effector, is complexed with hVPS45 and recruited to endosomes through a FYVE finger domain. J Cell Biol 151:601-612

Nielsen E, Severin F, Backer JM, Hyman AA, Zerial M (1999) Rab5 regulates motility of early endosomes on microtubules. Nat Cell Biol 1:376-382

Ortiz D, Medkova M, Walch-Solimena C, Novick P (2002) Ypt32 recruits the Sec4p guanine nucleotide exchange factor, Sec2p, to secretory vesicles; evidence for a Rab cascade in yeast. J Cell Biol 157:1005-1015

Ostermeier C, Brunger AT (1999) Structural basis of Rab effector specificity: crystal structure of the small G protein Rab3A complexed with the effector domain of rabphilin-3A. Cell 96:363-374

Otomo A, Hadano S, Okada T, Mizumura H, Kunita R, Nishijima H, Showguchi-Miyata J, Yanagisawa Y, Kohiki E, Suga E, Yasuda M, Osuga H, Nishimoto T, Narumiya S, Ikeda JE (2003) ALS2, a novel guanine nucleotide exchange factor for the small GTPase Rab5, is implicated in endosomal dynamics. Hum Mol Genet 12:1671-1687

Papini E, Satin B, Bucci C, de Bernard M, Telford JL, Manetti R, Rappuoli R, Zerial M, Montecucco C (1997) The small GTP binding protein rab7 is essential for cellular vacuolation induced by Helicobacter pylori cytotoxin. EMBO J 16:15-24

Pastural E, Barrat FJ, Dufourcq-Lagelouse R, Certain S, Sanal O, Jabado N, Seger R, Griscelli C, Fischer A, de Saint Basile G (1997) Griscelli disease maps to chromosome 15q21 and is associated with mutations in the myosin-Va gene. Nat Genet 16:289-292

Patki V, Lawe DC, Corvera S, Virbasius JV, Chawla A (1998) A functional PtdIns(3)P-binding motif. Nature 394:433-434

Pei L, Peng Y, Yang Y, Ling XB, Van Eyndhoven WG, Nguyen KC, Rubin M, Hoey T, Powers S, Li J (2002) PRC17, a novel oncogene encoding a Rab GTPase-activating protein, is amplified in prostate cancer. Cancer Res 62:5420-5424

Pereira-Leal JB, Seabra MC (2000) The mammalian Rab family of small GTPases: definition of family and subfamily sequence motifs suggests a mechanism for functional specificity in the Ras superfamily. J Mol Biol 301:1077-1087

Peterson MR, Burd CG, Emr SD (1999) Vac1p coordinates rab and phosphatidylinositol 3-kinase signaling in Vps45p-dependent vesicle docking/fusion at the endosome. Curr Biol 9:159-162

Pfeffer S (2003) Membrane domains in the secretory and endocytic pathways. Cell 112:507-517

Pfeffer SR (2001) Rab GTPases: specifying and deciphering organelle identity and function. Trends Cell Biol 11:487-491

Piddini E, Vincent JP (2003) Modulation of developmental signals by endocytosis: different means and many ends. Curr Opin Cell Biol 15:474-481

Provance DW, James TL, Mercer JA (2002) Melanophilin, the product of the leaden locus, is required for targeting of myosin-Va to melanosomes. Traffic 3:124-132

Rak A, Fedorov R, Alexandrov K, Albert S, Goody RS, Gallwitz D, Scheidig AJ (2000) Crystal structure of the GAP domain of Gyp1p: first insights into interaction with Ypt/Rab proteins. EMBO J 19:5105-5113

Rak A, Pylypenko O, Durek T, Watzke A, Kushnir S, Brunsveld L, Waldmann H, Goody RS, Alexandrov K (2003) Structure of Rab GDP-dissociation inhibitor in complex with prenylated YPT1 GTPase. Science 302:646-650

Ricard CS, Jakubowski JM, Verbsky JW, Barbieri MA, Lewis WM, Fernandez GE, Vogel M, Tsou C, Prasad V, Stahl PD, Waksman G, Cheney CM (2001) *Drosophila* rab GDI mutants disrupt development but have normal Rab membrane extraction. Genesis 31:17-29

Richardson PM, Zon LI (1995) Molecular cloning of a cDNA with a novel domain present in the tre-2 oncogene and the yeast cell cycle regulators BUB2 and cdc16. Oncogene 11:1139-1148

Rybin V, Ullrich O, Rubino M, Alexandrov K, Simon I, Seabra C, Goody R, Zerial M (1996) GTPase activity of Rab5 acts as a timer for endocytic membrane fusion. Nature 383:266-269

Sacher M, Barrowman J, Wang W, Horecka J, Zhang Y, Pypaert M, Ferro-Novick S (2001) TRAPP I implicated in the specificity of tethering in ER-to-Golgi transport. Mol Cell 7:433-442

Salminen A, Novick PJ (1987) A ras-like protein is required for a post-Golgi event in yeast secretion. Cell 49:527-538

Seabra MC, Brown MS, Goldstein JL (1993) Retinal degeneration in choroideremia: deficiency of rab geranylgeranyl transferase. Science 259:377-381

Seabra MC, Goldstein JL, Sudhof TC, Brown MS (1992) Rab geranylgeranyl transferase: a multisubunit enzyme that prenylates GTP-binding proteins terminating in Cys-X-Cys or Cys-Cys. J Biol Chem 267:14497-14503

Seachrist JL, Laporte SA, Dale LB, Babwah AV, Caron MG, Anborgh PH, Ferguson SS (2002) Rab5 association with the angiotensin II type 1A receptor promotes Rab5 GTP binding and vesicular fusion. J Biol Chem 277:679-685

Short B, Preisinger C, Korner R, Kopajtich R, Byron O, Barr FA (2001) A GRASP55-rab2 effector complex linking Golgi structure to membrane traffic. J Cell Biol 155:877-883

Short B, Preisinger C, Schaletzky J, Kopajtich R, Barr FA (2002) The Rab6 GTPase regulates recruitment of the dynactin complex to Golgi membranes. Curr Biol 12:1792-1795

Sincock PM, Ganley IG, Krise JP, Diederichs S, Sivars U, O'Connor B, Ding L, Pfeffer SR (2003) Self-assembly is important for TIP47 function in mannose 6-phosphate receptor transport. Traffic 4:18-25

Sivars U, Aivazian D, Pfeffer SR (2003) Yip3 catalyses the dissociation of endosomal Rab-GDI complexes. Nature 425:856-859

Soldati T, Rancano C, Geissler H, Pfeffer SR (1995) Rab7 and Rab9 are recruited onto late endosomes by biochemically distinguishable processes. J Biol Chem 270:25541-25548

Sonnichsen B, De Renzis S, Nielsen E, Rietdorf J, Zerial M (2000) Distinct membrane domains on endosomes in the recycling pathway visualized by multicolor imaging of Rab4, Rab5, and Rab11. J Cell Biol 149:901-914

Stein MP, Feng Y, Cooper KL, Welford AM, Wandinger-Ness A (2003) Human VPS34 and p150 are Rab7 interacting partners. Traffic 4:754-771

Stenmark H, Parton RG, Steele-Mortimer O, Lutcke A, Gruenberg J, Zerial M (1994) Inhibition of rab5 GTPase activity stimulates membrane fusion in endocytosis. EMBO J 13:1287-1296

Stinchcombe JC, Barral DC, Mules EH, Booth S, Hume AN, Machesky LM, Seabra MC, Griffiths GM (2001) Rab27a is required for regulated secretion in cytotoxic T lymphocytes. J Cell Biol 152:825-834

Strom M, Hume AN, Tarafder AK, Barkagianni E, Seabra MC (2002) A family of Rab27-binding proteins. Melanophilin links Rab27a and myosin Va function in melanosome transport. J Biol Chem 277:25423-25430

Stroupe C, Brunger AT (2000) Crystal structures of a Rab protein in its inactive and active conformations. J Mol Biol 304:585-598

Tall GG, Barbieri MA, Stahl PD, Horazdovsky BF (2001) Ras-activated endocytosis is mediated by the Rab5 guanine nucleotide exchange activity of RIN1. Dev Cell 1:73-82

Tisdale EJ (2003) Rab2 interacts directly with atypical protein kinase C (aPKC) iota/lambda and inhibits aPKCiota/lambda-dependent glyceraldehyde-3-phosphate dehydrogenase phosphorylation. J Biol Chem 278:52524-52530

Tisdale EJ, Balch WE (1996) Rab2 is essential for the maturation of pre-Golgi intermediates. J Biol Chem 271:29372-29379

Tisdale EJ, Jackson MR (1998) Rab2 protein enhances coatomer recruitment to pre-Golgi intermediates. J Biol Chem 273:17269-17277

Ullrich O, Stenmark H, Alexandrov K, Hubert L, Kaibuchi K, Sasaki T, Takai Y, Zerial M (1993) Rab GDP dissociation inhibitor as a general regulator for the membrane association of Rab proteins. J Biol Chem 268:18143-18150

van Ijzendoorn SC, Tuvim MJ, Weimbs T, Dickey BF, Mostov KE (2002) Direct interaction between Rab3b and the polymeric immunoglobulin receptor controls ligand-stimulated transcytosis in epithelial cells. Dev Cell 2:219-228

Verhoeven K, De Jonghe P, Coen K, Verpoorten N, Auer-Grumbach M, Kwon JM, Fitz-Patrick D, Schmedding E, De Vriendt E, Jacobs A, Van Gerwen V, Wagner K, Hartung HP, Timmerman V (2003) Mutations in the small GTP-ase late endosomal protein RAB7 cause Charcot-Marie-Tooth type 2B neuropathy. Am J Hum Genet 72:722-727

Vernoud V, Horton AC, Yang Z, Nielsen E (2003) Analysis of the small GTPase gene superfamily of Arabidopsis. Plant Physiol 131:1191-1208

Wang L, Merz AJ, Collins KM, Wickner W (2003) Hierarchy of protein assembly at the vertex ring domain for yeast vacuole docking and fusion. J Cell Biol 160:365-374

Wang W, Ferro-Novick S (2002) A ypt32p exchange factor is a putative effector of ypt1p. Mol Biol Cell 13:3336-3343

Wang W, Sacher M, Ferro-Novick S (2000) TRAPP stimulates guanine nucleotide exchange on Ypt1p. J Cell Biol 151:289-296

Waselle L, Coppola T, Fukuda M, Iezzi M, El-Amraoui A, Petit C, Regazzi R (2003) Involvement of the Rab27 binding protein Slac2c/MyRIP in insulin exocytosis. Mol Biol Cell 14:4103-4113

Waters MG, Pfeffer SR (1999) Membrane tethering in intracellular transport. Curr Opin Cell Biol 11:453-459

Whyte JR, Munro S (2002) Vesicle tethering complexes in membrane traffic. J Cell Sci 115:2627-2637

Wilson SM, Yip R, Swing DA, O'Sullivan TN, Zhang Y, Novak EK, Swank RT, Russell LB, Copeland NG, Jenkins NA (2000) A mutation in Rab27a causes the vesicle transport defects observed in ashen mice. Proc Natl Acad Sci USA 97:7933-7938

Wu G, Yussman MG, Barrett TJ, Hahn HS, Osinska H, Hilliard GM, Wang X, Toyokawa T, Yatani A, Lynch RA, Robbins J, Dorn GW, 2nd (2001) Increased myocardial Rab

GTPase expression: a consequence and cause of cardiomyopathy. Circ Res 89:1130-1137

Wu XS, Rao K, Zhang H, Wang F, Sellers JR, Matesic LE, Copeland NG, Jenkins NA, Hammer JA III (2002) Identification of an organelle receptor for myosin-Va. Nat Cell Biol 4:271-278

Wurmser AE, Sato TK, Emr SD (2000) New component of the vacuolar class C-Vps complex couples nucleotide exchange on the Ypt7 GTPase to SNARE-dependent docking and fusion. J Cell Biol 151:551-562

Yoshimura SI, Nakamura N, Barr FA, Misumi Y, Ikehara Y, Ohno H, Sakaguchi M, Mihara K (2001) Direct targeting of cis-Golgi matrix proteins to the Golgi apparatus. J Cell Sci 114:4105-4115

Zerial M, McBride H (2001) Rab proteins as membrane organizers. Nat Rev Mol Cell Biol 2:107-117

Zhou Y, Toth M, Hamman MS, Monahan SJ, Lodge PA, Boynton AL, Salgaller ML (2002) Serological cloning of PARIS-1: a new TBC domain-containing, immunogenic tumor antigen from a prostate cancer cell line. Biochem Biophys Res Commun 290:830-838

Zhu G, Liu J, Terzyan S, Zhai P, Li G, Zhang XC (2003) High resolution crystal structures of human Rab5a and five mutants with substitutions in the catalytically important phosphate-binding loop. J Biol Chem 278:2452-2460

Burd, Christopher G.
Department of Cell and Developmental Biology, University of Pennsylvania School of Medicine, 421 Curie Boulevard room 1010, Philadelphia, PA 19104-6158, USA
cburd@mail.med.upenn.edu

Collins, Ruth N.
Department of Molecular Medicine, C4-109, VMC, Cornell University, Ithaca NY 14853-6401, USA

Tethering proteins in membrane traffic

Sarah Chu and Wei Guo

Abstract

In eukaryotic cells, protein secretion and the transport of materials between membrane-bounded organelles require efficient and accurate delivery of carrier vesicles to their correct destination. Recent studies have revealed a large number of evolutionarily conserved proteins that may function in tethering vesicles to their target membrane in preparation for subsequent docking and fusion. Here, we review the proteins and protein complexes implicated in membrane tethering at various traffic stages.

1 Introduction

Membrane traffic between organelles is mediated by lipid-delimited carriers, which often take the form of vesicles. Elaborate machinery is needed to create the carriers from donor membranes, and to deliver them to the correct targeting membranes for docking and fusion. The fidelity of the transport process is vital for maintaining the integrity and function of distinctive membrane compartments in eukaryotic cells.

The interactions between the SNARE proteins are at the center of membrane docking and fusion (Rothman 1994). SNAREs are integral membrane proteins that reside on vesicles (v-SNARE) and target membranes (t-SNARE). The cognate v- and t-SNAREs form exceedingly stable complexes that bring the vesicles and target membranes to close apposition for fusion. The core of the highly stable SNARE complex consists of parallel four α-helix bundle contributed by membrane proximal SNARE motifs (for review see Ungar and Hughson 2003; Jahn et al. 2003). Cognate SNARE interaction was shown to be sufficient to drive the fusion of liposome vesicles (Weber et al. 1998; McNew et al. 2000). Furthermore, cells expressing the interaction domains of v- and t-SNAREs on the cell surface can spontaneously fuse with each other (Hu et al. 2003). Though the SNAREs are sufficient for membrane fusion, the intrinsic rate of *in vitro* SNARE assembly is slow. In the cells, there must be proteins that "catalyze" the SNARE assembly to its physiological rate.

The specific distribution of SNARE proteins to different trafficking compartments suggests that SNARE assembly confers targeting specificity. Even though the non-cognate SNAREs can form complexes *in vitro* comparable to the cognate pairs (Yang et al. 1999; Fasshauer et al. 1999; Tsui et al. 2000), *in vitro* lipid fu-

Topics in Current Genetics, Vol. 10
S. Keränen, J. Jäntti (Eds.): Regulatory Mechanisms of Intracellular Membrane Transport
DOI 10.1007/b98495 / Published online: 13 May 2004
© Springer-Verlag Berlin Heidelberg 2004

sion assays demonstrated a high degree of specificity for SNARE recognition (McNew et al. 2000; Parlati et al. 2000; Fukuda et al. 2000). However, SNARE pairing cannot be the sole determinant of targeting specificity in the cells (for review see Gotte and von Mollard 1998; Pelham 2001). First, some SNAREs function in multiple traffic steps and are found in several SNARE complexes (Fischer von Mollard et al. 1997, 1999; Sprang and Schekman 1998; Holthuis et al. 1998; Lupashin et al. 1997). Second, both the t- and v-SNAREs are present in both forward and recycling vesicles. Finally, in exocytosis of yeast, the t-SNARE proteins are evenly distributed along the entire plasma membrane (Brennwald et al. 1994), whereas, membrane fusion is restricted to limited subdomains of the plasma membrane. Thus, additional levels of control must be available to spatially restrict SNARE assembly.

Members of the Rab family of small GTP-binding proteins are localized to various membrane compartments and regulate different stages of traffic (Novick and Zerial 1997; Lazar et al. 1997; Schimmoller et al. 1998). Rab proteins are implicated in directing vesicles to their appropriate target membranes. However, Rab alone may not be sufficient. For example, the yeast Ypt1p appears to function in multiple steps early in the traffic pathway (Bacon et al. 1989; Jedd et al. 1995). A chimera of Ypt1p and Sec4p, a Rab protein functioning at the final stage of the yeast exocytic pathway, can fulfill the essential functions of either of the two proteins without mis-targeting cargo (Brennwald and Novick 1993; Dunn et al. 1993).

The term "tethering" has been used in the field to describe the physical association of transport intermediates to their target membranes preceding the formation of the *trans*-SNARE complex (for review see Pfeffer 1999; Waters and Pfeffer 1999; Waters and Hughson 2000; Guo et al. 2000; Whyte and Munro 2002). It represents the earliest molecular interaction event that eventually leads to membrane fusion. Tethering could be a loose and reversible interaction that leaves the two membranes at some distance while subsequent docking, on the other hand, is regarded as a more intimate interaction that brings the membranes to close apposition right before the fusion of the lipid bilayers. As the earliest interaction event, tethering represents a critical step for vesicle recognition and, therefore, targeting specificity.

The heterogeneity of the tethering complexes makes their categorization difficult, although it is possible to divide them to two broad classes: long, coiled-coil proteins such as EEA1 and p115, and large, multisubunit complexes such as the exocyst and the TRAPP complex. The long rod-like structure of the former class of molecules and the large sizes of multiprotein complexes allow them to tether vesicles to membranes at a distance or perhaps allow the membrane "sampling" before SNARE pairing and membrane fusion. Size (or length) matters here considering that the tethers may have the first contact with incoming vesicles. A visual clue for tethering molecules may come from examination of mammalian Golgi by freeze fracture electron microscopy (Orci et al. 1998), which showed fibrous elements associated with vesicles and Golgi stacks. An early biochemical demonstration of tethering came from the study of Uso1p, where Barlowe and his colleagues, in an *in vitro* reconstitution assay, revealed the "initial docking" of

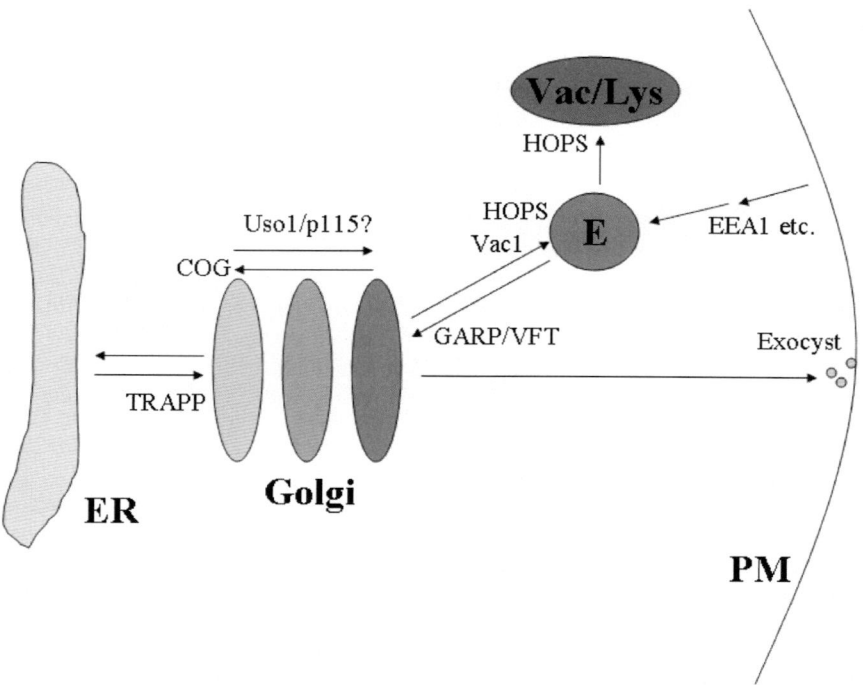

Fig. 1. Putative localizations of tethering proteins that function in various stages of membrane traffic. ER, endoplasmic reticulum; PM, plasma membrane; E, endosomes. For simplicity, endosomes are not subdivided into early and late endosomes in the figure. vac/lys, vacuole and lysosomes.

ER-derived vesicles to the Golgi that is independent of SNARE assembly (Cao et al. 1998).

In recent years, we have witnessed great progresses in our understanding of the molecular basis of vesicle tethering. Here, we only focus on several well-studied proteins or protein complexes that are postulated in tethering transport vesicles to their receptor membranes (Fig. 1). With rapid progresses in this field, we anticipate that in the near future new tethering proteins will be identified, or tethering functions will be assigned to some known proteins.

2 Tethering proteins in various traffic stages

2.1 TRAPP

The discovery of the TRAPP (*tr*ansport *p*rotein *p*article) complex started with the identification of one of its components, Bet3p. *BET3* gene was identified via its

genetic interactions with *BET1*, the gene product of which encodes a t-SNARE involved in ER-to-Golgi traffic (Rossi et al. 1995). Studies had shown that it tightly associated to the Golgi and was required for vesicle targeting but not fusion (Sacher et al. 1998). Epitope-tagged yeast Bet3p co-precipitated with nine other polypeptides in a complex that was given the name TRAPP (Sacher et al. 2000). Many of these subunits are highly conserved proteins, sharing between 29% and 54% sequence identity with their human counterparts, but having no sequence homology with other known proteins. TRAPP stably associates with the Golgi membrane and is resistant to conditions that normally disperse the Golgi apparatus (Barrowman et al. 2000). This static localization, in contrast to the mobility of SNAREs, suggested that TRAPP might function as a landmark for the delivery of vesicles to the Golgi.

It was later revealed that TRAPP represents two distinct complexes (Sacher et al. 2001). TRAPP I, approximately 300 kD, consists of seven subunits: Bet3p, Trs33p, Trs31p, Bet5p, Trs20p, Trs23p, and Trs85p. TRAPP II, approximately 1000 kD, contains three additional components: Trs65p, Trs130p, and Trs120p. Unlike the other seven subunits, the overexpression of these three subunits is unable to suppress the growth phenotype of a temperature-sensitive *bet3* mutant. TRAPP I and TRAPP II both co-fractionate with Golgi compartments of similar size and density but mediate different transport steps. TRAPP I, but not TRAPP II, is required for ER-to-Golgi transport *in vitro* and binds COPII vesicles derived from the ER (but not from post-Golgi or other membranes) (Sacher et al. 2001). Binding is blocked in the presence of GTPγS (Sacher et al. 2001), which is known to stabilize the COPII coat (Barlowe et al. 1994). This suggests that TRAPP I binds to vesicles after uncoating. Studies implicate TRAPP II in a later stage of Golgi traffic; a temperature-sensitive *trs130* mutant accumulates aberrant Golgi membranes at non-permissive temperatures, with structures markedly different from ER-to-Golgi blocked mutants. The *trs130* mutant also accumulates the Golgi forms of invertase and the vacuolar hydrolase carboxypeptidase Y (CPY) (Sacher et al. 2001). A possible role for TRAPP II is to bind retrograde vesicles moving from a later to an earlier Golgi compartment (Sacher et al. 2001).

Both TRAPP I and TRAPP II can bind and exchange nucleotides on the Rab protein Ypt1p (Jones et al. 2000; Wang et al. 2000; Sacher et al. 2001). This perhaps explains how Ypt1p can function at two different stages; Ypt1p is required for both ER-to-Golgi and intra-Golgi transport. Studies have also shown that TRAPP is a nucleotide exchanger for Ypt31/32 (Jones et al. 2000), a functional pair of Rabs necessary for transport from the *trans*-Golgi. This suggests that TRAPP is not only implicated in ER-to-Golgi and intra-Golgi vesicle transport, but may also function in transport from the *trans*-Golgi.

Uso1p and the COG complex are also implicated as tethers in ER-to-Golgi traffic (see below). However, binding of TRAPP I to COPII vesicles is specific and does not require Uso1p, COG or Ypt1p; COPII vesicles reconstituted with mutations in *USO1, YPT1, SEC34,* and *SEC35* bound to TRAPP I as efficiently as those in wild type situation (Sacher et al. 2001). In contrast, a deletion of *BET5* could not be bypassed by the overexpression of *YPT1, USO1,* or *SEC35* (Barrowman et al. 2000). These results suggest that TRAPP I acts upstream of the other

tethering factors, and that it binds COPII vesicles either directly or via other unidentified proteins. Sacher et al. have proposed a model in which TRAPP I binds COPII vesicles and activates Ypt1p, causing other tethering complexes to be recruited and complete the docking process prior to SNARE assembly and fusion (Sacher et al. 2001). This hypothesis is consistent with the observation that Uso1p-binding to membranes is dependent on Ypt1p (Cao et al. 1998). However, Uso1p was shown to be necessary to tether COPII vesicles to the Golgi *in vitro* (Cao et al. 1998) while TRAPP I alone can bind to COPII vesicles (Sacher et al. 2001). It will be interesting to further explore the relationship between the TRAPP and other factors in the tethering process.

2.2 p115

p115 seems to be involved in several stages of vesicular traffic. It was first identified as a necessary factor for intra-Golgi transport in mammalian cells, and later for the fusion of transcytotic vesicles to the plasma membrane (it is identical to the trancytosis-associated protein, TAP), as well as the docking and fusion of vesicular tubular clusters (VTCs) to the *cis*-Golgi (Waters et al. 1992; Barroso et al. 1995). It shares sequence homology to the yeast protein Uso1p, which functions in ER to Golgi transport, and is required for docking ER-derived COPII vesicles (see below). Both p115 and Uso1p have structures similar to myosin II, consisting of homodimers with two globular heads, an extensive coiled-coil domain and a short acidic tail (Sapperstein et al. 1995). p115 interacts with GTP-bound form of Rab1, and this interaction is important for the recruitment of p115 to the COPII vesicles budding from the endoplasmic reticulum (Allan et al. 2000). Furthermore, p115 crosslinks SNARE proteins syntaxin5, sly1, membrin, and rbet1. Inhibition of the interaction between p115 and the SNAREs blocks the COPII vesicle docking (Allan et al. 2000).

A role for p115 in tethering COPI-coated vesicles to the Golgi derives from studies of its involvement in Golgi fragmentation in mitosis. COPI vesicles transport molecules that need to be salvaged or recycled from the Golgi, and are also involved in anterograde transport and endocytosis. p115 binds to COPI vesicles via giantin, a peripheral membrane protein incorporated into budding COPI vesicles, and p115 and giantin are required for docking COPII vesicles to Golgi membranes *in vitro* (Sonnichsen et al. 1998). p115 also binds to the Golgi matrix protein GM130, which is tightly associated to the Golgi apparatus (particularly the *cis*-Golgi) via another protein, GRASP65 (Barr et al. 1997). GM130 is required for Golgi cisternal reassembly and stacking after mitosis. Like p115, both GM130 and giantin have long, fibrous, rod-like structures, consistent with their tethering functions. The p115-giantin and p115-GM130 interactions led to the proposal that p115 forms a ternary complex with its two membrane receptors, linking COPI vesicles to Golgi cisternae. At the onset of mitosis, GM130 phosphorylation by cdc2/cyclin B inhibits p115 binding to GM130 but not to giantin (Nakamura et al. 1997; Lowe et al. 1998). COPI vesicles continue to bud but are unable to tether to or fuse with the Golgi. *In vitro* studies indicate that up to two-thirds of the Golgi

membranes are converted into COPI vesicles in the absence of fusion, helping to account for the fragmentation of the Golgi apparatus and accumulation of COPI vesicles during mitosis. Pretreatment with antibodies against giantin precludes p115-giantin binding to the Golgi. Similarly, antibodies against GM130 inhibit p115 binding and vesicle docking to the Golgi, providing further support for the existence of the ternary complex. Recent studies, however, questioned the simple model wherein p115 simultaneously binds and, thus, bridges GM130 and giantin. Linstedt et al. have demonstrated that GM130 and p115 actually bind the same p115 domain in its C-terminus and compete for p115 binding (Linstedt 2000). These results suggest that p115 tethers GM130 and giantin in separate events, and that more complex mechanisms are involved in COPI docking to the Golgi than previously thought.

Downstream of its involvement in vesicle tethering, p115 promotes SNARE complex assembly in ER to Golgi and intra-Golgi transport (Shorter et al. 2002). In crude cellular extracts, a SNARE motif-related domain of p115 binds Golgi SNARE components including syntaxin-5 and GOS-28. p115, GM130, and giantin co-immunoprecipitate with both syntaxin-5 and GOS-28 but not syntaxin-1 involved in post-Golgi secretion. Importantly, p115 stimulates the assembly of the Golgi SNARE. The interaction of p115 with SNARE may represent an intermediate in the activation process of SNARE assembly. The ability of p115 to first mediate vesicle tethering to target membranes by binding to one set of proteins, and subsequently catalyze their fusion by binding to another set, suggests an efficient mechanism for coordinating target-specific vesicle tethering and fusion.

2.3 Uso1p

The yeast protein Uso1p is required for the transport and tethering of ER-derived vesicles to the Golgi (Nakajima et al. 1991; Cao et al. 1998). The yeast protein Uso1p, like its mammalian homolog p115, has a globular head domain and a long coiled coil extension (Yamakawa et al. 1996). However, it is almost twice the size of p115. Since the C-terminal region of Uso1p shares sequence similarity with GM130 in mammalian cells, it was suggested that Uso1p is a fusion of p115 and GM130 in yeast (Nakamura et al. 1997).

Barlowe and colleagues, using *in vitro* reconstitution experiments, demonstrated that Uso1p is required for the "initial docking" or "tethering" of ER-derived vesicles to the Golgi (Cao et al. 1998). This process also requires Ypt1p as GDI extraction of this Rab protein from the membrane affects the tethering process. On the other hand, the SNARE proteins did not affect tethering because mutant forms of Sed5p, Bet1p, Bos1p, and Sly1p, though blocking fusion, had no effect on the initial vesicle docking event. These experiments separated the tethering and subsequent fusion events and indicated the role of Uso1p and Ypt1p in the tethering step. This work is important because it, for the first time, demonstrated a "tethering" event separated from SNARE assembly by biochemical methods. The results were also consistent with genetic studies on Uso1p. The temperature-sensitive mutant *uso1-1* can be suppressed by overexpression of Ypt1p and the ER

to Golgi SNAREs such as Bet1p, Bos1p, Sec22p, and Ykt6p. Overexpression of Bet1p and Sec22p can even suppress *uso1* deletion, strongly suggesting that the SNAREs function downstream of Uso1p (Sapperstein et al. 1996). Consistent with these suppression data, the assembly of ER to Golgi SNARE complex is impaired in the *uso1-1* mutant (Sapperstein et al. 1996). Overall, both the strong genetic data and the reconstitution experiment clearly indicate that Uso1p is involved in the tethering of ER-derived vesicles to the Golgi. The remaining questions are how, exactly, Uso1p functions, and especially what is the relationship of Uso1p with other proteins including TRAPP and COG complexes in the tethering process.

2.4 The COG complex

The octameric COG complex was originally called Sec34/35p complex in yeast (Kim et al. 1999; Van Rheenen et al. 1999; Whyte and Munro 2001; Ram et al. 2002). Its mammalian counterpart is the GTC complex, which was purified from bovine brain cytosol based on its stimulatory effects on intra-Golgi transport (Walter et al. 1998; Suvorova et al. 2001; Ungar et al. 2002). Electron microscopy studies revealed that the mammalian COG complex has a 2-lobed structure (Ungar et al. 2002). Based on sequence analysis, it was shown that members of the COG complex share sequence homologies to other tethering complexes such as the exocyst and the VFT/GARP complex (Whyte and Munro 2001). These complexes were referred to as the "quatrefoil" tethering complexes as they all have a multiple of four subunits (Whyte and Munro 2001, 2002).

The COG complex has been proposed as a tether at the early Golgi compartments for recycling vesicles from multiple sources including later Golgi compartments and endosomes (Whyte and Munro 2001; Suvorova et al. 2002). A component of the complex, GRD20 (COG3) was identified as a protein required for the localization of TGN proteins (Spelbrink and Nothwehr 1999). Sec34p and Sec35p mutants accumulate the p1 form of CPY lacking Golgi-specific glycosylation (Wuestehube et al. 1996). Cells mutated for COG components were also shown to be defective in both N- and O-glycosylation associated with the Golgi function (Suvorova et al. 2002). Suvorova and colleagues further demonstrated that COG associates with a subset of Golgi SNAREs including Sed5p, Sec22p, Ykt6p, Gos1p, and Vti1p, all of which are implicated in the retrograde trafficking. In addition, COG interacts with the vesicle coat protein complex COPI. These interactions further suggest that COG may mediate the targeting and recruitment of retrograde intra-Golgi vesicles to cis-Golgi (Suvorova et al. 2002). The COG complex interacts with Ypt1p in its GTP-bound form.

2.5 The exocyst

Tethering of secretory vesicles at the plasma membrane is mediated by a multiprotein complex, the exocyst (Hsu et al. 2004). The exocyst complex consists of eight components: Sec3p, Sec5p, Sec6p, Sec8p, Sec10p, Sec15p, Exo70p, and Exo84p. They were first identified by genetic and biochemical methods in the budding yeast (TerBush and Novick 1995; TerBush et al. 1996; Guo et al. 1998), and were later found in mammalian cells based on sequence alignment searches and biochemical purification (Ting et al. 1995; Hsu et al. 1996; Kee et al. 1997; Guo et al. 1997; Matern et al. 2001). All the exocyst components are hydrophilic proteins. They interact with each other and are peripherally associated with the plasma membrane. The yeast exocyst complex was found to be 19.5S and the mammalian complex 17S in sizes based on gel filtration and velocity gradients fractionation (Bowser and Novick 1991; Ting et al. 1995).

In yeast, mutations in these genes lead to accumulation of secretory vesicles and blocks of exocytosis (Novick et al. 1980; Guo et al. 1999). The exocyst proteins are localized to regions of active surface expansion in yeast cells: the bud tip at the beginning of the cell cycle and the mother/daughter cell connection during cytokinesis (TerBush and Novick 1995; Finger et al. 1998; Guo et al. 1999). This localization pattern is in contrast to that of the t-SNAREs, which are evenly distributed along the entire plasma membrane (Brennwald et al. 1994). In epithelial cells, the Sec6p and Sec8p subunits were found to have both perinuclear and plasma membrane enrichment (Shin et al. 2000; Yeaman et al. 2001). In particular, the plasma membrane-localized exocyst staining was enriched at the tight junction or the lateral membranes (Grindstaff et al. 1998; Yeaman et al. 2001; Kreitzer et al. 2003) in MDCK cells. When MDCK cells were treated with the calcium chelator EGTA to disrupt the contacts between the cells, the plasma membrane-localized Sec8p was found to redistribute into the cytoplasm (Grindstaff et al. 1998). Thus, the plasma membrane localization of exocyst is dependent on cell-cell contact. In neuroendocrine PC12 cells the localization of the exocyst was dependent on the differentiation state of the cell (Vega and Hsu 2001). In undifferentiated PC12 cells, Exo70p displayed perinuclear enrichment. Upon the addition of nerve growth factor to promote neurite outgrowth, Exo70p is found distributing from the perinuclear region into the growing neurite and became enriched in the growth cone. In cultured hippocampal neurons, Sec6p and Sec8p were also found in the cell body, axons and dendrites (Hazuka et al. 1999).

Studies in yeast indicated that the exocyst subunit Sec3p is localized to sites of exocytosis independent of ongoing secretion (Finger et al. 1998). In a variety of secretion mutants blocked at or before the Golgi apparatus, Sec3-GFP is still localized at bud tips and mother/daughter cell connections. Furthermore, the localization of Sec3p appears to be independent of the other subunits of the exocyst. Based on these observations, it was proposed that Sec3p represents a spatial landmark for polarized secretion (Finger et al. 1998). While Sec3p is thought to be most proximal to the plasma membrane, another exocyst subunit, Sec15p, directly interacts with the GTP-bound form of the Rab protein, Sec4p, on the secretory vesicles (Guo et al. 1999). This interaction is specific for Sec4p but not for other

Rab proteins such as Ypt1p and Ypt51p. Replacing 4 amino acid residues in the effector loop of Sec4p with the corresponding residuals of Ypt1p completely abolished the interaction of the Rab protein with Sec15p. Sec4p promotes the protein-protein interactions among the exocyst components, which eventually link Sec15p to Sec3p (Guo et al. 1999). Therefore, the assembly of the exocyst complex may tether the secretory vesicles to specific sites on the plasma membrane for subsequent membrane fusion. Overexpression of SNARE proteins and Sec1p suppresses the temperature-sensitive growth defects of exocyst mutants, suggesting that the exocyst functions upstream of the SNAREs (Aalto et al. 1993; Brennwald et al. 1994). Consistent with these genetic data, SNARE assembly was found defective in the exocyst mutants (Grote et al. 2000). It was reported that the exocyst can co-immunoprecipitate with syntaxin in neuronal system (Hsu et al. 1996). However, a direct physical interaction between the exocyst and the SNAREs has not been found in yeast or mammals.

In addition to being an effector for the Rab proteins, the exocyst is also under the control of Rho family GTPases, which are master regulators of cell growth and morphogenesis. The polarized localization of the exocyst in the bud tip is lost in alleles of *rho1* and *cdc42* mutants (Guo et al. 2001; Zhang et al. 2001). Further studies indicate that the N-terminus of Sec3p interacts with both Rho1p and Cdc42p in their GTP-bound form. Truncation of this domain of Sec3p led to mislocalization of Sec3-GFP (Guo et al. 2001). Cdc42p is essential for the establishment of yeast polarity and Rho1p may be important for the maintenance of polarized growth. Their interactions with the exocyst may help to direct exocytosis to the site of polarized cell growth, where new membrane components are needed. Since Rho1p and Cdc42p compete with each other in their binding to Sec3p, their effect on exocyst localization may be exerted at different stages of cell growth. In addition to affecting the localization of the exocyst, mutations in Cdc42p also affect the kinetics of exocytosis (Adamo et al. 2001). Rho3p, another Rho protein in yeast binds to Exo70p in its GTP-bound form. Rho3p mutants exhibit depolarized actin as well as defects in exocytosis (Adamo et al. 1999; Robinson et al. 1999). However, there is no report of exocyst localization defects in these mutants. In mammalian cells, Exo70p was found to interact with TC10, a member of the Rho family of small GTPases that shares sequence similarity to Cdc42 (Inoue et al. 2003). Expression of the active form of TC10 in 3T3 adipocytes promoted the recruitment of Exo70 to the plasma membrane, where the exocyst carries out its physiological function.

The mammalian exocyst was also shown to be a downstream effector of Ral (Brymora et al. 2001; Moskalenko et al. 2002; Polzin et al. 2002; Sugihara et al. 2002; Moskalenko et al. 2003). Ral-GTP interacts with two components of the exocyst Sec5p and Exo84p. The disruption of Ral function has been shown to perturb the exocyst complex assembly (Moskalenko et al. 2002), an effect similar to that seen upon the loss of Sec4p function in yeast (Guo et al. 1999). In epithelial cells, expression of constitutively active RalA as well as the inhibition of either the function or synthesis of RalA had a similar effect on exocytosis. Proteins normally targeted to the basolateral surface of epithelial cells became randomly dis-

tributed between the apical and basolateral surfaces, while proteins normally targeted to the apical surface were unaffected (Moskalenko et al. 2002).

Recently, it was shown that the GTP-bound form of Arf6 interacts with Sec10p (Prigent et al. 2003). Arf6 regulates membrane recycling to regions of the plasma membrane via the endosomal pathway. In addition, Arf6 is involved in the generation of membrane protrusions concomitant with cortical actin remodeling. Prigent et al. showed that Sec10 is redistributed from internal membrane compartments to the plasma membrane ruffles and cells expression GTP-Arf6. Furthermore, dominant inhibition of Sec10 interferes with Arf6-induced cell spreading (Prigent et al. 2003).

While the exocyst were thought to be essential for the post-Golgi stage of membrane traffic, an interaction between the exocyst and translocon present at the endoplasmic reticulum was found in both yeast and mammalian cells (Toikkanen et al. 2003; Lipschutz et al. 2003). It seems surprising that the exocyst, which functions at the final stage of the secretory pathway interacts with translocon. However, an independent study revealed that *sec3* mutants are defective in the inheritance of cortical ER to the daughter cells (Wiederkehr et al. 2003). Sec3p may function in the "capture" of ER tubules at the bud tip, possibly through an association with translocon. Further support of the above connections came from genetic and functional studies. Overexpression of translocon components suppressed the exocyst mutants in yeast (Toikkanen et al. 2003). On the other hand, overexpression of Seb1p and Sec4p increases the production of secreted proteins in yeast (Toikkanen et al. 2003). In mammalian cells, overexpression of the exocyst component Sec10p leads to increased protein synthesis activity (Lipschutz et al. 2000; Lipschutz et al. 2003). The interaction between these proteins suggests that there is a circuit that functionally couples protein synthesis and exocytosis (reviewed by Guo and Novick 2004). Further investigation of this circuit will not only help us understand the regulatory mechanisms for protein synthesis and secretion, but also help understand human diseases such as autosomal dominant polycystic kidney disease (ADPKD), in which dramatic upregulation in protein synthesis and secretion results in gigantic cystic expansions, and ultimately kidney failure, in patients with ADPKD.

2.6 EEA1

EEA1 is believed to tether vesicles to the early endosome for both homotypic and heterotypic fusion (Rubino et al. 2000; Simonsen et al. 1998). EEA1 is not found on other endosome populations and is considered one of the most specific early endosome markers (Mu et al. 1995; Wilson et al. 2000). Immunofluorescence studies of EEA1 in several types of mammalian cells show that it is located on filamentous extension, in keeping with its role as a tether (Wilson et al. 2000). The protein contains 2 FYVE domains, a Rab5 binding domain on both the N-terminus and C-terminus, and an IQ domain. Normally localized on the cytosol, EEA1 can be recruited to the early endosome after Rab5 activation by the Rabaptin5/Rabex5 (GEF for Rab5) complex (Stenmark et al. 1996; Horiuchi et al. 1997). Although

EEA1 has a C-terminal Rab5 binding site, it is not believed to be used for attachment to the membrane (Lawe et al. 2002). Rab5 expression on the membrane is not enough to recruit EEA1 (Rubino et al. 2000). CCVs (clathrin-coated vesicles) were not able to recruit EEA1 in spite of activated Rab5. It was speculated that EEA1 specificity is determined by the concomitant expression of activated Rab5 and phosphatidylinositol 3-phosphate (Rubino et al. 2000), a phosphoinositol generated by another Rab5 effector hVPS34 (Christoforidis et al. 1999b; Nielsen et al. 2000). Structural studies of EEA1 have shown that it is a homodimer and Dumas et al. (2001) have determined that a bivalent thermodynamic model would account for the specificity of EEA1 binding to phosphatidylinositol 3-phosphate through its C-terminal FYVE domain. Binding in this conformation would orient EEA1 so that the rigid quaternary structure of its coiled-coil region extends into the cytoplasm where it would be capable of binding Rab5 on incoming vesicles (Dumas et al. 2001). Treatment of membrane with the phosphatidylinositol 3-phosphate inhibitor wortmannin causes dissociation of EEA1 from the membrane, showing that EEA1 attachment is dependent on phosphatidylinositol 3-phosphate (Simonsen et al. 1998). However, adding EEA1 as the only cytosolic factor bypasses the need for phosphatidylinositol 3-phosphate and allows fusion to occur independent of both phosphatidylinositol 3-phosphate and Rab5 (Christoforidis et al. 1999a). While EEA1 localization depends on phosphatidylinositol 3-phosphate binding, the later fusion is dependent on Rab5 binding to the EEA1 C-terminus (Lawe et al. 2002).

EEA1 acts upstream of SNARES in the tethering process (Christoforidis et al. 1999). Inhibition of SNARE priming by α-SNAP inhibited fusion but not tethering of vesicles (Christoforidis et al. 1999). Once EEA1 has tethered a vesicle, presumably by binding to its N-terminal Rab5 site (Rubino et al. 2000), it is believed to interact with NSF, SNARES and other Rab5 effectors to promote SNARE priming and vesicle fusion (McBride et al. 1999). Although Rab5 interacts with 22 other proteins as judged from a GST-Rab5-GTP column purification, EEA1 alone is sufficient and necessary to promote vesicle tethering and fusion in the absence of cytosol (Christoforidis et al. 1999). Addition of Rabaptin5 and Rabex5 increase this activity to normal levels (Christoforidis et al. 1999). The SNAREs syntaxin6 and syntaxin13 interact with EEA1 (McBride et al. 1999; Simonsen et al. 1999). Syntaxin6 competes with Rab5 for binding the C-terminus of EEA1 (Simonsen et al. 1999). Syntaxin6 is involved in TGN to endosome transport, therefore, it was speculated that EEA1 also regulates this fusion process by sequential interaction with Rab5 and syntaxin6 (Simonsen et al. 1999). Biosensor studies have shown a transient interaction between EEA1 and syntaxin13, a t-SNARE on the early endosome (McBride et al. 1999). Syntaxin13 interacts with the C-terminal FYVE domain of EEA1 (McBride et al. 1999). Inhibition of this interaction by addition of the EEA1 C-terminal FYVE peptide disrupted *in vitro* binding and stopped vesicle fusion, without causing EEA1 to dissociate from phosphatidylinositol 3-phosphate on the endosome membrane (McBride et al. 1999). EEA1 is found on oligomeric complexes only on the membrane with Raptain5, Rabex5, NSF, and syntaxin13 (McBride et al. 1999). McBride et al. propose that EEA1, the Rab effectors, and NSF form oligomers that act as membrane microdomains for vesicle

docking (McBride et al. 1999). Upon tethering, EEA1 interacts with syntaxin13 to form a fusion pore, akin to the mechanism of viral protein oligomers (McBride et al. 1999). NSF regulates this process by priming the SNAREs (McBride et al. 1999). EEA1, thus, provides a link between Rab5 and SNAREs to ensure specificity in docking of vesicles and promote fusion via interaction with SNAREs.

Besides EEA1, Rabenosyn-5 binds the GTP-bound form of both Rab4 and Rab5 (Nielsen et al. 2000). The early endosome is comprised of three distinct membrane domains defined by expression of Rab5, Rab4, or Rab11 (Sonnichsen et al. 2000). The bifunctional connection of Rabenosyn-5 with both Rab5 and Rab4 may physically connect entry and recycling sites of early endosomes, and play an important role in the microdomain organization of the endosome membrane (Zerial and McBride 2001; de Renzis et al. 2002; Pfeffer 2003).

In budding yeast, Vac1p shares similar sequence features to EEA1 and Rabenosyn-5. Like its mammalian counterpart, Vac1p interacts with phosphatidylinositol 3-phosphate and the GTP-bound form of Rab protein, Vps21p, and is required for Golgi-to-endosome transport (Tall et al. 1999; Peterson et al. 1999). In addition, Vac1p binds the Sec1p-like protein Vps45p. Rabenosyn-5 has also been shown to interact with mammalian Vps45p (Nielsen et al. 2000), suggesting functional conservation of these proteins.

2.7 HOPS

The HOPS (homotypic fusion and vacuole protein sorting) is a multiprotein complex consisting of Vps41p (Vam2p), Vps39p (Vam6p), and the class C Vps proteins including Vps11p (Vam1p), Vps16p (Vam9p), and Vps18p (Vam8p) (Reider and Emr 1997; Price et al. 2000a; Seals et al. 2000). It resides on the vacuolar membrane and mediates the tethering of multiple transport intermediates to the vacuole.

Biochemical analyses revealed sequential action of proteins involving the HOPS (Price et al. 2000b). First, proteins of the HOPS are associated with cis-paired SNAREs Vam3p and Nyv1p on the vacuoles as a 65S complex. This association is disrupted upon Sec18p (NSF)-driven priming. The HOPS then binds to GTP-bound Ypt7p, the Rab protein functioning in vacuole docking and fusion. While the complex is an effector complex for Ypt7p (Seals et al. 2000), Vam6p/Vps39p interacts with Ypt7p in its GDP-bound form and is a nucleotide exchange factor for Ypt7p (Wurmser et al. 2000). The HOPS complex may, therefore, couple Rab activation and Rab downstream activities promoting vacuole fusion.

HOPS was also shown to bind to the vacuolar t-SNARE Vam3p in its unpaired form (Sato et al. 2000). In addition, HOPS interacts with the Sec1p-homologue Vps33p (Reider and Emr 1997; Seals et al. 2000). The interaction between Vps33p with HOPS depends on Vps18p (Vam8p) (Sato et al. 2000). The HOPS is important for the SNARE assembly as temperature-sensitive mutants of HOPS components inhibited Vam3p-Vti1p-Vam7p pairing (Sato et al. 2000). On the

other hand, anti-SNARE antibodies that disrupt SNARE assembly did not disrupt vacuole tethering (Ungermann et al. 1998).

Interestingly, it was found that Cdc42p and Rho1p, members of Rho family of small GTPases, also affect tethering during homotypic vacuole fusion (Eitzen et al. 2001; Muller et al. 2001). These two small G-proteins function after the action of Ypt7p and before the SNARE pairing. The observed effects of Cdc42p and Rho1p is probably through their effects on actin. Indeed, G-actin co-purifies with vacuoles, and there is a G- to F-actin transition during the vacuole fusion reaction. Furthermore, actin-directed drugs latrunculin B and jasplakinolide affect the fusion reaction (Eitzen et al. 2002).

2.8 The GARP/VFT complex

The GARP (Golgi-associated retrograde protein)/VFT (Vps Fifty-Three) complex is required for the fusion of vesicles derived from both early and late endosomal populations with the *trans*-Golgi network (TGN). Now understood to be a tetrameric complex, GARP/VFT complex was originally identified as a tightly associated trimer consisting of Vps52p, Vps53p and Vps54p, all large coiled-coil proteins (Conibear and Stevens 2000). Later, another Vps protein, Vps51p, was found to be part of the complex by virtue of its association with the trimer and Tlg1p (Siniossoglou and Pelham 2002; Conibear et al. 2003). Loss of any of the four subunits causes Vps10p, which continuously recycles from the pre-vacuolar compartment to the TGN in wild type yeast, to be mislocalized to the vacuole. Under the same conditions, Snc1p, an exocytic v-SNARE that is normally recycled from the early endosome to the Golgi and then directed to the plasma membrane, is mislocalized intracellularly and colocalizes with the vacuole (Lewis et al. 2000).

The localization of this complex to the late Golgi is mediated by Ypt6p, which is required for the retrograde transport of late Golgi proteins from the endosome back to the Golgi. In *ypt6* mutants, Golgi proteins such as Kex2p and the Vps10p are mislocalized to the vacuole (Tsukada et al. 1999; Siniossoglou et al. 2000). Vps52p binds the Rab protein Ypt6p in its active form. Ypt6p itself is activated by the Ric1p-Rgp1p complex at the Golgi (Siniossoglou et al. 2000). Studies also showed that Vps51p interacts with the N-terminal domain of late Golgi t-SNARE Tlg1p, suggesting that it might function as a tethering molecule to couple Rab activity with SNARE complex formation (Siniossoglou and Pelham 2002; Conibear et al. 2003). The absence of Ypt6p causes Vps51p and Vps52p to be redistributed as shown by a dispersed, finely punctuate staining pattern, which may reflect their binding to endosome-derived transport vesicles rather than dispersion of the Golgi per se (Conibear et al. 2003). The proportion of Vps52p associated with membrane is unaffected, suggesting that after Ypt6p helps to localize GARP/VFT to the Golgi, another yet unidentified mechanism mediates the actual membrane association (Conibear et al. 2003). VFT/GARP may be recruited by activated Ypt6p. In addition, the binding of GARP/VFT to the N-terminal domain of Tlg1p may cause a conformational change that promotes the assembly of a SNARE fusion

complex. It has been observed that the removal of the N-terminal domain of syntaxin accelerates SNARE complex formation *in vitro* (Parlatti et al. 1999; Munson et al. 2000), and a similar principle might apply in which regulatory factors interact with the N-terminal domain of syntaxin-like t-SNAREs to modulate fusion. Another possible model is based on the fact that Tlg1p resides not only on the Golgi membrane but also on the vesicles themselves. Ric1p-Rgp1p would activate Ypt6p at the Golgi membrane, which would then recruit GARP/VFT and through it associate with endosome-derived vesicles bearing Tlg1p (Siniossoglou and Pelham 2001). Further studies are required to determine whether VFT/GARP directly modulates SNARE activity through Tlg1p.

3 Postulated functions of the tethering proteins

The tethering proteins may have at least two functions. First, protein-protein interactions linking incoming vesicles and targeting membrane may confer targeting specificity, ensuring the fidelity of membrane fusion. Although the targeting specificity could come from both tethering and docking, most of the specificity is likely to be conferred by the earlier tethering event in the sequence of the reactions. Second, tethering proteins may contribute in the kinetic aspects by physically bringing vesicles to the vicinity of the target membrane and then promote/catalyze cognate SNARE assembly, which is a relatively slow process by themselves (Weber et al. 1998). The "kinetic" and "targeting" aspects of tethering complex functions are not mutually exclusive. Kinetic acceleration of the assembly of correctly paired SNAREs may, at the end, be revealed as a specific targeting event.

Tethering proteins not only help confer specificities to cognate vesicle-target membrane recognition, they may also help to define the specific domains of target membrane for fusion. This can be regarded as another layer of targeting specificity. For example, the yeast exocyst is localized to the bud tip, sites of active exocytosis (TerBush and Novick 1995; Finger et al. 1998; Guo et al. 1999), whereas the t-SNARE components Sso1p, Sso2p and Sec9p is distributed to the entire plasma membrane in both mother and daughter cells (Brennwald et al. 1994). The exocyst is not only essential for secretion at the plasma membrane, but also defines the specific site of the plasma membrane for exocytosis in response to polarity cues such as Rho proteins (Guo et al. 2001; Zhang et al. 2001). In addition to the plasma membrane stage, membrane domain specification has also been revealed in other stages of traffic compartments such as endosomes. Several excellent papers have reviewed this area of research (Zerial and McBride 2001; Pfeffer 2003).

4 Emerging common features of the tethering proteins

The protein sequences of Rabs and the SNAREs functioning in various stages of traffic are conserved. How about the tethering proteins? Despite the evolutionary conservation across species, the sequence homologies among the tethering proteins at various traffic stages are not obvious. However, a careful comparison between components of the COG complex and components of other tethering complexes including the exocyst and GARP/VTF revealed homology at the N-terminal amphipathic helical regions, many of which have significant potential to form short coiled-coil structures (Whyte and Munro 2001, 2002). Since these complexes contain a multiple of four subunits, they were termed "quatrefoil" complexes (Whyte and Munro 2001, 2002). For tethering proteins that do not form complexes, many of them contain long coiled-coil structures and exist as parallel homodimers. These long coiled-coils may be important for the "sampling" of vesicles and "tethering" of cognate membranes. The large sizes of these proteins may also provide a scaffold for protein interactions (Gillingham and Munro 2003). From an evolutionary point of view, the tethering proteins are likely to have evolved at relatively late stages to enhance the functions of the core traffic machinery consisting of the SNAREs and the Rab proteins.

Common features are emerging for the tethering proteins. First, they tend to associate with the specific SNAREs at the target membranes. In many cases, direct physical associations were reported. Even if in some cases no direct binding has been revealed yet by biochemical methods (such as the exocyst and the plasma membrane t-SNARE Sso1, 2p), overwhelming genetic and functional data indicate a close relationship between SNAREs and the tethering proteins (Aalto et al. 1993; Brennwald et al. 1994). These interactions may be transient in nature. However, they may catalyze the activation of SNAREs for assembly. Shorter et al. (2002) investigated the interaction between p115 and Golgi SNAREs and suggested that the interaction of p115 with SNARE represents an intermediate in the activation process of SNARE assembly (Shorter et al. 2002). For yeast exocytosis, the rate of SNARE assembly is very slow due to a self-inhibitory conformation of the t-SNARE. *In vitro* studies indicate that removal of the N-terminal inhibitory region accelerates the core SNARE complex as much as 3000-fold (Nicholson et al. 1998; Munson et al. 2000). It is tempting to speculate that the exocyst or exocyst-associated protein(s) may transiently interact with the t-SNAREs Sso1, 2p and "open" this self-inhibition. Consistent with this speculation, it was found that the assembly of the SNAREs was defective in the exocyst mutants (Grote et al. 2000).

In addition to the t-SNAREs, some tethering proteins may also interact with Sec1p homologues. For example, HOPS was found to interact with Vps33p (Reider and Emr 1997; Seals et al. 2000). Vac1p binds Vps45p (Tall et al. 1999; Peterson et al. 1999), and its mammalian homologue Rabenosyn-5 has also been shown to interact with mammalian Vps45p (Nielsen et al. 2000). For exocytosis, overexpression of *SEC1* suppresses the exocyst mutations (Aalto et al. 1993). Since

Sec1p family proteins are important regulators of SNAREs, these interactions may functionally affect membrane fusion.

Another major feature for the tethering proteins is their close relationship to the Rab proteins. In many cases, these tethering proteins interact with the GTP-bound form of Rab proteins and function as their downstream effectors (e.g. Rab1-p115; Ypt1p-COG; Sec4p-exocyst; Rab5-EEA1; Ypt6p-GARP). In several cases, the tether proteins have guanine nucleotide exchange activity towards the Rabs (e.g. TRAPP-Ypt1p). In vacuole and endosomes tethering, both GEF and effectors were found in the same complex (e.g. HOPS and Rabaptin-5/Rabex-5). This type of molecular organization may provide a "positive-feedback" mechanism to create a membrane domain for active docking and fusion. The exact functional effects of Rabs on these tethering proteins are still unclear. In the case of endosomal trafficking, Rab5 may help to recruit tethering factors onto membranes in preparation for subsequent *trans*-SNARE assembly (McBride et al. 1999). In the case of exocytosis at the plasma membrane, at least one of the functions of Sec4p is to promote the assembly of the exocyst complex at the plasma membrane (Guo et al. 1999).

The tethering factors are not the only effectors for the Rab proteins. For example, in addition to EEA1, several proteins including phosphatidylinositol 3-kinases are found to be downstream effectors for Rab5 (Christoforidis et al. 1999b). EEA1 binds to the phosphatidylinositol 3-phosphate in the endosomal membrane through its "FYVE" domain (Dumas et al. 2001) and tethers endosomal membranes independent of trans-SNARE pairing (Christoforidis et al. 1999a). Rab5 may recruit phosphatidylinositol 3-kinases to the endosomal membranes and couple local phosphatidylinositol 3-phosphate production to EEA1 function. Rab proteins are also found to be intimately linked to cytoskeleton. For example, Rabkinesin-6, a kinesin-like motor protein associated with Golgi is a downstream effector of Rab6 (Echard et al. 1998). In mammalian cells, microtubule based cytoskeleton may direct Golgi vesicle trafficking. For exocytosis, polarized localization of Sec4p and components of the exocyst to the site of secretion requires actin "cable" and the motor protein Myo2p (Govindan et al. 1995; Walch-Solimena et al. 1997; Pruyne et al. 1998; Schott et al. 1999; Karpova et al. 2000). Rho GTPases including Cdc42p and Rho1p directly interact with the N-terminus of Sec3p in a GTP-dependent manner (Guo et al. 2001; Zhang et al. 2001). In addition to affecting the localization of the exocyst, mutations in Cdc42p also affect the kinetics of exocytosis (Adamo et al. 2001). Rho proteins may coordinate actin and exocytosis to ensure polarized membrane expansion and cell growth (Novick and Guo 2002; Lipschutz and Mostov 2002). For homotypic vacuole fusion in yeast, actin is involved (Eitzen et al. 2002) and Cdc42p and Rho1p are regulators for the reaction (Eitzen et al. 2001; Muller et al. 2001).

5 Regulations of the tethering proteins

Membrane traffic is under the control of a wide variety of cellular signaling events. Since the tethering events precede docking and fusion, the tethering proteins are ideal targets for cellular regulations. Therefore, besides studying the connections with Rab and the SNAREs, it will be important to search for molecules that interact with the tethering proteins. For exocytosis at the plasma membrane, different Rho proteins interact with members of the exocyst complex (Lipschutz and Mostov 2002; Novick and Guo 2002). The correct localization of the exocyst requires functional Rho proteins and polarized actin cytoskeleton. It is possible that different exocyst components communicate with cellular cues, and the assembly of the complex may help integrate various sources of cellular information to ensure that membrane fusion occurs at the right time and the right place. At the Golgi stage, phosphorylation of GM130 by Cdc2 kinase disrupts the interaction between p115 and GM130, which in turn inhibits the fusion of COPI vesicles to the Golgi. This may account for most of the observed Golgi fragmentation during mitosis (Nakamura et al. 1997; Lowe et al. 1998). At the endosomal stage, Rabaptin-5 can be cleaved by caspase-3 during apoptosis, which prevents the endosomal fusion and results in endosomal fragmentation (Cosulich et al. 1997).

6 Tethering proteins and sorting

A recent study has implicated the tethering proteins in sorting upon vesicle budding from the donor compartment (Morsomme and Riezman 2002). Using an *in vitro* assay, it was shown that Uso1p, along with Ypt1p and COG (but not the TRAPP complex), are necessary for the sorting of GPI-anchored proteins from other secretory proteins upon exit from the ER (Morsomme and Riezman 2002). The sorting and tethering functions are not interdependent. A model has been proposed in which Ypt1p recruits Uso1p and Sec34/35p at the ER where sorting and packaging occur. These factors would then be targeted to the Golgi membrane, possibly to the TRAPPI complex, which is statically localized to the *cis*-Golgi compartment and is the specific target of COPII-coated vesicles (Barrowman et al. 2000; Sacher et al. 2001). This result suggests that the traffic events at the donor membrane and the tethering events at the acceptor membrane are functionally coupled and Rab GTPases may coordinate this coupling. This may further ensure the targeting specificity throughout the various steps of trafficking. It will be interesting to investigate this coupling process in other trafficking stages such as protein sorting at TGN and targeting to the plasma membrane or endosomal compartments.

7 Future studies

With the players at various stages of traffic identified, it will be important to reconstitute the membrane tethering and fusion using minimal lipids and protein components. For example, with the SNAREpins in artificially synthesized membranes as the core (Weber et al. 1998; McNew et al. 2000), it will be interesting to apply various components of the tethering proteins individually or in combination to test if they affect the kinetics of SNAREpin formation.

Another important future direction will involve identification of proteins that interact with the tethering proteins. As mentioned above, an increasing number of proteins have been identified that regulate or work in concert with the tethering proteins. Studying these molecular interactions may reveal the cellular context of the traffic events. These studies will reveal how membrane traffic is functionally coupled to other essential cellular processes such as cell polarization, cell cycle progression, cell proliferation and apoptosis.

Acknowledgment

We thank Ms. Allison Zajac for her help in preparation of this manuscript. We also thank Dr. Elizabeth Conibear for her helpful comments. We apologize that many excellent papers were excluded from the review. W.G. is supported by National Institutes of Health and American Cancer Society. W.G. is a Pew Scholar of Biomedical Sciences.

References

Aalto MK, Ronne H, Keranen S (1993) Yeast syntaxins Sso1p and Sso2p belong to a family of related membrane proteins that function in vesicular transport. EMBO J 12:4095-4104

Adamo JE, Rossi G, Brennwald P (1999) The Rho GTPase Rho3 has a direct role in exocytosis that is distinct from its role in actin polarity. Mol Biol Cell 10:4121-4133

Allan BB, Moyer BD, Balch WE (2000) Rab1 recruitment of p115 into a cis-SNARE complex: programming budding COPII vesicles for fusion. Science 289:444-448

Aniento F, Gu F, Parton RG, Gruenberg J (1996) An endosomal beta COP is involved in the pH-dependent formation of transport vesicles destined for late endosomes. J Cell Biol 133:29-41

Bacon RA, Salminen A, Ruohola H, Novick P, Ferro-Novick S (1989) The GTP-binding protein Ypt1 is required for transport in vitro: the Golgi apparatus is defective in ypt1 mutants. J Cell Biol 109:1015-1022

Barlowe C, Orci L, Yeung T, Hosobuchi M, Hamamoto S, Salama N, Rexach MF, Ravazzola M, Amherdt M, Schekman R (1994) COPII: a membrane coat formed by Sec proteins that drive vesicle budding from the endoplasmic reticulum. Cell 77:895-907

Barr FA, Puype M, Vandekerckhove J, Warren G (1997) GRASP65, a protein involved in the stacking of Golgi cisternae. Cell 91:253-262

Barroso M, Nelson DS, Sztul E (1995) Transcytosis-associated protein (TAP)/p115 is a general fusion factor required for binding of vesicles to acceptor membranes. Proc Natl Acad Sci USA 92:527-531

Barrowman J, Sacher M, Ferro-Novick S (2000) TRAPP stable associates with the Golgi and is required for vesicle docking. EMBO J 19:862–869

Becherer KA, Rieder SE, Emr SD, Jones EW (1996) Novel syntaxin homologue, Pep12p, required for the sorting of lumenal hydrolases to the lysosome-like vacuole in yeast. Mol Biol Cell 7:579-594

Brennwald P, Novick P (1993) Interactions of three domains distinguishing the Ras-related GTP-binding proteins Ypt1 and Sec4. Nature 362:560-563

Brennwald P, Kearns B, Champion K, Keranen S, Bankaitis V, Novick P (1994) Sec9 is a SNAP-25-like component of a yeast SNARE complex that may be the effector of Sec4 function in exocytosis. Cell 79:245-258

Bowser R, Novick P (1991) Sec15 protein, an essential component of the exocytotic apparatus, is associated with the plasma membrane and with a soluble 19.5S particle. J Cell Biol 6:1117-1131

Burd CG, Peterson M, Cowles CR, Emr SD (1997) A novel Sec18p/NSF-dependent complex required for Golgi-to-endosome transport in yeast. Mol Biol Cell 8:1089-1104

Cao X, Ballew N, Barlowe C (1998) Initial docking of ER-derived vesicles requires Uso1p and Ypt1p but is independent of SNARE proteins. EMBO J 17:2156-2165

Conibear E, Stevens TH (2000) Vps52p, Vps53p, and Vps54p form a novel multisubunit complex required for protein sorting at the yeast late Golgi. Mol Biol Cell 11:305-323

Christoforidis S, McBride HM, Burgoyne RD, Zerial M (1999a) The Rab5 effector EEA1 is a core component of endosome docking. Nature 397:621-625

Christoforidis S, Miaczynska M, Ashman K, Wilm M, Zhao L, Yip SC, Waterfield MD, Backer JM, Zerial M (1999b) Phosphatidylinositol-3-OH kinases are Rab5 effectors. Nat Cell Biol 1:249-252

Conibear E, Cleck JN, Stevens TH (2003) Vps51p mediates the association of the GARP (Vps52/52/54) complex with the late Golgi t-SNARE Tlg1p. Mol Biol Cell 14:1610-1623

Cosulich SC, Horiuchi H, Zerial M, Clarke PR, Woodman PG (1997) Cleavage of rabaptin-5 blocks endosome fusion during apoptosis. EMBO J 16:6182-6191

Daro E, Sheff D, Gomez M, Kreis T, Mellman I (1997) Inhibition of endosome function in CHO cells bearing a temperature-sensitive defect in the coatomer (COPI) component epsilon-COP. J Cell Biol 139:1747-1759

Darsow T, Rieder SE, Emr SD (1997) A multispecificity syntaxin homologue, Vam3p, essential for autophagic and biosynthetic protein transport to the vacuole. J Cell Biol 138:517-529

de Renzis S, Sonnichsen B, Zerial M (2002) Divalent Rab effectors regulate the subcompartmental organization and sorting of early endosomes. Nat Cell Biol 4:124-133

Dumas JJ, Merithew E, Sudharshan E, Rajamani D, Hayes S, Lawe D, Corvera S, Lambright DG (2001) Multivalent endosome targeting by homodimeric EEA1. Mol Cell 8:947-958

Dunn B, Stearns T, Botstein D (1993) Specificity domains distinguish the Ras-related GTPases Ypt1 and Sec4. Nature 362:563-565

Echard A, Jollivet F, Martinez O, Lacapere JJ, Rousselet A, Janoueix-Lerosey I, Goud B (1998) Interaction of a Golgi-associated kinesin-like protein with Rab6. Science 279:580-585

Eitzen G, Thorngren N, Wickner W (2001) Rho1p and Cdc42p act after Ypt7p to regulate vacuole docking. EMBO J 20:5650-5656

Eitzen G, Wang L, Thorngren N, Wickner W (2002) Remodeling of organelle-bound actin is required for yeast vacuole fusion. J Cell Biol 158(4):669-679

Fasshauer D, Antonin W, Margittai M, Pabst S, Jahn R (1999) Mixed and non-cognate SNARE complexes. Characterization of assembly and biophysical properties. J Biol Chem 274:15440-15446

Fischer von Mollard G, Stevens TH (1999) The *Saccharomyces cerevisiae* v-SNARE Vti1p is required for multiple membrane transport pathways to the vacuole. Mol Biol Cell 10:1719-1732

Fukuda R, McNew JA, Weber T, Parlati F, Engel T, Nickel W, Rothman JE, Sollner TH (2000) Functional architecture of an intracellular membrane t-SNARE. Nature 407:198-202

Gasman S, Kalaidzidis Y, Zerial M (2003) RhoD regulates endosome dynamics through Diaphanous-related Formin and Src tyrosine kinase. Nat Cell Biol 5:195-204

Gillingham AK, Munro S (2003) Long coiled-coil proteins and membrane traffic. Biochim Biophys Acta 1641:71-85 Review

Grote E, Carr CM, Novick PJ (2000) Ordering the final events in yeast exocytosis. J Cell Biol 151:439-452

Gotte M, von Mollard GF (1998) A new beat for the SNARE drum. Trends Cell Biol 8:215-218 Review

Govindan B, Bowser R, Novick P (1995) The role of Myo2, a yeast class V myosin, in vesicular transport. J Cell Biol 128:1055-1068

Guo W, Sacher M, Barrowman J, Ferro-Novick S, Novick P (2000) Protein complexes in transport vesicle targeting. Trends Cell Biol 10:251-255 Review

Guo W, Roth D, Walch-Solimena C, Novick P (1999) The exocyst is an effector for Sec4p, targeting secretory vesicles to sites of exocytosis. EMBO J 18:1071-1080

Guo W, Grant A, Novick P (1999) Exo84p is an exocyst protein essential for secretion. J Biol Chem 274:23558-23564

Guo W, Tamanoi F, Novick P (2001) Spatial regulation of the exocyst complex by Rho1 GTPase. Nat Cell Biol 3:353-360

Guo W, Novick P (2004) The exocyst meets the translocon: a regulatory circuit for secretion and protein synthesis? Trend Cell Biol 14:61-63

Holthuis JC, Nichols BJ, Dhruvakumar S, Pelham HR (1998) Two syntaxin homologues in the TGN/endosomal system of yeast. EMBO J 17:113-126

Horiuchi H, Lippe R, McBride HM, Rubino M, Woodman P, Stenmark H, Rybin V, Wilm M, Ashman K, Mann M, Zerial M (1997) A novel Rab5 GDP/GTP exchange factor complexed to Rabaptin-5 links nucleotide exchange to effector recruitment and function. Cell 90:1149-1159

Hsu SC, TerBush D, Abraham M, Guo W (2004) The exocyst complex in polarized exocytosis. Int Rev Cytol 233:243-265

Hu C, Ahmed M, Melia TJ, Sollner TH, Mayer T, Rothman JE (2003) Fusion of cells by flipped SNAREs. Science 300:1745-1749

Jahn R, Lang T, Sudhof TC (2003) Membrane fusion. Cell 112:519-533 Review

Jedd G, Richardson C, Litt R, Segev N (1995) The Ypt1 GTPase is essential for the first two steps of the yeast secretory pathway. J Cell Biol 131:583-590

Jones S, Newman C, Liu F, Segev N (2000) The TRAPP complex is a nucleotide exchanger for Ypt1 and Ypt31/32. Mol Biol Cell 11:4403-4411

Karpova TS, Reck-Peterson SL, Elkind NB, Mooseker MS, Novick PJ, Cooper JA (2000) Role of actin and Myo2p in polarized secretion and growth of *Saccharomyces cerevisiae*. Mol Biol Cell 11:1727-1737

Lawe DC, Chawla A, Merithew E, Dumas J, Carrington W, Fogarty K, Lifshitz L, Tuft R, Lambright D, Corvera S (2002) Sequential roles for phosphatidylinositol 3-phosphate and Rab5 in tethering and fusion of early endosomes via their interaction with EEA1. J Biol Chem 277:8611-8617

Lazar T, Gotte M, Gallwitz D (1997) Vesicular transport: how many Ypt/Rab-GTPases make a eukaryotic cell? Trends Biochem Sci 22:468-472 Review

Levine TP, Rabouille C, Kieckbusch RH, Warren G (1996) Binding of the vesicle docking protein p115 to Golgi membranes is inhibited under mitotic conditions. J Biol Chem 271:17304-17311

Lewis MF, Nichols BJ, Prescianotto-Baschong C, Riezman H, Pelham HR (2000) Specific retrieval of the exocytic SNARE Snc1p from early yeast endosomes. Mol Biol Cell 11:23-38

Linstedt AD, HP Hauri (1993) Giantin, a novel conserved Golgi membrane protein containing a cytoplasmic domain of at least 350 kDa. Mol Biol Cell 4:679-693

Linstedt AD, Jesch SA, Mehta A, Lee TH, Garcia-Mata R, Nelson DS, Sztul E (2000) Binding relationships of membrane tethering components. The giantin N terminus and the GM130 N terminus compete for binding to the p115 C terminus. J Biol Chem 14:10196-10201

Lupashin VV, Pokrovskaya ID, McNew JA, Waters MG (1997) Characterization of a novel yeast SNARE protein implicated in Golgi retrograde traffic. Mol Biol Cell 8:2659-2676

Lipschutz JH, Lingappa VR, Mostov KE (2003) The exocyst affects protein synthesis by acting on the translocation machinery of the endoplasmic reticulum. J Biol Chem 278:20954-20960

Lipschutz JH, Mostov KE (2002) Exocytosis: the many masters of the exocyst. Curr Biol 12:R212-R214 Review

Lipschutz JH, Guo W, O'Brien LE, Nguyen YH, Novick P, Mostov KE (2000) Exocyst is involved in cystogenesis and tubulogenesis and acts by modulating synthesis and delivery of basolateral plasma membrane and secretory proteins. Mol Biol Cell 11:4259-4275

Lowe M (2000) Membrane transport: Tethers and TRAPPs. Curr Biol 10:R407-R409

Lowe M, Rabouille C, Nakamura N, Watson R, Jackman M, Jamsa E, Rahman D, Pappin DJ, Warren G (1998) Cdc2 kinase directly phosphorylates the cis-Golgi matrix protein GM130 and is required for Golgi fragmentation in mitosis. Cell 94:783-793

McBride HM, Rybin V, Murphy C, Giner A, Teasdale R, Zerial M (1999) Oligomeric complexes link Rab5 effectors with NSF and drive membrane fusion via interactions between EEA1 and syntaxin 13. Cell 98:377-386

McNew JA, Parlati F, Fukuda R, Johnston RJ, Paz K, Paumet F, Sollner TH, Rothman JE (2000) Compartmental specificity of cellular membrane fusion encoded in SNARE proteins. Nature 407:153-159

Misteli T, Warren G (1994) COP-coated vesicles are involved in the mitotic fragmentation of Golgi stacks in a cell-free system. J Cell Biol 125:269-282Morsommne P, Riezman H (2002) The Rab GTPase Ypt1p and tethering factors couple protein sorting at the ER to vesicle targeting to the Golgi apparatus. Dev Cell 2:307-317Moyer BD, Allan BB, Balch WE (2001) Rab1 interaction with a GM130 effector complex regulates COPII vesicle cis--Golgi tethering. Traffic 2:268-276

Mu FT, Callaghan JM, Steele-Mortimer O, Stenmark H, Parton RG, Campbell PL, McCluskey J, Yeo JP, Tock EP, Toh BH (1995) EEA1, an early endosome-associated protein. EEA1 is a conserved alpha-helical peripheral membrane protein flanked by cysteine "fingers" and contains a calmodulin-binding IQ motif. J Biol Chem 270:13503-13511

Muller O, Johnson DI, Mayer A (2001) Cdc42p functions at the docking stage of yeast vacuole membrane fusion. EMBO J 20:5657-5665

Munson M, Chen X, Cocina AE, Schultz SM, Hughson FM (2000) Interactions within the yeast t-SNARE Sso1p that control SNARE complex assembly. Nat Struct Biol 7:894-902

Nakajima H, Hirata A, Ogawa Y, Yonehara T, Yoda K, Yamasaki MA (1991) Cytoskeleton-related gene, uso1, is required for intracellular protein transport in *Saccharomyces cerevisiae*. J Cell Biol 113:245-260

Nakamura N, C Rabouille, R Watson, Nilsson T, Hui P, Slusarewicz N, Kreis TE, Warren G (1995) Characterization of a cis-Golgi matrix protein, GM130. J Cell Biol 131:1715-1726

Nakamura N, Lowe M, Levine TP, Rabouille C, Warren G (1997) The vesicle docking protein p115 binds GM130, a cis-Golgi matrix protein, in a mitotically regulated manner. Cell 89:445-455

Nelson DS, Alvarez C, Gao YS, Garcia-Mata R, Fialkowski E, Sztul E (1998) The membrane transport factor TAP/p115 cycles between the Golgi and earlier secretory compartments and contains distinct domains required for its localization and function. J Cell Biol 143:319-331

Nicholson KL, Munson M, Miller RB, Filip TJ, Fairman R, Hughson FM (1998) Regulation of SNARE complex assembly by an N-terminal domain of the t-SNARE Sso1p. Nat Struct Biol 5:793-802

Nielsen E, Christoforidis S, Uttenweiler-Joseph S, Miaczynska M, Dewitte F, Wilm M, Hoflack B, Zerial M (2000) Rabenosyn-5, a novel Rab5 effector, is complexed with hVPS45 and recruited to endosomes through a FYVE finger domain. J Cell Biol 151:601-612

Novick P, Guo W (2002) Ras family therapy: Rab, Rho and Ral talk to the exocyst. Trends Cell Biol 12:247-249

Novick P, Zerial M (1997) The diversity of Rab proteins in vesicle transport. (1997) Curr Opin Cell Biol 9:496-504 Review

Orci L, Perrelet A, Rothman JE (1998) Vesicles on strings: morphological evidence for processive transport within the Golgi stack. Proc Natl Acad Sci USA 95:2279-2283

Pelham HR (2001) SNAREs and the specificity of membrane fusion. Trends Cell Biol 11:99-101 Review

Parlati F, Weber T, McNew JA, Westermann B, Sollner TH, Rothman JE (1999) Rapid and efficient fusion of phospholipids vesicles by the alpha-helical core of a SNARE complex in the absence of an N-terminal regulatory domain. Proc Natl Acad Sci USA 96:12565-12570

Peterson MR, Burd CG, Emr SD (1999) Vac1p coordinates Rab and phosphatidylinositol 3-kinase signaling in Vps45p-dependent vesicle docking/fusion at the endosome. Curr Biol 9:159-162

Peterson MR, Emr SD (2001) The class C Vps complex functions at multiple stages of the vacuolar transport pathway. Traffic 2:476-486

Pfeffer SR (1999) Transport-vesicle targeting: tethers before SNAREs. Nat Cell Biol 1:E17-E22 Review

Pfeffer S (2003) Membrane domains in the secretory and endocytic pathways. Cell 112:507-517 Review

Price A, Seals D, Wickner W, Ungermann C (2000) The docking stage of yeast vacuole fusion requires the transfer of proteins from a cis-SNARE complex to a Rab/Ypt protein. J Cell Biol 148:1231-1238

Price A, Wickner W, Ungermann C (2000) Proteins needed for vesicle budding from the Golgi complex are also required for the docking step of homotypic vacuole fusion. J Cell Biol 148:1223-1229

Prigent M, Dubois T, Raposo G, Derrien V, Tenza D, Rosse C, Camonis J, Chavrier P (2003) ARF6 controls post-endocytic recycling through its downstream exocyst complex effector. J Cell Biol 163:1111-1121

Pruyne DW, Schott DH, Bretscher A (1998) Tropomyosin-containing actin cables direct the Myo2p-dependent polarized delivery of secretory vesicles in budding yeast. J Cell Biol 143:1931-1945

Ram RJ, Li B, Kaiser CA (2002) Identification of Sec36p, Sec37p, and Sec38p: components of yeast complex that contains Sec34p and Sec35p. Mol Biol Cell 13:1484-1500

Raymond CK, Howald-Stevenson I, Vater CA, Stevens TH (1992) Morphological classification of the yeast vacuolar protein sorting mutants: evidence for a prevacuolar compartment in class E vps mutants. Mol Biol Cell 3:1389-1402

Rieder SE, Emr SD (1997) A novel RING finger protein complex essential for a late step in protein transport to the yeast vacuole. Mol Biol Cell 8:2307-2327

Robinson NG, Guo L, Imai J, Toh-E A, Matsui Y, Tamanoi F (1999) Rho3 of *Saccharomyces cerevisiae*, which regulates the actin cytoskeleton and exocytosis, is a GTPase, which interacts with Myo2 and Exo70. Mol Cell Biol 19:3580-3587

Rothman JE (1994) Mechanisms of intracellular protein transport. Nature 372:55-63 Review

Rubino M, Miaczynska M, Lippe R, Zerial M (2000) Selective membrane recruitment of EEA1 suggests a role in directional transport of clathrin-coated vesicles to early endosomes. J Biol Chem 275:3745-3748

Rossi G, Kolstad K, Stone S, Palluault F, Ferro-Novick S (1995) BET3 encodes a novel hydrophilic protein that acts in conjunction with yeast SNAREs. Mol Biol Cell 6:1769–1780

Sacher M, Jiang Y, Barrowman J, Scarpa A, Burston J, Zhang L, Schieltz D, Yates JR 3rd, Abeliovich H, Ferro-Novick S (1998) TRAPP, a highly conserved novel complex on the cis-Golgi that mediates vesicle docking and fusion. EMBO J 17:2494–2503

Sacher M, Barrowman J, Schieltz D, Yates JR 3rd, Ferro-Novick S (2000) Identification, of five new subunits of TRAPP. Eur J Cell Biol 79:71–80

Sacher M, Barrowman J, Wang W, Horecka J, Zhang Y, Pypaert M, Ferro-Novick S (2001) TRAPP I implicated in the specificity of tethering in ER-to-Golgi transport. Mol Cell 7:433-442

Sapperstein SK, Walter DM, Grosvenor AR, Heuser JE, Waters MG (1995) p115 is a general vesicular transport factor related to the yeast endoplasmic reticulum to Golgi transport factor Uso1p. Proc Natl Acad Sci USA 92:522-526

Sapperstein SK, Lupashin VV, Schmitt HD, Waters MG (1996) Assembly of the ER to Golgi SNARE complex requires Uso1p. J Cell Biol 132:755-567

Sato TK, Rehling P, Peterson MR, Emr SD (2000) Class C Vps protein complex regulates vacuolar SNARE pairing and is required for vesicle docking/fusion. Mol Cell 6:661-671

Schimmoller F, Simon I, Pfeffer SR (1998) Rab GTPases, directors of vesicle docking. J Biol Chem 273:22161-22164 Review

Schott D, Ho J, Pruyne D, Bretscher A (1999) The COOH-terminal domain of Myo2p, a yeast myosin V, has a direct role in secretory vesicle targeting. J Cell Biol 147:791-808

Seals DF, Eitzen G, Margolis N, Wickner WT, Price A (2000) A Ypt/Rab effector complex containing the Sec1 homolog Vps33p is required for homotypic vacuole fusion. Proc Natl Acad Sci USA 97:9402-9407

Sevrioukov EA, He JP, Moghrabi N, Sunio A, Kramer H (1999) A role for the deep orange and carnation eye color genes in lysosomal delivery in *Drosophila*. Mol Cell 4:479-486

Shorter J, Beard MB, Seemann J, Dirac-Svejstrup AB, Warren G (2002) Sequential tethering of Golgins and catalysis of SNAREpin assembly by the vesicle-tethering protein p115. Cell Biol 1:45-62

Simonsen A, Lippe R, Christoforidis S, Gaullier JM, Brech A, Callaghan J, Toh BH, Murphy C, Zerial M, Stenmark H (1998) EEA1 links PI(3)K function to Rab5 regulation of endosome fusion. Nature 394:494-498

Simonsen A, Gaullier JM, D'Arrigo A, Stenmark H (1999) The Rab5 effector EEA1 interacts directly with syntaxin-6. J Biol Chem 274:28857-28860

Siniossoglou S, Peak-Chew SY, Pelham HR (2000) Ric1p and Rgp1p form a complex that catalyzes nucleotide exchange on Ypt6p. EMBO J 19:4886-4894

Siniossoglou S, Pelham HR (2001) An effector of Ypt6p binds the SNARE Tlg1p and mediates selective fusion of vesicles with late Golgi membranes. EMBO J 20:5991-5998

Sonnichsen B, De Renzis S, Nielsen E, Rietdorf J, Zerial M (2000) Distinct membrane domains on endosomes in the recycling pathway visualized by multicolor imaging of Rab4, Rab5, and Rab11. J Cell Biol 149:901-914

Sonnichsen B, Lowe M, Levine T, Jamsa E, Dirac-Svejstrup B, Warren G (1998) A role for giantin in docking COPI vesicles to Golgi membranes. J Cell Biol 140:1013-1021

Spang A, Schekman R (1998) Reconstitution of retrograde transport from the Golgi to the ER *in vitro*. J Cell Biol 143:589-599

Stenmark H, Aasland R, Toh BH, D'Arrigo A (1996) Endosomal localization of the autoantigen EEA1 is mediated by a zinc-binding FYVE finger. J Biol Chem 271:24048-24054

Suvorova ES, Kurten RC, Lupashin VV (2001) Identification of a human orthologue of Sec34p as a component of the cis-Golgi vesicle tethering machinery. J Biol Chem 276:22810-22818

Suvorova ES, Duden R, Lupashin VV (2002) The Sec34/Sec35p complex, a Ypt1p effector required for retrograde intra-Golgi trafficking, interacts with Golgi SNAREs and COPI vesicle coat proteins. J Cell Biol 157:631-643

Tall GG, Hama H, DeWald DB, Horazdovsky BF (1999) The phosphatidylinositol 3-phosphate binding protein Vac1p interacts with a Rab GTPase and a Sec1p homologue to facilitate vesicle-mediated vacuolar protein sorting. Mol Biol Cell 10:1873-1889

Ting AE, Hazuka CD, Hsu SC, Kirk MD, Bean AJ, Scheller RH (1995) rSec6 and rSec8, mammalian homologs of yeast proteins essential for secretion. Proc Natl Acad Sci USA 21:9613-9617

Toikkanen JH, Miller KJ, Soderlund H, Jantti J, Keranen S (2003) The beta subunit of the Sec61p endoplasmic reticulum translocon interacts with the exocyst complex in *Saccharomyces cerevisiae*. J Biol Chem 278:20946-20953

Tsui MM, Banfield DK (2000) Yeast Golgi SNARE interactions are promiscuous. J Cell Sci 113(Pt 1):145-152

Tsukada M, Will E, Gallwitz D (1999) Structural and functional analysis of a novel coiled-coil protein involved in Ypt6 GTPase-regulated protein transport in yeast. Mol Biol Cell 10:63-75

Ungar D, Oka T, Brittle EE, Vasile E, Lupashin VV, Chatterton JE, Heuser JE, Krieger M, Waters MG (2002) Characterization of a mammalian Golgi-localized protein complex, COG, that is required for normal Golgi morphology and function. J Cell Biol 157:405-415

Ungar D, Hughson FM (2003) SNARE protein structure and function. Annu Rev Cell Dev Biol 19:493-517

Ungermann C, Sato K, Wickner W (1998) Defining the functions of trans-SNARE pairs. Nature 396:543-548

Ungermann C, Price A, Wickner W (2000) A new role for a SNARE protein as a regulator of the Ypt7/Rab-dependent stage of docking. Proc Natl Acad Sci USA 97:8889-8891

VanRheenen SM, Cao X, Sapperstein SK, Chiang EC, Lupashin VV, Barlowe C, Waters MG (1999) Sec34p, a protein required for vesicle tethering to the yeast Golgi apparatus, is in a complex with Sec35p. J Cell Biol 147:729-742

VanRheenen SM, Cao X, Lupashin VV, Barlowe C, Waters MG (1998) Sec35p, a novel peripheral membrane protein, is required for ER to Golgi vesicle docking. J Cell Biol 141:1107-1119

Vida TA, Huyer G, Emr SD (1993) Yeast vacuolar proenzymes are sorted in the late Golgi complex and transported to the vacuole via a prevacuolar endosome-like compartment. J Cell Biol 121:1245-1256

Walch-Solimena C, Collins RN, Novick PJ (1997) Sec2p mediates nucleotide exchange on Sec4p and is involved in polarized delivery of post-Golgi vesicles. J Cell Biol 137:1495-1509

Waters MG, Clary DO, Rothman JE (1992) A novel 115kD peripheral membrance protein is required for interciternal transport in the Golgi stack. J Cell Biol 118:1015-1026

Waters MG, Pfeffer SR (1999) Membrane tethering in intracellular transport. Curr Opin Cell Biol. 11:453-459 Review

Waters MG, Hughson FM (2000) Membrane tethering and fusion in the secretory and endocytic pathways. Traffic 1:588-597 Review

Wang W, Sacher M, Ferro-Novick S (2000) TRAPP stimulates guanine nucleotide exchange on Ypt1p. J Cell Biol 151:289-296

Weber T, Zemelman BV, McNew JA, Westermann B, Gmachl M, Parlati F, Sollner TH, Rothman JE (1998) SNAREpins: minimal machinery for membrane fusion. Cell 92:759-772

Whyte JR, Munro S (2001) The Sec34/35 Golgi transport complex is related to the exocyst, defining a family of complexes involved in multiple steps of membrane traffic. Dev Cell 1:527-537

Whyte JR, Munro S (2002) Vesicle tethering complexes in membrane traffic. J Cell Sci 115(Pt 13):2627-2637 Review

Wiederkehr A, Du Y, Pypaert M, Ferro-Novick S, Novick P (2003) Sec3p is needed for the spatial regulation of secretion and for the inheritance of the cortical endoplasmic reticulum. Mol Biol Cell 14:4770-4782

Wilson JM, de Hoop M, Zorzi N, Toh BH, Dotti CG, Parton RG (2000) EEA1, a tethering protein of the early sorting endosome, shows a polarized distribution in hippocampal neurons, epithelial cells, and fibroblasts. Mol Biol Cell 11:2657-2671

Wuestehube LJ, Duden R, Eun A, Hamamoto S, Korn P, Ram R, Schekman R (1996) New mutants of *Saccharomyces cerevisiae* affected in the transport of proteins from the endoplasmic reticulum to the Golgi complex. Genetics 142:393-406

Wurmser AE, Sato TK, Emr SD (2000) New component of the vacuolar class C-Vps complex couples nucleotide exchange on the Ypt7 GTPase to SNARE-dependent docking and fusion. J Cell Biol 151:551-562

Yamakawa H, Seog DH, Yoda K, Yamasaki M, Wakabayashi T (1996) Uso1 protein is a dimer with 2 globular heads and a long coiled-coil tail. J Struct Biol 116:356-365

Yang B, Gonzalez L Jr, Prekeris R, Steegmaier M, Advani RJ, Scheller RH (1999) SNARE interactions are not selective. Implications for membrane fusion specificity. J Biol Chem 274:5649-5653

Zerial M, McBride H (2001) Rab proteins as membrane organizers. Nat Rev Mol Cell Biol 2:107-117

Zhang X, Bi E, Novick P, Du L, Kozminski KG, Lipschutz JH, Guo W (2001) Cdc42 interacts with the exocyst and regulates polarized secretion. J Biol Chem 276:46745-46750

Chu, Sarah
   University of Pennsylvania, Department of Biology, Philadelphia, PA19104-6018, USA

Guo, Wei
   University of Pennsylvania, Department of Biology, Philadelphia, PA19104-6018, USA
   guowei@sas.upenn.edu

The function of Sec1/Munc18 proteins - Solution of the mystery in sight?

Maria Kauppi, Jussi Jäntti, and Vesa M. Olkkonen

Abstract

Members of the Sec1/Munc18 (SM) protein family form a central part of the machinery responsible for transport vesicle docking and fusion. Their hallmark property is interaction with SNARE proteins belonging to the syntaxin family. A wealth of evidence proposes an essential positive role for SM proteins in the different intracellular trafficking events in all cell types and organisms studied. However, the exact nature of their function is not understood. The existing data suggests functions as chaperones of syntaxins, in transport vesicle docking, promotion of SNARE complex formation, and in fusion pore opening. Intriguingly, all SM proteins do not bind their cognate syntaxin partners in a similar manner, but four different binding modes can be distinguished. This suggests that the unifying factor in SM function is not an intimate tight-fitting contact with syntaxins. We propose here a model, based on increasing experimental evidence that the key function of SM proteins is to bridge the Rab-GTPase dependent machineries of transport vesicle tethering and the SNARE apparatus responsible for vesicle docking and fusion.

1 Introduction

Intracellular membrane trafficking in eukaryotic cells employs vesicular carriers that bud from one membrane compartment and fuse with another. This process is dependent on compartment-specific membrane anchored proteins denoted collectively as SNAREs [soluble N-ethyl maleimide sensitive fusion protein (NSF) attachment protein receptors] (Rothman 1994; Chen and Scheller 2001; Jahn et al. 2003). The SNARE proteins present on the transport vesicles (v-SNAREs; related to the neuronal synaptobrevin/VAMP proteins) and the target membranes (t-SNAREs; homologues of the neuronal syntaxin and SNAP-25 proteins) are characterized by the presence of one or two "SNARE motifs", α-helical sequences capable of forming coiled coil helix bundles which bridge vesicle and target membranes (*trans*-SNARE complexes). A prerequisite for membrane fusion is the formation of SNARE complexes usually composed of one v-SNARE and two t-SNAREs. In the formed complexes, the four α-helical segments are in register so that in a highly conserved layer of amino acid residues

Topics in Current Genetics, Vol. 10
S. Keränen, J. Jäntti (Eds.): Regulatory Mechanisms of Intracellular Membrane Transport
DOI 10.1007/b97172 / Published online: 11 March 2004
© Springer-Verlag Berlin Heidelberg 2004

there are three glutamine (Q) residues and an arginine (R). Based on this structural definition, the SNARE proteins have also been classified as Q- and R-SNAREs (Fasshauer et al. 1998). Assembly of *trans*-SNARE complexes results in a close apposition of the vesicle and target membranes, and is suggested to indirectly or directly cause fusion of the membrane bilayers (Ungermann et al. 1998; Weber et al. 1998). After fusion, the SNAREs remain entangled but are now all anchored in the target membrane. For re-use of the SNAREs, these *cis*-SNARE complexes are disassembled by the ATPase NSF, which binds to SNARE complexes via soluble NSF attachment proteins, SNAPs (Jahn et al. 2003).

Syntaxins comprise a large family of t-SNAREs/Q-SNAREs (Bock et al. 2001) that have a central role in SNARE complex assembly. They are type II membrane proteins anchored to the bilayer by a C-terminal transmembrane segment (Bennett et al. 1993). Syntaxins have an N-terminal regulatory domain (Habc) folded as a three-helix bundle (helices a, b, and c), a linker region, and a membrane-proximal helix 3 (H3) that engages in coiled-coil SNARE complexes (Fernandez et al. 1998; Dulubova et al. 1999; Fiebig et al. 1999) (Fig.1C and D).

For the fidelity and temporal control of SNARE complex formation and activity extensive regulatory processes are required. This is especially important since the interactions of the SNARE proteins as such show marked promiscuity (Fasshauer et al. 1999; Yang et al. 1999; Tsui and Banfield 2000). Several families of accessory proteins are known to regulate and facilitate the SNARE interactions. Of these, the most extensively characterized are the small Rab GTPases (Pfeffer 2001; Stenmark and Olkkonen 2001; Zerial and McBride 2001) and the Sec1/Munc18-related (SM) proteins (Gallwitz and Jahn 2003; Toonen and Verhage 2003) (Table 1). The hallmark features of the SM proteins are the ability to interact with syntaxin-related t-SNAREs/Q-SNAREs and functional effects on membrane trafficking at the vesicle docking/fusion stages. In the present review, we focus on the SM proteins and summarize the current perception of their role in vesicle transport.

Fig. 1 (overleaf). Structure of Munc18a and syntaxin 1A. A. A schematic presentation of Munc18a topology. Domain 1 is shown in blue, domain 2 in green, and domain 3 in yellow. Helices are shown as cylinders and β-strands as arrows. Asterisks indicate breaks in the structure. Dashed outlines denote domains 3a and 3b. B. A ribbon representation of Munc18a, coloured as in A. C. Topology diagram of syntaxin 1A. The Habc domain is shown in red, the linker region in orange, and H3 in purple. D. A ribbon representation of syntaxin 1A, with the same colour coding as in C. E, F. Ribbon representations of the Munc18a-syntaxin 1A complex. Two different projections are shown. E is a view looking down the long syntaxin 1A helices and D is rotated about the vertical axis by 90°. Syntaxin 1A is in the complex located in the cleft between Munc18a domains 1 and 3a (reproduced with permission from Misura et al. 2000; http://www.nature.com/).

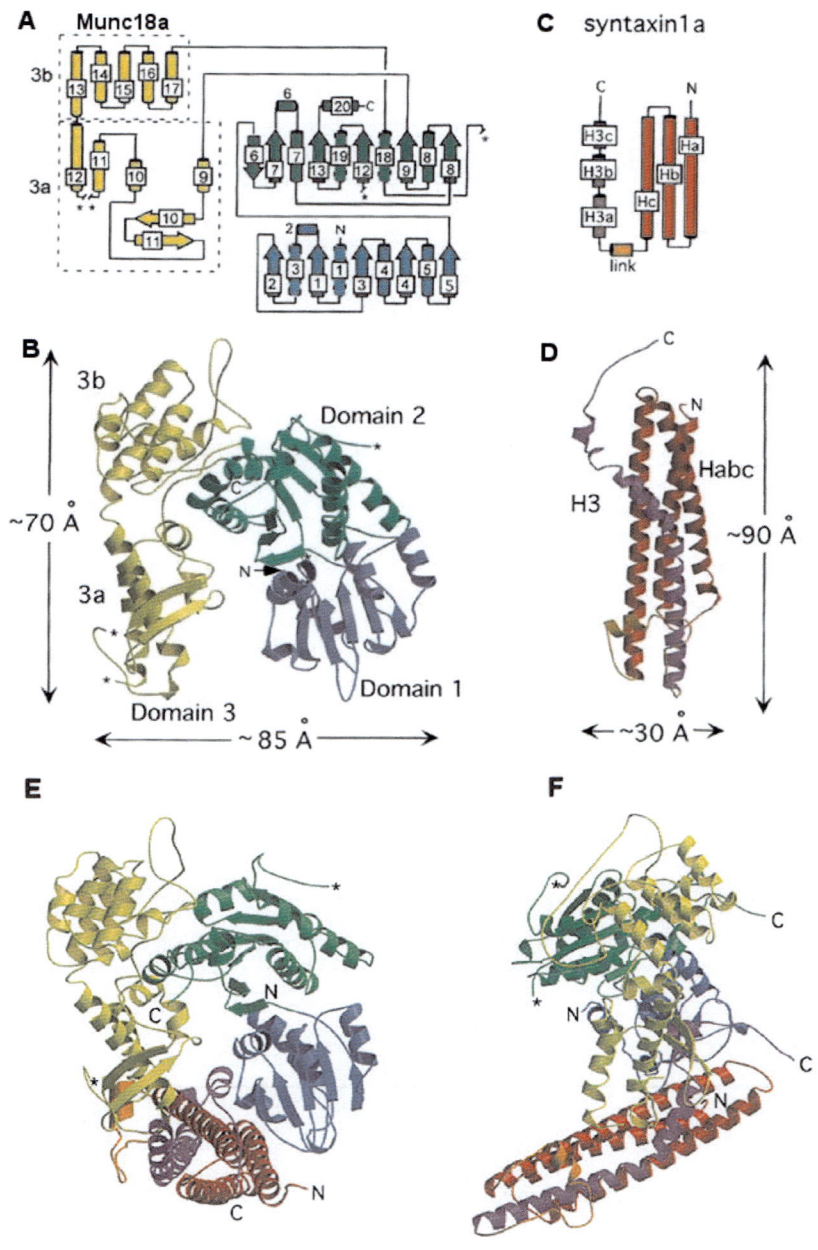

2 Genetic studies indicate an essential positive function for SM proteins in vesicle transport

In a pioneering study Novick and Schekman (1979) isolated the temperature sensitive *Saccharomyces cerevisiae sec1-1* mutant, which stopped dividing and accumulated secretory vesicles at the restrictive temperature. This initial study that introduced the *SEC1* genes into the field of transport biology provided evidence for an essential positive role of yeast Sec1p in the secretion process. The first indication of functional interaction between SM proteins and syntaxins was obtained in yeast when the syntaxin homologues *SSO1* and *SSO2* were isolated as multicopy suppressors of the temperature sensitive *sec1-1* mutation (Aalto et al. 1993). Similar observations have been made in several different organisms and for a number of different SM proteins (Table 1). In *S. cerevisiae*, four Sec1-like proteins exist, Sly1p, Vps33p, Vps45p, and Sec1p (Aalto et al. 1992; Cowles et al. 1994; Piper et al. 1994). Mutations in the *SEC1* homologue *VPS45* were found to lead to vacuolar protein sorting defects and intracellular accumulation of vesicles, which represent transport intermediates between the Golgi apparatus and the vacuole (Cowles et al. 1994; Piper et al. 1994). Furthermore, mutant alleles of the *VPS33* complementation group lead to defects in the sorting of vacuolar hydrolases and a vacuolar membrane enzyme marker; some alleles also caused a temperature sensitive growth phenotype (Robinson et al. 1988). *VPS33* is the same gene as *SLP1*, shown to be essential for vacuolar morphogenesis and function (Wada et al. 1990). Vps45p apparently has functions in more than one trafficking step between the TGN and the vacuole; functional Vps45p is required for the formation of complexes of the SNAREs Tlg2p (a syntaxin homologue), Vti1p, and Tlg1p (Bryant and James 2001), but it also associates (indirectly) with the syntaxin protein Pep12p (Abeliovich et al. 1999; Dulubova et al. 2002). Vps33p is part of a so-called class-C Vps protein complex that is involved in the late trafficking steps of the vacuolar transport pathway and interacts with the vacuolar syntaxin homologue Vam3p (Sato et al. 2000; Seals et al. 2000). In *vps33* null mutant cells, Vam3p is unable to form a complex with its cognate SNARE partners Vti1p and Vam7p (Sato et al. 2000). The fourth *S. cerevisiae SEC1* homologue, *SLY1*, was originally isolated as a gene whose allele *SLY1-20* was able to suppress the loss of Ypt1p, an essential secretory pathway small GTPase (Dascher et al. 1991; Ossig et al. 1991). *Sly1* null mutations are lethal, and the protein, which interacts with the syntaxin homologues Sed5p and Ufe1p, contributes to the specificity of ER-Golgi SNARE complex assembly (Peng and Gallwitz 2002). The conclusion arising from work on the four yeast SM genes/proteins is that the products of these genes exert important positive functions in distinct vesicular trafficking events. These functions involve regulation of SNARE complex assembly and transport vesicle docking/fusion at the target membranes.

Studies on mutations of the SM genes in several non-vertebrates provide evidence consistent with the above data on yeast SM genes. Mutants of the *Caenorhabditis elegans* SM gene *unc-18* show uncoordinated movements, retardation of

Table 1. Central characteristics of the SM proteins in different organisms.

Species	SM Protein	Syntaxin interactions	Syntaxin binding mode[a]	Proposed function
S. cerevisiae	Sec1p	Sso1p, Sso2p	Core complex	Vesicular transport to plasma membrane
	Sly1p	Sed5p, Ufe1p	N-terminus	ER to Golgi transport
	Vps33p	Vam3p	Indirect	Endosome and vacuole membrane trafficking
	Vps45p	Tgl2p, Pep12p	N-terminus	Golgi to late endosome/vacuole transport
D. melanogaster	ROP	Syntaxin	N.D.[b]	Synaptic vesicle release and general secretion
	Sly1	N.D.	N.D.	N.D.
	Vps45	N.D.	N.D.	N.D.
	Vps33/carnation	N.D.	N.D.	Transport to lysosomes and pigment granules
C. elegans	UNC-18	Syntaxin (UNC-64)	Closed conformation	Synaptic vesicle release
	5 other genes in G.D.[c]	N.D.	N.D.	N.D.
Mammals	Munc18a	Syntaxin-1,-2,-3	Closed conformation	Synaptic vesicle release, chromaffin granule exocytosis
	Munc18b	Syntaxin-1,-2,-3	N.D.	Apical trafficking in epithelial cells, mast cell granule secretion
	Munc18c	Syntaxin-2,-4	N.D.	GLUT4 translocation, platelet granule exocytosis
	mVPS45	Syntaxin-4,-6,-13,-16	N-terminus	N.D.
	mVPS33A, B	N.D.	N.D.	N.D.
	mSly1	Syntaxin-5,-18	N.D.	ER to Golgi transport

[a] The binding modes are illustrated in Fig. 2; closed conformation=A, core complex=B, N-terminal=C, Indirect interaction=D
[b] N.D.= Not determined
[c] G.D.= Genome database (www.wormbase.org)

post-embryonic development and defects in the maintenance of normal acetylcholine levels (Hosono et al. 1992; Gengyo-Ando et al. 1993). In *Drosophila*, muta-

tions in the Sec1-related *Rop* gene were found to result in loss of normal synaptic responses to light stimulus and defects in general secretion (Harrison et al. 1994). Later, detailed characterization of the phenotypic effects of *Rop* point mutations and changes in *Rop* gene dosage was carried out (Wu et al. 1998). This work revealed that while some mutations caused a dramatic reduction of both spontaneous and evoked neurotransmitter release, others resulted in an increase of evoked transmission. Furthermore, half the amount of wild type ROP resulted in a marked reduction of neurotransmission, indicating that ROP is a rate-limiting regulator of exocytosis and that optimal ROP concentration is required for proper neurotransmission. The *carnation* (*car*) gene in *Drosophila* encodes a homologue of Vps33p, and mutations in this gene lead to a pigmentation defect and to disturbed protein delivery to lysosomes (Sevrioukov et al. 1999; Sriram et al. 2003). The *Arabidopsis* mutants of the gene KEULE that encodes an SM protein, display defects in cytokinesis and accumulation of vesicles which are apparently unable to fuse at the equator of dividing cells (Waizenegger et al. 2000; Assaad et al. 2001).

Evidence in agreement with the positive role of SM proteins in membrane fusion has also been obtained in a mammalian system: null mutant mice for the mammalian neuronal Munc18 isoform, Munc18a (also denoted as Munc18-1, n-Sec1, or rbSec1; (Hata et al. 1993; Garcia et al. 1994; Pevsner et al. 1994b)), develop a normal nervous system but display complete loss of neurotransmission and die immediately after birth (Verhage et al. 2000). Further, it was recently reported that the organellar biogenesis mutant *bf* results from a mutation in the mouse *Vps33a* gene. The mutant phenotype includes defective pigmentation, aberrant targeting of lysosomal hydrolases, prolonged bleeding, and immunodeficiency (Suzuki et al. 2003).

3 The structure of SM proteins

The unifying functional feature of the SM proteins is that they all, directly or indirectly, interact with Q-SNAREs of the syntaxin protein family. However, different SM proteins have been shown to bind their cognate SNARE molecules in different ways. Detailed analysis of the different interaction modes is greatly hampered because of the limited structural information on SM-syntaxin complexes. X-ray crystallographic structures are only available for a complex between the rat nervous system SM protein Munc18a and the cytosolic domain of syntaxin 1A (Misura et al. 2000), and for the *S. cerevisiae* SM protein Sly1p complexed with a short N-terminal peptide of the yeast syntaxin Sed5p (Bracher and Weissenhorn 2002). In addition, crystal structure of an uncomplexed squid neuronal Sec1p has been resolved (Bracher et al. 2000). All currently available SM protein structures, free or complexed with syntaxin, reveal a similar arch shaped organisation, composed of three domains 1, 2, and 3 (Fig. 1B). Domain 3 in the mammalian Munc18a has been further divided into domains 3a and b (Misura et al. 2000). Based on comparison of the uncomplexed and syntaxin bound neuronal SM structures, it appears that only small structural changes in the SM protein occur upon binding to syn-

taxin (Bracher et al. 2000). Analysis of different crystal forms of squid neuronal Sec1 suggests a hinge region between domains 1 and 2, which allows domain 1 to rotate. This movement could be associated with release of bound syntaxin, possibly upon interaction with yet unspecified Sec1 effector molecules (Bracher and Weissenhorn 2001). Munc18a interacts both with the H3 and the Habc regions of syntaxin1A in the closed conformation (Misura et al. 2000). This binding is largely mediated by interaction of the cavity between Munc18 domains 1 and 3a with amino acids in the syntaxin1 Habc region. This is in marked contrast to the binding mode of the yeast Sly1p with Sed5p, although yeast Sly1p shares an overall fold very similar to the mammalian SM proteins (Bracher and Weissenhorn 2002). Unlike the SM interaction with the closed conformation of the syntaxin1A, a short, N-terminal peptide sequence preceding the Habc domain of Sed5p interacts predominantly with domain 1 of Sly1p. This interaction is not dependent on the conformational state of Sed5p. Notably, the N-terminal peptide is not present in all syntaxins. Sly1p domains 1 and 2 superimpose well with the corresponding domains in mammalian SM proteins. However, domain 3 is structurally and sequence-wise more diverse from the other SM and even Sly1 proteins. Although the binding of Munc18a and Sly1p to syntaxins appear to occur through different mechanisms, Bracher and Weissenhorn (2002) proposed that also Munc18a could possess a peptide binding function in the same position as Sly1p. However, the recent study by Dulubova et al. (2003) did not support this idea. It may be that the different binding modes observed for Munc18 and Sly1p by structural methods may nevertheless allow rather similar functions *in vivo*.

Domain 2 of SM proteins is formed by assembly of polypeptide chains that emerge from domains 1 and 3 (Fig. 1A). This feature has greatly hampered the mutational analysis of Sec1 proteins as deletions in the peptide backbone are likely to result in serious misfolding of the protein and in compromised expression or biological activity. Interestingly, the point mutations resulting in temperature sensitive behaviour of the *S. cerevisiae* Sec1p localise to domain 3 (Brummer et al. 2001). In addition, the dominant *SLY1-20* mutation, which results in rescue of the loss of the small GTPase Ypt1p, localises to the linker, connecting domain 3 to domain 2. These results imply that domain 3 may be functionally important for binding of effector molecules. Another potential effector binding site is located in the cavity between domains 1 and 2 in Munc18a. Identification of the binding partners for these sites is an important future challenge.

4 SM proteins display different binding modes to SNARE molecules

The structural information, together with a number of molecular biological/biochemical studies, suggests four different modes of SM protein-syntaxin binding (Fig. 2).

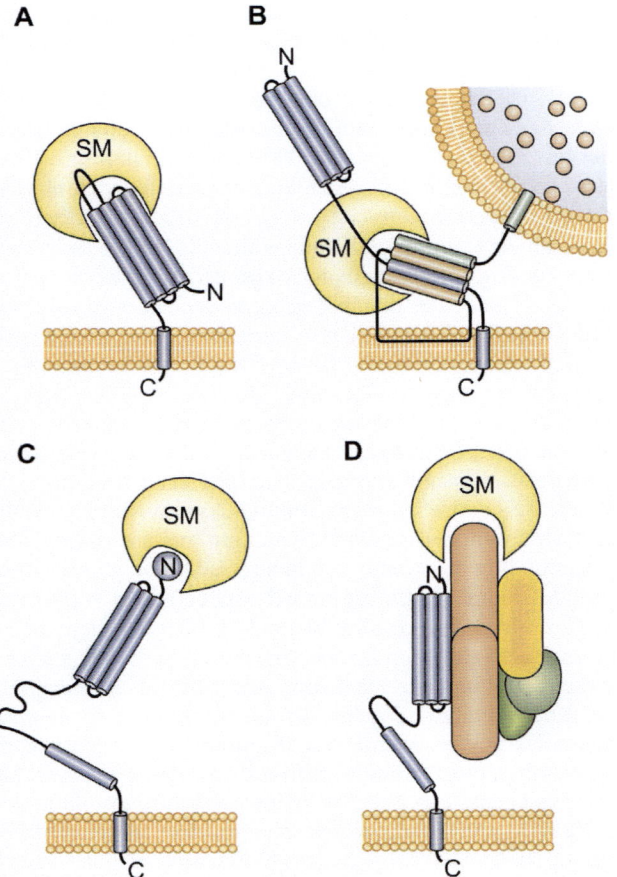

Fig. 2. The different modes of interaction of SM proteins (yellow) with syntaxins (blue). A. Binding to a monomeric syntaxin in a closed conformation (Munc18a-syntaxin 1A complex). B. Binding to a SNARE complex without affinity for monomeric syntaxin (Interaction of the *S. cerevisiae* Sec1p with a complex of Ssop, Sncp, and Sec9p). C. Binding to an N-terminal peptide motif of syntaxin (Sly1p and Vps45p interactions with Sed5p and Tlg2p, respectively). This binding mode allows for association with syntaxin both in monomeric form and within SNARE complexes. D. Indirect association with syntaxin mediated by protein complexes (Binding of Vps33p to Vam3p). Even though not clear from the images, the syntaxin binding surface on the SM protein may vary for the different binding modes.

A. Binding of SM with the closed conformation of syntaxin. The crystal structure of Munc18a in complex with syntaxin 1A revealed that the syntaxin bound to the cleft of an arch-shaped Munc18a protein displayed a "closed" conformation where the N-terminal Habc domain had folded onto the C-terminal H3 motif

(Misura et al. 2000; see also Dulubova et al. 1999; Yang et al. 2000). Compared to the syntaxin 1A bound Munc18a, free neuronal squid Sec1 displays a very similar arch shaped overall organisation (Bracher et al. 2000). The structural data implies that only limited sterical changes are required for binding of Munc18a to syntaxin and its release from the SNARE, and proposes a framework for other molecules to interact with the SM protein.

  B. Binding of SM proteins to assembled SNARE complexes. Yeast Sec1p has been shown to bind assembled tripartite SNARE complexes and not to interact with free Sso1p (a syntaxin homologue), which nevertheless can resume a closed conformation when not in SNARE complexes (Carr et al. 1999; Munson et al. 2000). The actual binding mode of Sec1p with the assembled Sso1/2p-Snc1/2p-Sec9p SNARE complex is unknown. Also a direct interaction between Sec1p and the syntaxin homologue Sso2p has been detected in the yeast two-hybrid system (Brummer et al. 2001).

  C. Binding of SM to a short motif at the very N-terminal end of syntaxin. Recently, both yeast and mammalian SM proteins Sly1p and Vps45p were shown to interact with their respective syntaxins through a short, conserved peptide sequence (Dulubova et al. 2002; Yamaguchi et al. 2002). Because this binding is apparently insensitive to the conformation of syntaxin, it could allow a positive and active function for SM proteins in SNARE complex regulation. Importantly, the syntaxin binding surface on the SM proteins is different in this binding mode and mode A (Misura et al. 2000; Bracher and Weissenhorn 2002; Dulubova et al. 2003).

  D. Binding of SM proteins with SNAREs is mediated by a multiprotein complex. Binding of Vps33p in yeast to its cognate syntaxin Vam3p involves a six member multiprotein complex, the homotypic fusion and vacuole protein sorting (HOPS) complex (Seals et al. 2000). The actual physical interactions required for the action of this complex in conjunction with Vam3p are poorly understood. Importantly, HOPS interacts also with the small Rab GTPase Ypt7p, thus, intimately linking Rab proteins with the regulation of SNARE interactions.

5 Are the data suggesting an inhibitory role for SM proteins physiologically relevant?

Pevsner et al. (1994a) showed in their *in vitro* study that Munc18a that binds syntaxins 1, 2, and 3 with high affinity was absent from the 7S (consisting of VAMP, synaptotagmin, SNAP-25 and syntaxin) and 20S (consisting of VAMP, SNAP-25, syntaxin, NSF and αSNAP) particles, and inhibited VAMP or SNAP-25 binding to syntaxin. This suggests that interaction of Munc18a with syntaxin precedes formation of the 7S and 20S SNARE complexes and may exert regulatory constraints on their assembly. If one considers those SM proteins that are able to bind syntaxins in monomeric form and whose binding interface covers the Helix 3 part of the syntaxins, one can envision that an excess amount of the SM protein may prevent participation of the syntaxin in SNARE complex formation, resulting in an

inhibitory effect on the transport event(s) controlled by the syntaxin. Findings consistent with this idea have been made in cells overexpressing SM proteins: an excess of Munc18b, also called Munc18-2 (Hata and Sudhof 1995; Katagiri et al. 1995; Tellam et al. 1995; Riento et al. 1996) in epithelial cells was reported to inhibit formation of syntaxin 3-based SNARE complexes and apical transport of a marker protein (Riento et al. 1998, 2000; Kauppi et al. 2002). Further, increased expression of Munc18b, its domain 3, or a specific 'effector loop' included in domain 3 were reported to inhibit the IgE-triggered exocytosis of mast cell granules (Martin-Verdeaux et al. 2003), and Munc18b-syntaxin 3 complexes in this cell type were found to be spatially separated from syntaxin 3 containing SNARE complexes involved in granule exocytosis (Pombo et al. 2003). Similarly, overexpression of Munc18c or microinjection of a Munc18c 'effector loop' peptide inhibited GLUT4 translocation to the plasma membrane in 3T3L1 adipocytes (Tamori et al. 1998; Thurmond et al. 1998, 2000). Moreover, corresponding peptides derived from either Munc18a or Munc18c were recently found to inhibit granule exocytosis in permeabilised human platelets (Schraw et al. 2003). In the same cell system, inhibition of Munc18c-syntaxin 4 association by Munc18c antibody or peptides corresponding to a syntaxin 4 binding region were found to enhance Ca^{2+}- dependent platelet granule exocytosis (Houng et al. 2003). Both the peptides used by Houng et al. (2003) and the 'effector loop' peptides (Dresbach et al. 1998; Thurmond et al. 2000) have been reported to inhibit SM protein-syntaxin interactions. The opposite effects obtained with the two types of peptides in the platelet system are, therefore, somewhat puzzling but may be due to the fact that the anticipated primary targets of the peptides are different. Overexpression of a Munc18c transgene in a mouse model resulted in an inhibition of insulin secretion by pancreatic islets and reduction in glucose uptake by skeletal muscle and white adipose tissue, implying a defect in GLUT4 translocation (Spurlin et al. 2003). Microinjection of the squid SM protein s-Sec1 into giant squid axons was reported to inhibit evoked neurotransmitter release (Dresbach et al. 1998). Furthermore, overexpression of ROP in Drosophila had an inhibitory effect on neurotransmission (Schulze et al. 1994; Wu et al. 1998). However, inhibitory effects are not observed in all cases of SM protein overexpression: excess Munc18a in primary chromaffin cells was shown to promote secretory vesicle fusion rather than to inhibit it (Voets et al. 2001). Further, Graham et al. (1997) observed no effect of Munc18a addition in secretion assays based on permeabilised chromaffin or PC12 cells. Importantly, the inhibitory effect of ROP overexpression in Drosophila was overcome by simultaneous expression of syntaxin (Wu et al. 1998), and similarly, co-injection of an N-terminal cytosolic fragment of syntaxin into squid giant axons together with s-Sec1 abolished the inhibitory effect of s-Sec1 (Dresbach et al. 1998).

The conclusion arising from this data is that an appropriately balanced stoichiometric ratio between an SM protein and its cognate syntaxin partners is crucial for normal membrane trafficking. An excess amount of SM protein can in many cases lead to inhibition of specific vesicle transport events, probably due to hindrance of syntaxin participation in the assembly of SNARE core complexes (Pevsner et al. 1994a; Araki et al. 1997; Tellam et al. 1997; Riento et al. 1998;

Tamori et al. 1998; Thurmond et al. 2000). The inhibitory effects may arise upon overexpression of SM proteins that are capable of binding monomeric syntaxins and perhaps stabilising their closed conformation. However, the neuronal Munc18a appears to form an exception here, the reason for which is not well understood. It is possible that since Munc18a functions in highly specialized, rapid Ca^{2+}-triggered secretory events, the functional properties of this protein are significantly different from those of the related SM proteins involved in constitutive membrane trafficking processes. Alternatively, it is possible that SM protein overexpression can lead to out-titration of accessory factors, such as GTPases, kinases and phosphatases, which are likely to participate in the membrane fusion event. We tend to believe that the inhibitory effects of SM proteins do not reflect an intrinsically negative regulatory role of these proteins but rather an experimental situation, which, even though it has provided useful insight into the function of SM proteins, is unlikely to occur *in vivo*.

6 At what stage of the vesicle tethering-docking-fusion process do SM proteins exert their effects?

Recently, *unc-18* mutants were reported to display reduction of docked vesicles at the synaptic active zones, providing evidence for a function of the protein as a facilitator of vesicle docking (Weimer et al. 2003). Similar interpretations can be made from work on *S. cerevisiae sec1* (Novick and Schekman 1979) and *vps45* (Cowles et al. 1994; Piper et al. 1994) mutants, which accumulate apparently undocked transport vesicles in the cytoplasm. Furthermore, a marked reduction in the number of docked large dense-core vesicles was detected in the chromaffin cells of *munc18a* null mutant mice (Voets et al. 2001). Contradictory evidence can also be found: *S. cerevisiae sly1* mutants display a defect in the fusion of ER-derived vesicles with the Golgi apparatus but docking of the transport vesicles seems not to be affected (Cao et al. 1998). Further, a normal amount of docked synaptic vesicles can be observed in the *munc18a* null mutant mouse embryos (Verhage et al. 2000). However, there is quite convincing evidence that the function of SM proteins is required for successful assembly of SNARE complexes: depletion of *S. cerevisiae* Vps45p or Vps33p resulted in reduced amounts of the endosomal and vacuolar SNARE complexes, respectively (Sato et al. 2000; Bryant and James 2001). Furthermore, Kosodo et al. (2002) reported that binding of recombinant Sly1p to the syntaxin Sed5p promoted the formation of yeast early Golgi *trans*-SNARE complexes *in vitro*. Consistent with this data, suppression analyses in yeast indicate location of SM protein function between the vesicle tethering apparatus and the SNARE machinery (Dascher et al. 1991; Aalto et al. 1993; Sapperstein et al. 1996).

The high-resolution structure of a complex between the neuronal Munc18a and its cognate partner t-SNARE syntaxin 1A (Misura et al. 2000) (Fig. 1) revealed that the SM protein engulfs large parts of the syntaxin in its closed conformation, including the N-terminal Habc domain and the membrane-proximal H3. The struc-

ture also suggested a mechanism by which Munc18a could facilitate formation of SNARE core complexes by acting as a platform on which the α-helical parts of the neuronal SNAREs could wind together in a zipper-like fashion. If this were the case, one could envision that Munc18a and possibly other SM proteins could have strong regulatory effects on the assembly of SNARE complexes, including the specificity of SNARE interactions. In fact, this type of function would be possible even for SM proteins that do not form as intimate contacts with their syntaxin partners as Munc18a does. Is there any evidence that SM proteins would increase the fidelity of SNARE interactions? Peng and Gallwitz (2002) studied the assembly of SNARE complexes containing the syntaxin homologue Sed5p. When the complexes assembled in the absence of the Sed5p partner SM protein Sly1p, several presumably non-physiological SNARE complexes were generated, while they were not observed when Sed5p was first bound to Sly1p. The results demonstrate that at least *in vitro* the SM protein Sly1p can prevent the formation of non-cognate SNARE complexes, thus, promoting the specificity of SNARE interactions.

SM proteins themselves do display a high degree of functional specificity: different SM proteins from the same species cannot functionally replace each other, as shown, for example, by multicopy suppressor studies in yeast showing that *SEC1* overexpression cannot rescue growth of *sly1* or *vps33* mutants under restrictive conditions (Aalto et al. 1993). Furthermore, the *munc18a* null mutant mice which do have normal Munc18b and c genes show a lethal phenotype (Verhage et al. 2000). The latter observation is quite remarkable considering the fact the Munc18a and Munc18b show the same syntaxin specificity, and still Munc18b is unable to compensate for the loss of Munc18a. Murine Munc18a but not Munc18c is capable of complementing the *unc-18* null mutant phenotype in *C. elegans* (Gengyo-Ando et al. 1996). Also, Munc18b is not capable of replacing Munc18c in GLUT4 translocation to the transverse tubules in skeletal muscle (Khan et al. 2001), and Munc18c does not inhibit mast cell degranulation as Munc18b does (Martin-Verdeaux et al. 2003). These latter findings can, however, be explained by the different syntaxin binding specificity of these two SM proteins: Munc18a and Munc18b bind syntaxins 1, 2, and 3, while Munc18c binds syntaxins 2 and 4 (Hata and Sudhof 1995; Riento et al. 1996; Tellam et al. 1997).

Fisher et al. (2001) reported the unexpected finding that overexpression of a Munc18a mutant protein in chromaffin cells affected the very final step in exocytotic fusion, opening of the fusion pore. This suggested that the SM protein might exert a function downstream of SNARE complex assembly, while most of the literature rather supports functional roles of the SM proteins upstream of the process. The relevance of this finding, however, has remained unclear, since there are discrepant reports on the topic. In the chromaffin cells of *munc18a* null mutant mice, the biophysical characteristics of single fusion events seemed to be unaffected (Voets et al. 2001), while Barclay et al. (2003) recently reported that expression of Munc18a with phosphomimetic mutations in chromaffin cells did modify single vesicle release kinetics.

Finally, Kosodo et al. (2002) reported that disassembly of *cis*-SNARE complexes containing the syntaxin Sed5p was retarded in a temperature-sensitive yeast

4 Protein phosphorylation in the early secretory pathway 160
5 Conclusions 161
Acknowledgments 162
References 162

Phosphoinositides and membrane traffic in health and disease 171
Anna Godi, Antonella Di Campli, and Maria Antonietta De Matteis 171
Abstract 171
1 PIs metabolism 171
2 The PI kinases: their subcellular localization and role in the secretory
pathway 173
2.1 The PtdIns 3-kinases 173
2.2 The PtdIns 4-kinases 175
2.3 The PtdIns monophosphate kinases 177
3 The PI phosphatases: localization and function in the secretory
pathway 179
3.1 The PI 3-phosphatases 179
3.2 The PI 4-phosphatases 181
3.3 The PI 5-phosphatases 181
4 PI metabolism and disease 182
4.1 PTEN and human cancers 182
4.2 Myotubular myopathy and Charcot-Marie-Tooth disease 183
4.3 Lowe syndrome 184
5 Conclusions 185
Acknowledgements 185
References 185

Regulation of exocytotic events by centrosome-analogous structures 193
Christof Taxis and Michael Knop 193
Abstract 193
1 MTOCs: a brief overview 193
2 Spindle pole bodies and plasma membrane biogenesis in yeast
meiosis 195
2.1 PSM initiation 196
2.2 Growth and shaping of the PSM 199
2.3 Closure of the PSM 200
3 The IMC in *Apicomplexa* 200
4 Centrioles, basal bodies and plasma membrane 204
5 Conclusion 204
Acknowledgement 205
References 205

Index 209

8 Protein phosphorylation regulates interactions of SM proteins with syntaxins

Munc18a possesses predicted consensus sequences for phosphorylation by protein kinase A, tyrosine kinase, protein kinase C (PKC), casein kinase II and two sites for cyclin dependent kinase 5 (Cdk5) (Kennelly and Krebs 1991; Pevsner et al. 1994a, 1994b). PKC stimulates Ca^{2+}-dependent exocytosis in various secretory cell types and phosphorylates Munc18a in a cell free system. This phosphorylation inhibits Munc18a interaction with syntaxin 1A but PKC does not phosphorylate assembled Munc18a-syntaxin complex (Fujita et al. 1996) or cause disassembly of the complex (Fletcher et al. 1999). In isolated rat nerve terminals, Munc18a was reported to be almost completely non-phosphorylated due to the action of endogenous phosphatases. K^+-evoked depolarization or pharmacologic PKC activation leads to an increase of Munc18a phosphorylation, making the phosphorylation dynamics of Munc18a a prominent candidate to account for PKC-mediated enhancement of secretion (de Vries et al. 2000; Craig et al. 2003). Similar results were reported in a chromaffin cell model (Craig et al. 2003). Analysis of the effects of a phosphomimetic mutant of Munc18a (resembling a PKC-phosphorylated form of the protein) in chromaffin cells (Barclay et al. 2003) did not reveal changes in the number of exocytotic events. However, the single vesicle release kinetics was affected by the mutant in a similar fashion as phorbol ester (PKC activator) treatment, supporting the conclusions of the above studies. Interestingly, also phosphorylation of syntaxins may regulate the SM protein-syntaxin interactions: syntaxin 1A is subject to phosphorylation by a synaptic protein kinase called DAP (death-associated protein), and modification of this t-SNARE by DAP inhibits dramatically the binding of Munc18a to the syntaxin (Tian et al. 2003).

Reed et al. (1999) reported that Munc18c is *in vitro* a substrate for PKC, and is specifically phosporylated by PKC in permeabilised platelets after cellular stimulation by phorbol esters or thrombin. In agreement with the data on Munc18a, the phosphorylation of Munc18c was reported to inhibit its interaction with syntaxin 4. Also the *C. elegans* UNC-18 is subject to phosphorylation by PKC (Sassa et al. 1996), and the membrane association of yeast Vps45p was reported to be regulated by protein phosphatase 1 (Bryant and James 2003). These results collectively indicate a likely *in vivo* regulation of SM proteins through phosphorylation/dephosphorylation cycles. In a recent whole *S. cerevisiae* phosphoproteome analysis, Sec1p was found to be phosphorylated at the C-terminal amino acids Ser^{687} ja Ser^{691} (Ficarro et al. 2002). However, the relevance of these phosphoamino acids for Sec1p function has not been established.

Cdk5 is a member of the Cdc2 protein kinase family and has been demonstrated to act as a proline-directed serine/threonine kinase, phosphorylating neurofilaments (Lew et al. 1992; Shetty et al. 1993). Cdk5 is one of the key factors determining axonal shape and structure, and was reported to form a stable complex with Munc18a. This SM protein, which in addition to the nerve terminals is also distributed along the axons, was reported to stimulate Cdk5 kinase activity and to co-precipitate with Cdk5 and neuronal cytoskeletal components (Shetty et al.

1995; Bhaskar et al. 2004). Munc18a is phosphorylated by Cdk5 in its C-terminal part (Thr^{574}), and this reduces the affinity of the SM protein for syntaxin 1A (Shuang et al. 1998; Fletcher et al. 1999). Importantly, unlike PKC, Cdk5 seems to be able to phosphorylate Munc18a in a preformed Munc18a-syntaxin heterodimer, resulting in disassembly of the complex (Fletcher et al. 1999). These findings suggest that Munc18a may, in addition to controlling of events in the presynaptic plasma membrane region, be involved in the dynamic regulation of cytoskeletal elements in the axon.

9 Non-syntaxin interaction partners of SM proteins

9.1 Interactions of SM proteins with transport vesicle tethering complexes

The *S. cerevisiae SEC1* interacts genetically with several genes involved in the late secretory pathway function. Overexpression of *SEC1* suppresses defects in most of the exocyst mutants (Aalto et al. 1993; Brummer et al. 2001), and *sec1* mutant alleles display various synthetic phenotypes when combined with mutations in exocyst complex components (Finger and Novick 2000). *sec1* mutants form harmful or lethal combinations with *sec4*, as well as *sec2* and *sec19* implicated in the GTP/GDP cycle of the small GTPase Sec4p and interestingly, in *sec4-8* mutant cells GFP-tagged Sec1p loses its focal localisation at the sites of exocytosis and the Sec1p SNARE interaction is impaired (Carr et al. 1999). However, the only known non-syntaxin Sec1 binding protein in yeast is Mso1p. The *MSO1* gene was isolated as a specific high copy suppressor of the temperature sensitive *sec1-1* mutation (Aalto et al. 1997). *MSO1* does not suppress mutations in any other *SEC1*-related or late acting *SEC* genes. Mso1p is a small hydrophilic protein that is most abundant in the microsomal membrane fraction. Deletion of *MSO1* results in accumulation of secretory vesicles in the bud, yet the cells remain viable (Aalto et al. 1997). Loss of *MSO1* shows synthetic interactions with most of the genes encoding components of the exocyst complex that interacts directly with the Rab GTPase Sec4p and plays a central role in secretory vesicle tethering at the plasma membrane (TerBush et al. 1996; Guo et al. 1999). Importantly, loss of Mso1p in combination with *sec1*, *sec4* or the Sec4p GEF *sec2* mutants is lethal. Mso1p, thus, provides an interesting link between a membrane tethering apparatus (exocyst), small Rab GTPase, SM proteins and SNARE complexes.

   A number of genetic studies have identified interaction partners for *S. cerevisiae SLY1* among genes involved in the early secretory pathway function. However, there is no evidence for direct binding of these partners to Sly1p. A dominant allele of *SLY1*, termed *SLY1-20*, suppresses mutations in *YPT1* and *USO1* (Dascher et al. 1991; Sapperstein et al. 1996). Ypt1p belongs to the Rab GTPase family which plays a central role in different trafficking steps (Ferro-Novick and Novick 1993; Zerial and McBride 2001), and *USO1* encodes a protein with a suggested tethering function in ER to Golgi membrane trafficking (Cao et al. 1998).

Both of these proteins are essential for formation of SNARE complexes functional in ER to Golgi membrane trafficking (Sapperstein et al. 1996). The genetic data suggests that Sly1p functions downstream of Ypt1p and Uso1p, consistent with the idea that, while Ypt1p and Uso1p control the tethering process, Sly1p is more directly involved in SNARE complex formation, docking and fusion. The third gene that interacts genetically with *SLY1* is *SEC35* encoding a small cytosolic protein that associates peripherally with membranes (VanRheenen et al. 1999). Sec35p is part of a large protein complex that, like Ypt1p and Uso1p, has been suggested to mediate tethering of ER derived transport intermediates in the *cis*-Golgi (Kim et al. 1999; VanRheenen et al. 1999; Whyte and Munro 2001).

The *S. cerevisiae* Vps33p is a component of a hetero-oligomeric protein complex containing the class-C Vps proteins. Mutations in class-C *VPS* genes result in severe defects of vacuolar protein sorting and morphology. The complex consists of four proteins, Vps11p, Vps16p, Vps18p, and Vps33p that interact genetically and physically (Rieder and Emr 1997), and associate with the vacuolar syntaxin homologue Vam3p (Sato et al. 2000). Importantly, loss-of-function mutations in the class-C *VPS* genes lead to defects in the formation of Vam3p-based SNARE complexes (Sato et al. 2000). The class-C Vps complex is part of a larger 38S complex also called HOPS. This is an assembly of six proteins that binds Ypt7p to initiate vesicle tethering/docking. The HOPS complex is, prior to vesicle priming by Sec18p/NSF, found within a 65S complex containing SNARE proteins in a *cis* arrangement. After priming, the HOPS complex is released and, thereby, gains the capacity to associate with the GTP-bound form of the Rab protein Ypt7p (Seals et al. 2000). The class-C Vps complex is, thus, involved in ordered tethering and SNARE pairing in vesicle transport to the vacuole and homotypic vacuole-vacuole fusion (Price et al. 2000; Sato et al. 2000; Seals et al. 2000; Wurmser et al. 2000). The function of the HOPS complex and the SM protein Vps33p therein forms a clear link between the Rab-based tethering apparatus and the SNARE-based docking/fusion machinery. In *Drosophila*, there are several genes involved in eye colour development. Two of them are related to the *S. cerevisiae* class-C *VPS* genes. The Dor (*deep-orange*) protein shares similarity with Vps18p and the Car (*carnation*) protein with Vps33p. Both of these proteins are part of a larger protein complex that is required for the biogenesis of pigment granules and also for the normal delivery of Golgi-derived vesicles to lysosomes (Sevrioukov et al. 1999; Sriram et al. 2003). This protein complex may represent a functional homologue of the yeast class-C Vps or HOPS complex.

The *S. cerevisiae* Vps45p associates directly with the zinc-binding FYVE finger protein Vac1p, which regulates trafficking between the Golgi and endosomes, the FYVE motif binding the signalling lipid PtdIns(3)P on endosomal membranes. Vac1p also associates with the Rab GTPase Vps21p and, thus, forms a physical connection between a Rab and an SM protein (Peterson et al. 1999; Tall et al. 1999). The mammalian counterparts of Vac1p are apparently the early endosomal antigen-1 (EEA1) and Rabenosyn-5, which both interact with the small GTPase Rab5 and bind PtsIns(3)P via a FYVE finger motif (Simonsen et al. 1998; Nielsen et al. 2000). Of these, Rabenosyn-5 interacts directly with mammalian VPS45 (Nielsen et al. 2000), while EEA1 was shown to associate with syntaxins 6 and 13

(McBride et al. 1999; Simonsen et al. 1999). These Rab5 effectors represent long rod-shaped polypeptides that mediate the tethering of TGN or plasma membrane derived vesicles with early endosomes and homotypic endosome-endosome interactions. The pancreatic β-cell protein granuphilin interacts with both the GTP bound form of Rab3 and Munc18a (Coppola et al. 2002). Granuphilin is a member of synaptotagmin-like (Slp) protein family and is associated with insulin-containing secretory granules. This protein, like other Slp proteins, has C2-calcium-phospholipid binding domains with a Ca^{2+} sensor function. In addition, it shares structural similarities with other Rab effectors, such as a Zn^{2+}-finger motif involved in Rab binding. A number of Rab effector proteins, thus, appear to bridge their Rab partners and the SM protein/SNARE machinery involved in membrane docking/fusion.

9.2 Interaction partners that regulate the SM protein-syntaxin association

Tomosyn was the first mammalian protein found to displace Munc18a from a Munc18a/syntaxin1A heterodimer (Fujita et al. 1998). The neuronal-specific 130 kDa isoform of tomosyn contains a carboxyl-terminal VAMP-2-like domain that mediates binding to syntaxin1A (Masuda et al. 1998). In adipocytes, the b-tomosyn isoform binds with high affinity to syntaxin 4 with the same domain. In contrast to the situation in neurons, tomosyn does not affect Munc18c/syntaxin 4 binding: Munc18c was reported to associate similarly with both the syntaxin 4/b-tomosyn binary complex and syntaxin 4 containing SNARE complexes (Widberg et al. 2003). These findings indicate that there may be fundamental differences between the association of Munc18a with syntaxin 1A and, on the other hand, Munc18c with syntaxin 4. The authors suggest that tomosyn and Munc18c operate at a similar stage of the SNARE assembly cycle, and their function perhaps involves priming of syntaxin 4 for entry into the ternary SNARE complex. The data of Widberg et al. (2003) is, however, somewhat discrepant with a number of studies suggesting that binding of Munc18c to syntaxin 4 inhibits its association with the cognate SNARE partners (Tellam et al. 1997; Reed et al. 1999; Thurmond et al. 2000).

The *C. elegans* SM protein UNC-18 interacts transiently with UNC-13 *in vitro*. UNC-13 is a vesicle transport regulator that contains two calcium phospholipid-binding domains (C2) and a potential phorbol ester binding domain (C1) (Ahmed et al. 1992; Kazanietz et al. 1995). UNC-13 functions at a post-docking step of exocytosis and has the capacity to displace UNC-18 from the UNC-18-syntaxin complex without forming a stable ternary complex (Sassa et al. 1999). The protein plays an essential role in synaptic vesicle priming possibly by promoting the open conformation of syntaxin that participates in SNARE complex assembly (Richmond et al. 2001).

Jantti, Jussi
    VTT Biotechnology, Tietotie 2, Espoo, P.O. Box 1500, FIN-02044 VTT, Finland
    jussi.jantti@vtt.fi

Kauppi, Maria
    Department of Molecular Medicine, National Public Health Institute, Biomedicum, P.O.Box 104, FIN-00251, Helsinki, Finland
    maria.kauppi@ktl.fi

Keränen, Sirkka
    VTT Biotechnology, Tietotie 2, Espoo, P.O. Box 1500, FIN-02044 VTT, Finland
    sirkka.keranen@vtt.fi

Knop, Michael
    EMBL, Cell Biology and Biophysics Programme, Meyerhofstr. 1, 69117 Heidelberg
    knop@embl.de

McPherson, Peter S.
    Montreal Neurological Institute, McGill University, 3801 University, Montreal, QC, H3A 2B4 Canada
    peter.mcpherson@mcgill.ca

Olkkonen, Vesa M.
    Department of Molecular Medicine, National Public Health Institute, Biomedicum, P.O.Box 104, FIN-00251, Helsinki, Finland

Ritter, Brigitte
    Montreal Neurological Institute, McGill University, 3801 University, Montreal, QC, H3A 2B4 Canada

Taxis, Christof
    EMBL, Cell Biology and Biophysics Programme, Meyerhofstr. 1, 69117 Heidelberg

Weinberger, Adina
    Department of Molecular Genetics, Weizmann Institute of Science, Rehovot 76100, Israel

Fig. 3 (overleaf). Model for a common unifying function of SM proteins. A. A transport vesicle is tethered at the target membrane via the action of a Rab protein in GTP-bound form (GTP) and a tethering complex (T) recruited to membranes by the small GTPase. The SM protein binds to factor X (or possibly several factors) and then to syntaxin (alternative a), or binds to syntaxin first and then recruits factor X (alternative b). SNARE proteins are marked with S; syntaxin is shown in blue. B. Factor X links the SM protein and the complexed syntaxin to the tethering apparatus. C. The SM protein facilitates the assembly of a trans-SNARE complex in a spatially and temporally controlled manner. The tethering complex dissociates; the Rab in GDP-bound form (GDP), the SM protein, and factor X are released. In some cases, the SM protein may remain bound to the SNARE complex and regulate events during the following fusion process.

No physical interaction has been detected between Munc18a and Munc13-1 (Brose et al. 1995), the mammalian homologues of *C. elegans* UNC-18 and UNC-13. Munc13-1 in mammalian cells was reported to associate with DOC2A, and also with the open conformation of syntaxins, but not with Munc18a (Mochida et al. 1998). However, Munc18a interacts directly with members of the DOC2 protein family. The DOC2 family consists of two members, the neuronal DOC2A and the ubiquitously expressed DOC2B. They are enriched on synaptic vesicles and contain a double C2 domain capable of binding to Ca^{2+} and phospholipids and have a role in Ca^{2+}-evoked neurotransmitter release. Practically the entire amino acid sequence of Munc18a is required for the interaction with the DOC2 family proteins (Verhage et al. 1997). Interestingly, syntaxins and DOC2 proteins compete for Munc18a binding, suggesting that the DOC2 proteins may be involved in the regulation of Munc18-syntaxin association.

9.3 The Mint proteins

Another family of components interacting directly with Munc18a are the Mint (Munc18 interacting) proteins (Okamoto and Sudhof 1997). There are three members in the Mint family: Mint1 and -2 are brain specific whereas Mint3 is ubiquitously expressed. Mint1, also known as X11α, was originally characterised as a protein interacting with the Alzheimer's disease-linked β-amyloid precursor protein (APP) (Borg et al. 1996). The Mint proteins have a phosphotyrosine binding domain (PTB) in the C-terminal region and two PDZ domains. In addition to phosphotyrosines, PTB domains bind also other phosphopeptides and phosphatidylinositol phosphates (PIPs). Mint protein PTB domains mediate the association with APP and they affect the distribution and turnover of APP (Borg et al. 1998; Sastre et al. 1998). Munc18a binds a motif in the N-terminal part of Mint1 and -2. This N-terminal part is missing from Mint3 and this protein seems to be unable to bind Munc18a or Munc18b (Okamoto and Sudhof 1997, 1998). Mint1 was found to participate in a tripartite complex also containing CASK and Veli1, -2, or - 3 proteins, which was suggested to play a role in the coupling between synaptic vesicle exocytosis, signal transduction, and cell adhesion in the brain (Butz et al. 1998). The interaction of Munc18a with Mint1 in this model would create a link

of the tripartite complex to the synaptic vesicle docking/fusion machinery. Mints also associate directly with the small GTPase Arf that controls vesicle coat formation in the Golgi, and are actually suggested to represent a novel class of clathrin adaptors in this organelle (Hill et al. 2003).

10 Do the SM proteins share a conserved function despite the different modes of interaction with SNAREs?

Different SM proteins interact with the SNARE machinery in highly different ways, ranging from extensive contacts with the Habc and H3 portions of monomeric syntaxins to indirect contacts with these t-SNAREs mediated through large protein complexes (Fig. 2). On the other hand, the proteins show a relatively high degree of conservation of primary sequence and structure (Bracher and Weissenhorn 2002; Dulubova et al. 2003; Gallwitz and Jahn 2003). The latter fact speaks for functional conservation, but the former one makes one wonder how a common function could be exerted by proteins forming such diverse interactions with their SNARE partners. Even though several SM proteins clearly form intimate contacts with their cognate syntaxins and may play important roles in the stability of these t-SNAREs or as platforms for SNARE complex assembly, this is unlikely be the unifying function of the protein family. A possible common function that would be compatible with the different interaction modes would be to act as carriers or bridging molecules that recruit accessory factors to the site of SNARE complex assembly/function. We have previously shown that Munc18b mutants that show no or drastically reduced affinity for syntaxin behave in different ways in an apical transport assay carried out in epithelial cells. Some of the mutants seem functionally inert while others have inhibitory effects comparable to those of the wild type protein (Riento et al. 2000). Our results are consistent with a model in which Munc18b with a low affinity for syntaxin is still capable of performing its essential function by associating with the t-SNARE at a high turn-over rate and recruiting a yet unknown essential factor to the SNARE complex assembly site (Fig. 3). The reduced affinity would compensate for the increased SM protein concentration due to overexpression. Therefore, a low affinity mutant would not inhibit transport like excess wild type protein does. However, an overexpressed mutant protein with totally abolished syntaxin binding could still interact with the unknown cytosolic or membrane-associated factor but not bring it to the SNARE complex assembly site. This sequestering effect would lead to inhibition of transport.

A tempting possibility is that the factors that SM proteins bridge to the SNARE complex assembly site are components of the membrane tethering apparatus controlled by the small Rab GTPases. Creating temporal and spatial control between the tethering and docking/fusion steps is an absolutely essential but yet poorly understood process. Indeed, there is increasing evidence for this type of role of SM proteins (see paragraph 9.1). To fully understand the role of SM proteins in vesicle transport we now need to investigate in great detail the function and the non-

syntaxin interaction partners of different SM proteins with distinct modes of syntaxin binding. Furthermore, we believe that phosphorylation/dephosphorylation cycles of SM proteins and their binding partners will turn out to play major roles in the regulation of vesicle docking and fusion. Grasping the common functional principle of these proteins is a major task for the future and will undoubtedly represent a major leap in our comprehension of how cells control the essential but highly complex process of vesicle transport.

Acknowledgements

The work in the authors' laboratories has been supported by the Finnish Cultural Foundation, Instrumentariumin Tiedesäätiö, Farmoksen Tutkimus- ja Tiedesäätiö, the Sigrid Juselius Foundation, the Academy of Finland (grants 42160, 49894, 49987, 50641, and 54301), and the research program "VTT Industrial Biotechnology" (Academy of Finland; Finnish Centre of Excellence program, 2000-2005, project no. 64330). The authors thank Mikko Jalanko for helping in preparation of the figures.

References

Aalto MK, Jäntti J, Östling J, Keränen S, Ronne H (1997) Mso1p: a yeast protein that functions in secretion and interacts physically and genetically with Sec1p. Proc Natl Acad Sci USA 94:7331-7336

Aalto MK, Keränen S, Ronne H (1992) A family of proteins involved in intracellular transport. Cell 68:181-182

Aalto MK, Ronne H, Keränen S (1993) Yeast syntaxins Sso1p and Sso2p belong to a family of related membrane proteins that function in vesicular transport. EMBO J 12:4095-4104

Abeliovich H, Darsow T, Emr SD (1999) Cytoplasm to vacuole trafficking of aminopeptidase I requires a t-SNARE-Sec1p complex composed of Tlg2p and Vps45p. EMBO J 18:6005-6016

Ahmed S, Maruyama IN, Kozma R, Lee J, Brenner S, Lim L (1992) The *Caenorhabditis elegans* unc-13 gene product is a phospholipid-dependent high-affinity phorbol ester receptor. Biochem J 287 Pt 3:995-999

Araki S, Tamori Y, Kawanishi M, Shinoda H, Masugi J, Mori H, Niki T, Okazawa H, Kubota T, Kasuga M (1997) Inhibition of the binding of SNAP-23 to syntaxin 4 by Munc18c. Biochem Biophys Res Commun 234:257-262

Assaad FF, Huet Y, Mayer U, Jurgens G (2001) The cytokinesis gene KEULE encodes a Sec1 protein that binds the syntaxin KNOLLE. J Cell Biol 152:531-543

Barclay JW, Craig TJ, Fisher RJ, Ciufo LF, Evans GJ, Morgan A, Burgoyne RD (2003) Phosphorylation of Munc18 by protein kinase C regulates the kinetics of exocytosis. J Biol Chem 278:10538-10545

Bennett MK, Garcia-Arraras JE, Elferink LA, Peterson K, Fleming AM, Hazuka CD, Scheller RH (1993) The syntaxin family of vesicular transport receptors. Cell 74:863-873

Bhaskar K, Shareef MM, Sharma VM, Shetty AP, Ramamohan Y, Pant HC, Raju TR, Shetty KT (2004) Co-purification and localization of Munc18-1 (p67) and Cdk5 with neuronal cytoskeletal proteins. Neurochem Int 44:35-44

Bock JB, Matern HT, Peden AA, Scheller RH (2001) A genomic perspective on membrane compartment organization. Nature 409:839-841

Borg JP, Ooi J, Levy E, Margolis B (1996) The phosphotyrosine interaction domains of X11 and FE65 bind to distinct sites on the YENPTY motif of amyloid precursor protein. Mol Cell Biol 16:6229-6241

Borg JP, Yang Y, De Taddeo-Borg M, Margolis B, Turner RS (1998) The X11alpha protein slows cellular amyloid precursor protein processing and reduces Abeta40 and Abeta42 secretion. J Biol Chem 273:14761-14766

Bracher A, Perrakis A, Dresbach T, Betz H, Weissenhorn W (2000) The X-ray crystal structure of neuronal Sec1 from squid sheds new light on the role of this protein in exocytosis. Structure Fold Des 8:685-694

Bracher A, Weissenhorn W (2001) Crystal structures of neuronal squid Sec1 implicate inter-domain hinge movement in the release of t-SNAREs. J Mol Biol 306:7-13

Bracher A, Weissenhorn W (2002) Structural basis for the Golgi membrane recruitment of Sly1p by Sed5p. EMBO J 21:6114-6124

Brose N, Hofmann K, Hata Y, Südhof TC (1995) Mammalian homologues of *Caenorhabditis elegans* unc-13 gene define novel family of C2-domain proteins. J Biol Chem 270:25273-25280

Brummer MH, Kivinen KJ, Jäntti J, Toikkanen J, Söderlund H, Keränen S (2001) Characterization of the sec1-1 and sec1-11 mutations. Yeast 18:1525-1536

Bryant NJ, James DE (2001) Vps45p stabilizes the syntaxin homologue Tlg2p and positively regulates SNARE complex formation. EMBO J 20:3380-3388

Bryant NJ, James DE (2003) The Sec1p/Munc18 (SM) protein, Vps45p, cycles on and off membranes during vesicle transport. J Cell Biol 161:691-696

Butz S, Okamoto M, Südhof TC (1998) A tripartite protein complex with the potential to couple synaptic vesicle exocytosis to cell adhesion in brain. Cell 94:773-782

Cao X, Ballew N, Barlowe C (1998) Initial docking of ER-derived vesicles requires Uso1p and Ypt1p but is independent of SNARE proteins. EMBO J 17:2156-2165

Carr CM, Grote E, Munson M, Hughson FM, Novick PJ (1999) Sec1p binds to SNARE complexes and concentrates at sites of secretion. J Cell Biol 146:333-344

Chen YA, Scheller RH (2001) SNARE-mediated membrane fusion. Nat Rev Mol Cell Biol 2:98-106

Coppola T, Frantz C, Perret-Menoud V, Gattesco S, Hirling H, Regazzi R (2002) Pancreatic beta-cell protein granuphilin binds Rab3 and Munc-18 and controls exocytosis. Mol Biol Cell 13:1906-1915

Cowles CR, Emr SD, Horazdovsky BF (1994) Mutations in the VPS45 gene, a SEC1 homologue, result in vacuolar protein sorting defects and accumulation of membrane vesicles. J Cell Sci 107 Pt 12:3449-3459

Craig TJ, Evans GJ, Morgan A (2003) Physiological regulation of Munc18/nSec1 phosphorylation on serine-313. J Neurochem 86:1450-1457

Dascher C, Ossig R, Gallwitz D, Schmitt HD (1991) Identification and structure of four yeast genes (SLY) that are able to suppress the functional loss of YPT1, a member of the RAS superfamily. Mol Cell Biol 11:872-885

de Vries KJ, Geijtenbeek A, Brian EC, de Graan PN, Ghijsen WE, Verhage M (2000) Dynamics of munc18-1 phosphorylation/dephosphorylation in rat brain nerve terminals. Eur J Neurosci 12:385-390

Dresbach T, Burns ME, O'Connor V, DeBello WM, Betz H, Augustine GJ (1998) A neuronal Sec1 homolog regulates neurotransmitter release at the squid giant synapse. J Neurosci 18:2923-2932

Dulubova I, Sugita S, Hill S, Hosaka M, Fernandez I, Südhof TC, Rizo J (1999) A conformational switch in syntaxin during exocytosis: role of munc18. EMBO J 18:4372-4382

Dulubova I, Yamaguchi T, Arac D, Li H, Huryeva I, Min SW, Rizo J, Südhof TC (2003) Convergence and divergence in the mechanism of SNARE binding by Sec1/Munc18-like proteins. Proc Natl Acad Sci USA 100:32-37

Dulubova I, Yamaguchi T, Gao Y, Min SW, Huryeva I, Südhof TC, Rizo J (2002) How Tlg2p/syntaxin 16 'snares' Vps45. EMBO J 21:3620-3631

Fasshauer D, Antonin W, Margittai M, Pabst S, Jahn R (1999) Mixed and non-cognate SNARE complexes. Characterization of assembly and biophysical properties. J Biol Chem 274:15440-15446

Fasshauer D, Sutton RB, Brunger AT, Jahn R (1998) Conserved structural features of the synaptic fusion complex: SNARE proteins reclassified as Q- and R-SNAREs. Proc Natl Acad Sci USA 95:15781-15786

Fernandez I, Ubach J, Dulubova I, Zhang X, Südhof TC, Rizo J (1998) Three-dimensional structure of an evolutionarily conserved N-terminal domain of syntaxin 1A. Cell 94:841-849

Ferro-Novick S, Novick P (1993) The role of GTP-binding proteins in transport along the exocytic pathway. Annu Rev Cell Biol 9:575-599

Ficarro SB, McCleland ML, Stukenberg PT, Burke DJ, Ross MM, Shabanowitz J, Hunt DF, White FM (2002) Phosphoproteome analysis by mass spectrometry and its application to *Saccharomyces cerevisiae*. Nat Biotechnol 20:301-305

Fiebig KM, Rice LM, Pollock E, Brunger AT (1999) Folding intermediates of SNARE complex assembly. Nat Struct Biol 6:117-123

Finger FP, Novick P (2000) Synthetic interactions of the post-Golgi sec mutations of *Saccharomyces cerevisiae*. Genetics 156:943-951

Fisher RJ, Pevsner J, Burgoyne RD (2001) Control of fusion pore dynamics during exocytosis by Munc18. Science 291:875-878

Fletcher AI, Shuang R, Giovannucci DR, Zhang L, Bittner MA, Stuenkel EL (1999) Regulation of exocytosis by cyclin-dependent kinase 5 via phosphorylation of Munc18. J Biol Chem 274:4027-4035

Fujita Y, Sasaki T, Fukui K, Kotani H, Kimura T, Hata Y, Südhof TC, Scheller RH, Takai Y (1996) Phosphorylation of Munc-18/n-Sec1/rbSec1 by protein kinase C: its implication in regulating the interaction of Munc-18/n-Sec1/rbSec1 with syntaxin. J Biol Chem 271:7265-7268

Fujita Y, Shirataki H, Sakisaka T, Asakura T, Ohya T, Kotani H, Yokoyama S, Nishioka H, Matsuura Y, Mizoguchi A, Scheller RH, Takai Y (1998) Tomosyn: a syntaxin-1-binding protein that forms a novel complex in the neurotransmitter release process. Neuron 20:905-915

Gallwitz D, Jahn R (2003) The riddle of the Sec1/Munc-18 proteins - new twists added to their interactions with SNAREs. Trends Biochem Sci 28:113-116

Garcia EP, Gatti E, Butler M, Burton J, De Camilli P (1994) A rat brain Sec1 homologue related to Rop and UNC18 interacts with syntaxin. Proc Natl Acad Sci USA 91:2003-2007

Gengyo-Ando K, Kamiya Y, Yamakawa A, Kodaira K, Nishiwaki K, Miwa J, Hori I, Hosono R (1993) The C. elegans unc-18 gene encodes a protein expressed in motor neurons. Neuron 11:703-711

Gengyo-Ando K, Kitayama H, Mukaida M, Ikawa Y (1996) A murine neural-specific homolog corrects cholinergic defects in Caenorhabditis elegans unc-18 mutants. J Neurosci 16:6695-6702

Graham ME, Sudlow AW, Burgoyne RD (1997) Evidence against an acute inhibitory role of nSec-1 (munc-18) in late steps of regulated exocytosis in chromaffin and PC12 cells. J Neurochem 69:2369-2377

Guo W, Roth D, Walch-Solimena C, Novick P (1999) The exocyst is an effector for Sec4p, targeting secretory vesicles to sites of exocytosis. EMBO J 18:1071-1080

Harrison SD, Broadie K, van de Goor J, Rubin GM (1994) Mutations in the Drosophila Rop gene suggest a function in general secretion and synaptic transmission. Neuron 13:555-566

Hata Y, Slaughter CA, Südhof TC (1993) Synaptic vesicle fusion complex contains unc-18 homologue bound to syntaxin. Nature 366:347-351

Hata Y, Südhof TC (1995) A novel ubiquitous form of Munc-18 interacts with multiple syntaxins. Use of the yeast two-hybrid system to study interactions between proteins involved in membrane traffic. J Biol Chem 270:13022-13028

Hill K, Li Y, Bennett M, McKay M, Zhu X, Shern J, Torre E, Lah JJ, Levey AI, Kahn RA (2003) Munc18 interacting proteins: ADP-ribosylation factor-dependent coat proteins that regulate the traffic of beta-Alzheimer's precursor protein. J Biol Chem 278:36032-36040

Hosono R, Hekimi S, Kamiya Y, Sassa T, Murakami S, Nishiwaki K, Miwa J, Taketo A, Kodaira KI (1992) The unc-18 gene encodes a novel protein affecting the kinetics of acetylcholine metabolism in the nematode Caenorhabditis elegans. J Neurochem 58:1517-1525

Houng A, Polgar J, Reed GL (2003) Munc18-syntaxin complexes and exocytosis in human platelets. J Biol Chem 278:19627-19633

Jahn R, Lang T, Südhof TC (2003) Membrane fusion. Cell 112:519-533

Katagiri H, Terasaki J, Murata T, Ishihara H, Ogihara T, Inukai K, Fukushima Y, Anai M, Kikuchi M, Miyazaki J, et al. (1995) A novel isoform of syntaxin-binding protein homologous to yeast Sec1 expressed ubiquitously in mammalian cells. J Biol Chem 270:4963-4966

Kauppi M, Wohlfahrt G, Olkkonen VM (2002) Analysis of the Munc18b-syntaxin binding interface. Use of a mutant Munc18b to dissect the functions of syntaxins 2 and 3. J Biol Chem 277:43973-43979

Kazanietz MG, Lewin NE, Bruns JD, Blumberg PM (1995) Characterization of the cysteine-rich region of the Caenorhabditis elegans protein Unc-13 as a high affinity phorbol ester receptor. Analysis of ligand-binding interactions, lipid cofactor requirements, and inhibitor sensitivity. J Biol Chem 270:10777-10783

Kennelly PJ, Krebs EG (1991) Consensus sequences as substrate specificity determinants for protein kinases and protein phosphatases. J Biol Chem 266:15555-15558

Khan AH, Thurmond DC, Yang C, Ceresa BP, Sigmund CD, Pessin JE (2001) Munc18c regulates insulin-stimulated glut4 translocation to the transverse tubules in skeletal muscle. J Biol Chem 276:4063-4069

Kim DW, Sacher M, Scarpa A, Quinn AM, Ferro-Novick S (1999) High-copy suppressor analysis reveals a physical interaction between Sec34p and Sec35p, a protein implicated in vesicle docking. Mol Biol Cell 10:3317-3329

Kosodo Y, Noda Y, Adachi H, Yoda K (2002) Binding of Sly1 to Sed5 enhances formation of the yeast early Golgi SNARE complex. J Cell Sci 115:3683-3691

Lew J, Winkfein RJ, Paudel HK, Wang JH (1992) Brain proline-directed protein kinase is a neurofilament kinase which displays high sequence homology to p34cdc2. J Biol Chem 267:25922-25926

Martin-Verdeaux S, Pombo I, Iannascoli B, Roa M, Varin-Blank N, Rivera J, Blank U (2003) Evidence of a role for Munc18-2 and microtubules in mast cell granule exocytosis. J Cell Sci 116:325-334

Masuda ES, Huang BC, Fisher JM, Luo Y, Scheller RH (1998) Tomosyn binds t-SNARE proteins via a VAMP-like coiled coil. Neuron 21:479-480

McBride HM, Rybin V, Murphy C, Giner A, Teasdale R, Zerial M (1999) Oligomeric complexes link Rab5 effectors with NSF and drive membrane fusion via interactions between EEA1 and syntaxin 13. Cell 98:377-386

Misura KM, Scheller RH, Weis WI (2000) Three-dimensional structure of the neuronal-Sec1-syntaxin 1a complex. Nature 404:355-362

Mochida S, Orita S, Sakaguchi G, Sasaki T, Takai Y (1998) Role of the Doc2 alpha-Munc13-1 interaction in the neurotransmitter release process. Proc Natl Acad Sci USA 95:11418-11422

Munson M, Chen X, Cocina AE, Schultz SM, Hughson FM (2000) Interactions within the yeast t-SNARE Sso1p that control SNARE complex assembly. Nat Struct Biol 7:894-902

Nichols BJ, Holthuis JC, Pelham HR (1998) The Sec1p homologue Vps45p binds to the syntaxin Tlg2p. Eur J Cell Biol 77:263-268

Nielsen E, Christoforidis S, Uttenweiler-Joseph S, Miaczynska M, Dewitte F, Wilm M, Hoflack B, Zerial M (2000) Rabenosyn-5, a novel Rab5 effector, is complexed with hVPS45 and recruited to endosomes through a FYVE finger domain. J Cell Biol 151:601-612

Novick P, Schekman R (1979) Secretion and cell-surface growth are blocked in a temperature-sensitive mutant of *Saccharomyces cerevisiae*. Proc Natl Acad Sci USA 76:1858-1862

Okamoto M, Südhof TC (1997) Mints, Munc18-interacting proteins in synaptic vesicle exocytosis. J Biol Chem 272:31459-31464

Okamoto M, Südhof TC (1998) Mint 3: a ubiquitous mint isoform that does not bind to munc18-1 or -2. Eur J Cell Biol 77:161-165

Ossig R, Dascher C, Trepte HH, Schmitt HD, Gallwitz D (1991) The yeast SLY gene products, suppressors of defects in the essential GTP-binding Ypt1 protein, may act in endoplasmic reticulum-to-Golgi transport. Mol Cell Biol 11:2980-2993

Peng R, Gallwitz D (2002) Sly1 protein bound to Golgi syntaxin Sed5p allows assembly and contributes to specificity of SNARE fusion complexes. J Cell Biol 157:645-655

Peterson MR, Burd CG, Emr SD (1999) Vac1p coordinates Rab and phosphatidylinositol 3-kinase signaling in Vps45p-dependent vesicle docking/fusion at the endosome. Curr Biol 9:159-162

Pevsner J, Hsu SC, Braun JE, Calakos N, Ting AE, Bennett MK, Scheller RH (1994a) Specificity and regulation of a synaptic vesicle docking complex. Neuron 13:353-361

Pevsner J, Hsu SC, Scheller RH (1994b) n-Sec1: a neural-specific syntaxin-binding protein. Proc Natl Acad Sci USA 91:1445-1449

Pfeffer SR (2001) Rab GTPases: specifying and deciphering organelle identity and function. Trends Cell Biol 11:487-491

Piper RC, Whitters EA, Stevens TH (1994) Yeast Vps45p is a Sec1p-like protein required for the consumption of vacuole-targeted, post-Golgi transport vesicles. Eur J Cell Biol 65:305-318

Pombo I, Rivera J, Blank U (2003) Munc18-2/syntaxin3 complexes are spatially separated from syntaxin3-containing SNARE complexes. FEBS Lett 550:144-148

Price A, Seals D, Wickner W, Ungermann C (2000) The docking stage of yeast vacuole fusion requires the transfer of proteins from a cis-SNARE complex to a Rab/Ypt protein. J Cell Biol 148:1231-1238

Reed GL, Houng AK, Fitzgerald ML (1999) Human platelets contain SNARE proteins and a Sec1p homologue that interacts with syntaxin 4 and is phosphorylated after thrombin activation: implications for platelet secretion. Blood 93:2617-2626

Richmond JE, Weimer RM, Jorgensen EM (2001) An open form of syntaxin bypasses the requirement for UNC-13 in vesicle priming. Nature 412:338-341

Rieder SE, Emr SD (1997) A novel RING finger protein complex essential for a late step in protein transport to the yeast vacuole. Mol Biol Cell 8:2307-2327

Riento K, Galli T, Jansson S, Ehnholm C, Lehtonen E, Olkkonen VM (1998) Interaction of Munc-18-2 with syntaxin 3 controls the association of apical SNAREs in epithelial cells. J Cell Sci 111 Pt 17:2681-2688

Riento K, Jäntti J, Jansson S, Hielm S, Lehtonen E, Ehnholm C, Keränen S, Olkkonen VM (1996) A sec1-related vesicle-transport protein that is expressed predominantly in epithelial cells. Eur J Biochem 239:638-646

Riento K, Kauppi M, Keränen S, Olkkonen VM (2000) Munc18-2, a functional partner of syntaxin 3, controls apical membrane trafficking in epithelial cells. J Biol Chem 275:13476-13483

Robinson JS, Klionsky DJ, Banta LM, Emr SD (1988) Protein sorting in *Saccharomyces cerevisiae*: isolation of mutants defective in the delivery and processing of multiple vacuolar hydrolases. Mol Cell Biol 8:4936-4948

Rothman JE (1994) Mechanisms of intracellular protein transport. Nature 372:55-63

Rowe J, Calegari F, Taverna E, Longhi R, Rosa P (2001) Syntaxin 1A is delivered to the apical and basolateral domains of epithelial cells: the role of munc-18 proteins. J Cell Sci 114:3323-3332

Rowe J, Corradi N, Malosio ML, Taverna E, Halban P, Meldolesi J, Rosa P (1999) Blockade of membrane transport and disassembly of the Golgi complex by expression of syntaxin 1A in neurosecretion-incompetent cells: prevention by rbSEC1. J Cell Sci 112 Pt 12:1865-1877

Sapperstein SK, Lupashin VV, Schmitt HD, Waters MG (1996) Assembly of the ER to Golgi SNARE complex requires Uso1p. J Cell Biol 132:755-767

Sassa T, Harada S, Ogawa H, Rand JB, Maruyama IN, Hosono R (1999) Regulation of the UNC-18-*Caenorhabditis elegans* syntaxin complex by UNC-13. J Neurosci 19:4772-4777

Sassa T, Ogawa H, Kimoto M, Hosono R (1996) The synaptic protein UNC-18 is phosphorylated by protein kinase C. Neurochem Int 29:543-552

Sastre M, Turner RS, Levy E (1998) X11 interaction with beta-amyloid precursor protein modulates its cellular stabilization and reduces amyloid beta-protein secretion. J Biol Chem 273:22351-22357

Sato TK, Rehling P, Peterson MR, Emr SD (2000) Class C Vps protein complex regulates vacuolar SNARE pairing and is required for vesicle docking/fusion. Mol Cell 6:661-671

Schraw TD, Lemons PP, Dean WL, Whiteheart SW (2003) A role for Sec1/Munc18 proteins in platelet exocytosis. Biochem J 374:207-217

Schulze KL, Littleton JT, Salzberg A, Halachmi N, Stern M, Lev Z, Bellen HJ (1994) rop, a *Drosophila* homolog of yeast Sec1 and vertebrate n-Sec1/Munc-18 proteins, is a negative regulator of neurotransmitter release *in vivo*. Neuron 13:1099-1108

Seals DF, Eitzen G, Margolis N, Wickner WT, Price A (2000) A Ypt/Rab effector complex containing the Sec1 homolog Vps33p is required for homotypic vacuole fusion. Proc Natl Acad Sci USA 97:9402-9407

Sevrioukov EA, He JP, Moghrabi N, Sunio A, Kramer H (1999) A role for the deep orange and *carnation* eye color genes in lysosomal delivery in *Drosophila*. Mol Cell 4:479-486

Shetty KT, Kaech S, Link WT, Jaffe H, Flores CM, Wray S, Pant HC, Beushausen S (1995) Molecular characterization of a neuronal-specific protein that stimulates the activity of Cdk5. J Neurochem 64:1988-1995

Shetty KT, Link WT, Pant HC (1993) cdc2-like kinase from rat spinal cord specifically phosphorylates KSPXK motifs in neurofilament proteins: isolation and characterization. Proc Natl Acad Sci USA 90:6844-6848

Shuang R, Zhang L, Fletcher A, Groblewski GE, Pevsner J, Stuenkel EL (1998) Regulation of Munc-18/syntaxin 1A interaction by cyclin-dependent kinase 5 in nerve endings. J Biol Chem 273:4957-4966

Simonsen A, Gaullier JM, D'Arrigo A, Stenmark H (1999) The Rab5 effector EEA1 interacts directly with syntaxin-6. J Biol Chem 274:28857-28860

Simonsen A, Lippe R, Christoforidis S, Gaullier JM, Brech A, Callaghan J, Toh BH, Murphy C, Zerial M, Stenmark H (1998) EEA1 links PI(3)K function to Rab5 regulation of endosome fusion. Nature 394:494-498

Spurlin BA, Thomas RM, Nevins AK, Kim HJ, Kim YJ, Noh HL, Shulman GI, Kim JK, Thurmond DC (2003) Insulin resistance in tetracycline-repressible Munc18c transgenic mice. Diabetes 52:1910-1917

Sriram V, Krishnan KS, Mayor S (2003) deep-orange and *carnation* define distinct stages in late endosomal biogenesis in *Drosophila melanogaster*. J Cell Biol 161:593-607

Stenmark H, Olkkonen VM (2001) The Rab GTPase family. Genome Biol 2:REVIEWS3007

Suzuki T, Oiso N, Gautam R, Novak EK, Panthier JJ, Suprabha PG, Vida T, Swank RT, Spritz RA (2003) The mouse organellar biogenesis mutant buff results from a mutation in Vps33a, a homologue of yeast vps33 and *Drosophila carnation*. Proc Natl Acad Sci USA 100:1146-1150

Tall GG, Hama H, DeWald DB, Horazdovsky BF (1999) The phosphatidylinositol 3-phosphate binding protein Vac1p interacts with a Rab GTPase and a Sec1p homologue to facilitate vesicle-mediated vacuolar protein sorting. Mol Biol Cell 10:1873-1889

Tamori Y, Kawanishi M, Niki T, Shinoda H, Araki S, Okazawa H, Kasuga M (1998) Inhibition of insulin-induced GLUT4 translocation by Munc18c through interaction with syntaxin4 in 3T3-L1 adipocytes. J Biol Chem 273:19740-19746

Tellam JT, Macaulay SL, McIntosh S, Hewish DR, Ward CW, James DE (1997) Characterization of Munc-18c and syntaxin-4 in 3T3-L1 adipocytes. Putative role in insulindependent movement of GLUT-4. J Biol Chem 272:6179-6186

Tellam JT, McIntosh S, James DE (1995) Molecular identification of two novel Munc-18 isoforms expressed in non-neuronal tissues. J Biol Chem 270:5857-5863

TerBush DR, Maurice T, Roth D, Novick P (1996) The exocyst is a multiprotein complex required for exocytosis in *Saccharomyces cerevisiae*. EMBO J 15:6483-6494

Thurmond DC, Ceresa BP, Okada S, Elmendorf JS, Coker K, Pessin JE (1998) Regulation of insulin-stimulated GLUT4 translocation by Munc18c in 3T3L1 adipocytes. J Biol Chem 273:33876-33883

Thurmond DC, Kanzaki M, Khan AH, Pessin JE (2000) Munc18c function is required for insulin-stimulated plasma membrane fusion of GLUT4 and insulin-responsive amino peptidase storage vesicles. Mol Cell Biol 20:379-388

Tian JH, Das S, Sheng ZH (2003) Ca2+-dependent phosphorylation of syntaxin-1A by the death-associated protein (DAP) kinase regulates its interaction with Munc18. J Biol Chem 278:26265-26274

Toonen RF, Verhage M (2003) Vesicle trafficking: pleasure and pain from SM genes. Trends Cell Biol 13:177-186

Tsui MM, Banfield DK (2000) Yeast Golgi SNARE interactions are promiscuous. J Cell Sci 113 Pt 1:145-152

Ungermann C, Sato K, Wickner W (1998) Defining the functions of trans-SNARE pairs. Nature 396:543-548

Wada Y, Kitamoto K, Kanbe T, Tanaka K, Anraku Y (1990) The SLP1 gene of *Saccharomyces cerevisiae* is essential for vacuolar morphogenesis and function. Mol Cell Biol 10:2214-2223

Waizenegger I, Lukowitz W, Assaad F, Schwarz H, Jurgens G, Mayer U (2000) The *Arabidopsis* KNOLLE and KEULE genes interact to promote vesicle fusion during cytokinesis. Curr Biol 10:1371-1374

VanRheenen SM, Cao X, Sapperstein SK, Chiang EC, Lupashin VV, Barlowe C, Waters MG (1999) Sec34p, a protein required for vesicle tethering to the yeast Golgi apparatus, is in a complex with Sec35p. J Cell Biol 147:729-742

Weber T, Zemelman BV, McNew JA, Westermann B, Gmachl M, Parlati F, Söllner TH, Rothman JE (1998) SNAREpins: minimal machinery for membrane fusion. Cell 92:759-772

Weimer RM, Richmond JE, Davis WS, Hadwiger G, Nonet ML, Jorgensen EM (2003) Defects in synaptic vesicle docking in unc-18 mutants. Nat Neurosci 6:1023-1030

Verhage M, de Vries KJ, Roshol H, Burbach JP, Gispen WH, Sudhof TC (1997) DOC2 proteins in rat brain: complementary distribution and proposed function as vesicular adapter proteins in early stages of secretion. Neuron 18:453-461

Verhage M, Maia AS, Plomp JJ, Brussaard AB, Heeroma JH, Vermeer H, Toonen RF, Hammer RE, van den Berg TK, Missler M, Geuze HJ, Südhof TC (2000) Synaptic assembly of the brain in the absence of neurotransmitter secretion. Science 287:864-869

Whyte JR, Munro S (2001) The Sec34/35 Golgi transport complex is related to the exocyst, defining a family of complexes involved in multiple steps of membrane traffic. Dev Cell 1:527-537

Widberg CH, Bryant NJ, Girotti M, Rea S, James DE (2003) Tomosyn interacts with the t-SNAREs syntaxin4 and SNAP23 and plays a role in insulin-stimulated GLUT4 translocation. J Biol Chem 278:35093-35101

Voets T, Toonen RF, Brian EC, de Wit H, Moser T, Rettig J, Südhof TC, Neher E, Verhage M (2001) Munc18-1 promotes large dense-core vesicle docking. Neuron 31:581-591

Wu MN, Littleton JT, Bhat MA, Prokop A, Bellen HJ (1998) ROP, the *Drosophila* Sec1 homolog, interacts with syntaxin and regulates neurotransmitter release in a dosage-dependent manner. EMBO J 17:127-139

Wurmser AE, Sato TK, Emr SD (2000) New component of the vacuolar class C-Vps complex couples nucleotide exchange on the Ypt7 GTPase to SNARE-dependent docking and fusion. J Cell Biol 151:551-562

Yamaguchi T, Dulubova I, Min SW, Chen X, Rizo J, Südhof TC (2002) Sly1 binds to Golgi and ER syntaxins via a conserved N-terminal peptide motif. Dev Cell 2:295-305

Yang B, Gonzalez L Jr, Prekeris R, Steegmaier M, Advani RJ, Scheller RH (1999) SNARE interactions are not selective. Implications for membrane fusion specificity. J Biol Chem 274:5649-5653

Yang B, Steegmaier M, Gonzalez LC Jr, Scheller RH (2000) nSec1 binds a closed conformation of syntaxin1A. J Cell Biol 148:247-252

Yang C, Coker KJ, Kim JK, Mora S, Thurmond DC, Davis AC, Yang B, Williamson RA, Shulman GI, Pessin JE (2001) Syntaxin 4 heterozygous knockout mice develop muscle insulin resistance. J Clin Invest 107:1311-1318

Zerial M, McBride H (2001) Rab proteins as membrane organizers. Nat Rev Mol Cell Biol 2:107-117

Jäntti, Jussi
    VTT Biotechnology, P.O.Box 1500, FIN-02044, Espoo, Finland

Kauppi, Maria
    Department of Molecular Medicine, National Public Health Institute, Biomedicum, P.O.Box 104, FIN-00251, Helsinki, Finland
    maria.kauppi@ktl.fi

Olkkonen, Vesa M.
    Department of Molecular Medicine, National Public Health Institute, Biomedicum, P.O.Box 104, FIN-00251, Helsinki, Finland

Regulation of SNARE assembly by protein phosphorylation

Adina Weinberger and Jeffrey E. Gerst

Abstract

Protein phosphorylation is emerging as an important regulatory mechanism that controls the secretory pathway. It enables coupling between the vesicular transport machinery and signaling cascades that are activated by both internal and external stimuli. Proteins that mediate the last steps of membrane fusion such as SNAREs, Sec1/Munc18, Rab proteins, and others undergo post-translational modification by phosphorylation to control membrane fusion. Control of fusion by protein phosphorylation is operant in both constitutive and regulated exocytic processes, and appears to regulate transport throughout the secretory pathway. The kinases involved are diverse and represent various signaling paths, including those of protein kinase A (PKA), PKC, casein kinase, death-associated protein kinase (DAPK), calcium/calmodulin-dependent protein kinase II (CaMKII), and cyclin-dependent kinase (Cdk). The phosphatases include protein phosphatase 1 (PP1) and ceramide-activated protein phosphatase (CAPP). Here, we present an overview on most recent information available concerning phosphorylative control of SNARE assembly leading to membrane fusion.

1 Introduction

The secretory pathway in eukaryotes consists of distinct membrane-bound intracellular compartments that transfer proteins and lipids from one to the other. This occurs via transport vesicles or tubulovesicular structures that undergo fusion with subsequent compartments. The final steps in vesicular transport comprise vesicle tethering, docking and, ultimately, bilayer fusion. The proteins needed for these different steps include: tethering proteins (i.e. p115), SNARE assembly factors [small GTPases (i.e. Rabs), *N*-ethylmaleimide sensitive fusion protein (NSF), SNAP (soluble NSF attachment protein), and Sec1/Munc18 (SM) proteins], and SNAREs (SNAP receptors), which act as the membrane fusogens. Once a secretory vesicle arrives at its target membrane, it is tethered and docked via a process mediated by a consortium of proteins located on the vesicle, such as p115 and the Rab proteins, as well as on the target membranes, such as the Exocyst complex (Pfeffer 1999). Upon tethering, the SNAREs are thought to be in close proximity in which to interact *in trans* between donor and acceptor membranes. All SNAREs

Topics in Current Genetics, Vol. 10
S. Keränen, J. Jäntti (Eds.): Regulatory Mechanisms of Intracellular Membrane Transport
DOI 10.1007/b98496 / Published online: 13 May 2004
© Springer-Verlag Berlin Heidelberg 2004

are membrane-associated proteins and share a common feature termed the "SNARE motif", a short coiled-coil stretch of residues that can assemble into a helical bundle with other SNARE motifs (Chen and Scheller 2001). The SNAREs have been divided classically into the v-SNAREs, which reside on vesicles, and the t-SNAREs, which reside on the target membrane. Both sequence analysis and X-ray crystallography studies of the SNAREs have allowed for their classification into Q-SNAREs and R-SNAREs (Fasshauer et al. 1998), depending upon the amino acid, either glutamine or arginine, present at the ionic layer of the SNARE complex (Sutton et al. 1998). A t-SNARE complex composed of syntaxin and SNAP-25/light chain family members, assembled into three-helix bundle and arrayed *in cis*, is prerequisite for binding to the SNARE motif donated by the v-SNARE (Pevsner et al. 1994; Sutton et al. 1998). Once tethered, the SNAREs from apposed membranes interact productively to fold into detergent-resistant *trans* SNARE complexes (Sollner et al. 1993, Sutton et al. 1998; Chen and Scheller 2001) that reduce the distance between membranes, extrude interposed water molecules and, ultimately, catalyze bilayer fusion (Mayer 2001). After fusion, the AAA-ATPase, NSF, together with α–SNAP unwind the *cis* SNARE complex to separate the v- and t-SNAREs, and to allow for subsequent rounds of SNARE assembly upon their recycling to transport vesicles (Mayer et al. 1996). Importantly, SNAREs alone are able to catalyze the fusion of liposomes *in vitro* (Weber et al. 1998), and even cells when presented artificially on the cell surface (Hu et al. 2003). Thus, SNAREs are the prime candidates for membrane fusogens. While distinct sets of SNARE molecules mediate specific membrane fusion events throughout the secretory system, some SNAREs operate in multiple trafficking events (Chen and Scheller 2001). Thus, while, SNAREs are central elements of the fusion machinery, they alone are insufficient *in vivo* and additional proteins are required to confer vesicular transport. For example, membrane fusion between intracellular compartments appears dependent on SM proteins, which are required in the final steps leading to fusion and probably act to promote SNARE complex assembly. Yet, despite the substantial efforts made towards understanding the function of SM proteins, their mode of action still remains elusive (Gallwitz and Jahn 2003).

There are two major types of exocytosis that occur in eukaryotic cells. One is the constitutive transport of vesicles to the plasma membrane and to other membranous organelles, and the second is the regulated exocytosis of membranal and secreted proteins, as well as small molecules. The former allows for cell growth and the maintenance of cell polarity. The latter (also referred to as stimulus-coupled secretion), allows for secretion in response to specific external stimuli and is cell-type specific (i.e. neurotransmitter release in neurons or hormone secretion in adrenal chromaffin cells). Constitutive vesicular transport, being responsible for the maintenance of the endomembrane system in the living cell, should also have an intrinsic responsiveness to external growth signals, such as nutrient availability or stress conditions. Surprisingly, although cell cycle control has been well studied, little is known about the interface between cell cycle regulation and the control of vesicular transport and, in particular, the regulation of SNAREs and proteins that affect SNARE complex assembly.

Many proteins involved in membrane fusion have been identified as phospho-proteins, while studies on secretion, endocytosis, and vacuolar fusion in yeast have provided good evidence for the involvement of protein phosphorylation in the regulation of the secretory pathway (Gerst 2003). Moreover, extensive work on the mitotic fragmentation and reassembly of the Golgi apparatus has also provided important insights into mechanisms that allow signaling cascades to regulate membrane fusion. In this review, we wish to focus on phosphorylation of the pro-teins required for the final steps in vesicular transport as a means of controlling both the constitutive and the regulated secretory pathways. This type of regulation by post-translational modification appears to play a role not only in the control of SNARE-SNARE interactions, but also to affect both SNARE localization and or-ganelle morphology. Proteins involved in membrane trafficking and fusion that are known to be phosphorylated, as well as the proposed outcome of phosphoryla-tion, are summarized in Table 1. A proposed general model for the possible role of kinases and phosphatases in regulating SNARE assembly is shown in Figure 1.

2 Phosphorylation and control of the exocytic machinery

2.1 Mammalian syntaxins

Syntaxin family members are t-SNAREs that were first identified in neurons (Bennett et al. 1992) and the neuronal Syntaxin 1 was the first SNARE shown to be modified by phosphorylation. Syntaxins share three common features: an NH_2-terminal regulatory domain composed of three α-helixes called the Habc domain, a SNARE motif, and a COOH-terminal trans-membrane anchor. In those syntax-ins involved in exocytosis, the Habc domain folds back on the central helix of the SNARE motif to generate a closed and inactive conformation (Dulubova et al. 1999; Munson et al. 2000). The formation of this four-helix bundle has not been shown for all syntaxins, as syntaxins from the yeast endosomal and vacuolar com-partments, Tlg2, Pep12, and Vam3 (Dulubova et al. 2001, 2002) do not appear to undergo this conformation. Nevertheless, the fact that these syntaxin homologs possess an Habc domain suggests the existence of alternative ternary structures that may reflect different types of regulation, either by signaling or by protein-protein interactions within their specific compartments.

Syntaxin 1. While Syntaxin 1 was one of the first SNAREs shown to be phos-phorylated the actual pathway leading to phosphorylation, and the exact purpose for the modification, are still not understood. Casein kinase II (CKII), a constitu-tively active, second messenger-independent, serine/threonine kinase involved in a wide range of cellular functions, has been shown repeatedly to act upon Syntaxin 1 *in vitro* (Bennett et al. 1993; Hirling and Scheller 1996; Foster et al. 1998; Ris-inger and Bennett 1999; Foletti et al. 2000; Dubois et al. 2002). Using an anti-body, which recognizes a specific phosphorylated form of Syntaxin 1, it was pos-sible to demonstrate that serine-14 is the site for phosphorylation by CKII *in vitro* (Foletti et al. 2000). This CKII phosphorylation site was also confirmed by mass

Table 1. Phosphorylated SNAREs and SNARE regulators

Protein	Kinase	Position	Function	Reference
Syntaxin 1	CKII	serine-14	1. Enhances binding to synaptotagmin 2. Reduced binding to SNAP-25 3. Change in subcellular localization	Bennett et al. 1993; Foster et al. 1998; Risinger and Bennett 1999; Foletti et al. 2000; Dubois et al. 2002
	CKI	threonine-21	ND	Dubois et al. 2002
	CaMKII	ND	ND	Risinger and Bennett 1999, Hirling and Scheller 1996
	PKC	ND	ND	Risinger and Bennett 1999
Syntaxin 4	DAPK	serine-188	Reduced binding to Munc18	Tian et al. 2003
	PKC	ND	Reduced binding to SNAP-23	Chung et al. 2000
	PKA	ND	Reduced binding to SNAP-23	Foster et al. 1998
	CKII	ND	Reduced binding to SNAP-23	Foster et al. 1998
AtSyp122	ND	serine-6 and -8	ND	Nühse et al. 2003
Sso1/2	PKA	serine-79 in Sso1	ND	Marash and Gerst 2001, 2003
Tlg1/2	PKA	serine-90 in Tlg2	1. Reduced binding to Sec9 2. Increased binding to Vsm1	Gurunathan et al. 2002
SNAP-25	PKA	threonine-138	Reduced binding of Tlg1 to Tlg2 No effect on SNARE complex assembly or syn- aptotagmin binding	Risinger and Bennett 1999
SNAP-25	PKC	serine-187	1. Reduced binding to Syntaxin 1 2. Change in subcellular localization	Shimazaki et al. 1996; Kataoka et al. 2000; Gonelle- Gispert et al. 2002
SNAP-23	PKC	ND	Reduced binding to Syntaxin 4	Polgar et al. 2003
	SNAK	ND	Reduced binding to Syntaxin 4	Cabaniols et al. 1999
VAMP	CaMKII	serine-61	ND	Hirling and Scheller 1996
VAMP4	CKII	serine-30	Enhanced binding to adaptor proteins that affect sorting of VAMP4	Hinners et al. 2003
Rab4	Cdc2	serine-196	Induce the release from endosomal membranes	van der Sluijs et al. 1992b
Munc18a	PKC	serine-306 and -313	Reduced binding to Syntaxin 1	Fujita et al. 1996; Barclay et al. 2003
	CDK5	threonine-574	Reduced binding to Syntaxin 1	Shuang et al. 1998; Fletcher et al. 1999
Munc18c	PKC	ND	Reduced binding to Syntaxin 4	Schraw et al. 2003
Synapto- tagmin I	CKII	threonine-125, -128	Reduced binding to Syntaxin 2 and 4	Davletov et al. 1993; Hilfiker et al. 1999
	CaMKII	threonine-112	ND	Popoli et al. 1993; Hilfiker et al. 1999
	PKC	threonine-112 (serine-123 in Aplysia)	ND	Hilfiker et al. 1999; Nakhost et al. 2003

ND: not determined

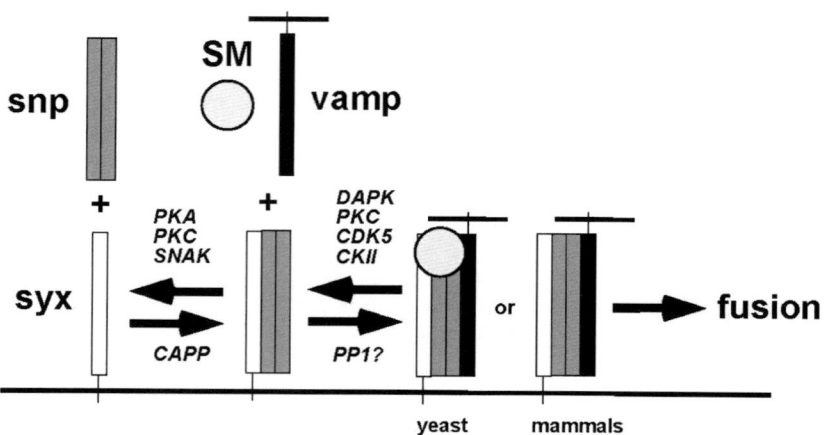

Fig. 1. A general model for the role of kinase and phosphatase activity in the regulation of SNARE assembly and exocytosis. The assembly of the SNAP-25 (snp) and syntaxin (syx) t-SNAREs into a t-SNARE complex is inhibited by kinase (i.e. PKA, PKC, SNAK) activity, but stimulated (at least in yeast) by phosphatase (i.e. CAPP) action. Assembly with the VAMP v-SNARE into a *trans* v-t SNARE complex is facilitated by the binding of a SNARE regulator (i.e. SM), which may either be promoted by phosphatase action (i.e. PP1) or inhibited by kinases (i.e. DAPK, PKC, CDK5, CKII). Thus, the local regulation of phosphatase and kinase activity may be an important mechanism for the control of membrane fusion. The presence or absence of the SM protein from the fusion complex may be dependent upon the system, as SM proteins facilitate SNARE assembly and tend to remain associated with functional SNARE complexes in yeast, but may undergo more transient interactions with the mammalian exocytic SNARE complex. This model does not take into consideration the likelihood of kinase action on multiple targets operating at the different steps leading vesicle fusion, nor differential regulation of the various pools of vesicles as they become secretion-competent. Given the complex nature of regulated exocytosis, the combinatorial effects of kinase action may result in an enhancement of secretion, while inhibiting SNARE assembly *per se*.

spectrometry analysis of the phosphorylated peptides derived from Syntaxin 1 after CKII phosphorylation (Dubois et al. 2002). Studies performed with an anti-phospho-Syntaxin 1 antibody showed three significant findings. First, immunohistochemical studies using rat brain slices revealed that phosphorylated Syntaxin 1 exists only in a subset of axons, while a non-phosphospecific anti-Syntaxin antibody gave a homogenous staining of axons in the rat cortex (Foletti et al. 2000). Second, phosphorylated Syntaxin 1 did not co-localize with a secretory vesicle marker, SV2A (Foletti et al. 2000), suggesting that phosphorylation could restrict Syntaxin 1 from the active zone of the synapse. Third, immunoprecipitation of the SNAP-25 t-SNARE from rat brain extracts revealed a higher affinity for the phosphorylated form of Syntaxin 1 than to the non-phosphorylated form (Foletti et al. 2000). Thus, phosphorylation appears to modulate both Syntaxin 1 localization as well as its ability to associate with SNARE partners. Interestingly, in an earlier

work the phosphorylation of Syntaxin 1 by CKII was shown to cause a significant increase in the binding to the SNARE regulator, synaptotagmin, however, the phosphorylation site was not mapped (Risinger and Bennett 1999). The two Syntaxin 1 phosphorylation events mediated by CKII may be distinct from one another as synaptotagmin, a synaptic vesicle protein proposed to act as a calcium sensor to modulate Ca^{2+}-dependent secretion, is present in the active zone of neurons (Yoshihara et al. 2003). Given that CKII-phosphorylated Syntaxin 1 is excluded from the active zone and, presumably, from its interaction with synaptotagmin (Foletti et al. 2000) the result of synaptotagmin binding to the phosphorylated form of Syntaxin 1 is somewhat paradoxical and requires more extensive analysis.

Other questions regarding the phosphorylation of Syntaxin 1 have yet to be answered. First, it is not known how CKII phosphorylation of Syntaxin 1 affects secretion. Second, it is not known if phosphorylation occurs on the SNARE in its monomeric form or when assembled into t- or v-t SNARE complexes. Thus, it is not possible to ascertain whether phosphorylation by CKII directly controls Ca^{2+}-triggered secretion as shown, for example, for synaptotagmin phosphorylation by CaMKII (Popoli et al. 1997). That said, CKII has been shown to be associated with transport vesicles and whose activity is modulated by its association with vesicle coat proteins (Korolchuk and Banting 2002). Thus, CKII is likely to play a role in vesicle trafficking.

The phosphorylation of Syntaxin 1 by a death-associated protein kinase (DAPK) was demonstrated recently (Tian et al. 2003). DAPK is a calcium/calmodulin-dependent protein kinase (Cohen et al. 1997) shown to be a potential mediator of cell death by a variety of signals (Cohen and Kimchi 2001). The death promoting effects of DAPK depend on its catalytic activity, its correct intracellular localization, and a COOH-terminal death domain (Cohen and Kimchi 2001). Interestingly, the DAPK COOH-terminal domain was isolated in a two-hybrid screen for Syntaxin 1-interacting proteins and full-length DAPK protein was shown to phosphorylate Syntaxin 1 *in vitro* in a Ca^{2+}/CaM-dependent manner (Tian et al. 2003). This phosphorylation site, mapped to serine-188, was not found to affect SNARE assembly *in vitro*, but binding to the Munc18 SNARE regulator was significantly reduced (Tian et al. 2003). Thus, DAPK activity could, potentially, regulate exocytic events (Tian et al. 2003). That the minimal region of DAPK required for Syntaxin 1 binding includes the death domain raises the possibility for a role of the exocytic machinery in regulated cell death. Apoptotic cells are usually regarded as cells that only passively transmit signals to their surroundings. It is possible that certain apoptotic signals may also stimulate exocytosis to generate autocrine and paracrine effects. For example, it was shown that apoptotic eosinophils are capable of exocytosis upon stimulation with phorbol ester (Seton et al. 2003), demonstrating that stimulus-coupled exocytic machinery is still intact and active during cell death.

Interestingly, an autophosphorylated form of CaMKII was found to bind directly to the region linking the Habc and the SNARE motifs of Syntaxin 1 (Ohyama et al. 2002). CaMKII binding appeared to be inhibited in the presence of Munc18, a SNARE regulator proposed to maintain mammalian Syntaxin 1 in a

closed conformation. Although Syntaxin 1 was not demonstrated to be phosphory-lated by CaMKII, CaMKII binding may act to regulate the open and active form of t-SNARE, allowing for SNARE assembly.

Syntaxin 4. Syntaxin 4, a non-neuronal homolog of Syntaxin 1 is expressed in platelets and a variety of other tissues, and forms a SNARE complex with VAMP2 and SNAP-23 (Foster et al. 1998) that mediates constitutive exocytosis. Syntaxin 4 was shown to be phosphorylated *in vitro* by several kinases such as CKII (Risinger and Bennett 1999; Foster et al. 1998), PKC (Foster et al. 1998; Chung et al. 2000), and PKA (Foster et al. 1998). Kinase activity present in Rab3D-containing immunoprecipitates from mast cells was responsible for the specific phosphorylation of Syntaxin 4 *in vitro* and this activity was downregulated after stimulation in a calcium-dependent manner (Pombo et al. 2001). Phosphorylation of the NH_2 regulatory domain of the t-SNARE was found to reduce the binding of Syntaxin 4 to SNAP-23 (Pombo et al. 2001). PKA phosphorylation of Syntaxin 4 was also found to reduce the binding of SNAP-23, but not that of VAMP2 *in vitro* (Foster et al. 1998). This contrasts with the results obtained with phosphorylation of Syntaxin 1 by CKII which caused an increase in binding to SNAP-25 (Foletti et al. 2000). More surprising is that VAMP2 v-SNARE binding was not affected, although according to the accepted mechanism for SNARE assembly, it should bind only to the Syntaxin 4-SNAP-23 t-SNARE complex. In any event, these examples clearly demonstrate that phosphorylation affects formation of the binary t-SNARE complex and, thus, are indicative of the control of SNARE assembly.

Platelets provide a good system for studying the coupling of signaling pathways to exocytosis. They are highly specialized secretory cells that are activated for secretion upon extracellular stimulation. What separates platelets from neurons, which also secrete upon external stimulation, is their simplicity. Platelets have no nucleus, almost no Golgi, and their intracellular trafficking is dedicated mostly to secretion. The cascade of signaling events that leads to membrane fusion in an injured blood vessel is well known and involves both an increase in intracellular Ca^{2+} and the activation of PKC. Both events synergistically amplify secretory granule exocytosis in activated platelets (Blockmans et al. 1995). The phosphorylation of Syntaxin 4 was examined in platelets after treating cells with thrombin, a natural agonist for platelet activation. Thrombin-induced PKC-dependent phosphorylation of Syntaxin 4 led to a mild decrease in SNAP-23 binding (Chung et al. 2000). However, *in vitro* phosphorylation of Syntaxin 4 by PKC caused a more significant reduction (~70%) in binding to SNAP-23 (Chung et al. 2000). Together, these studies indicate a role for PKC-dependent phosphorylation in the control of binary t-SNARE complex formation, which may play a role in modulating exocytosis.

2.2 Yeast syntaxins

Sso1/2. A better understood example of syntaxin phosphorylation and its regulation of SNARE assembly and exocytosis comes from the yeast *Saccharomyces cerevisiae*. Budding yeast grow in a polarized fashion similar to neurons, however,

exocytosis in yeast does not depend on the influx of calcium ions or on any external stimulus. Nevertheless, the Sso1 and Sso2 exocytic t-SNAREs, which are yeast syntaxin orthologs, are phosphorylated both *in vitro* and *in vivo* by PKA (Marash and Gerst 2001), a kinase that senses external nutrient availability as transduced through changes in the level of intracellular cyclic-AMP (Griffioen and Thevelein 2002). Phosphorylation of Sso1 occurs at position 79 of the Habc domain and reduces the affinity of Sso to Sec9, the yeast SNAP-25 ortholog, to lower SNARE assembly. The dephosphorylation of Sso is known to be mediated by a ceramide-activated protein phosphatase (CAPP) that is involved in cell cycle control in both yeast and mammals (Perry and Hannun 1998). Addition of a ceramide analog to growing cells or the introduction of a serine-to-alanine substitution in the relevant phosphorylation site on Sso increases the binding of Sso1 to Sec9 and rescues cells bearing mutations in their v- or t-SNAREs (Marash and Gerst 2001). Interestingly, more recent work has shown that PKA phosphorylation of the Sso t-SNAREs greatly enhances their association with a SNARE regulator, Vsm1, at the expense of Sec9 (Marash and Gerst 2003). Since the Sso t-SNAREs are distributed evenly over the cell surface (Brennwald et al. 1994) and are not exclusive to the site of exocytosis, PKA phosphorylation may restrict their functioning at the plasma membrane. It is expected then that local changes in CAPP regulation act to dephosphorylate and activate Sso at the site of exocytosis. As the secretory pathway is well-conserved between yeast and mammals, the paradigm of Sso phosphorylation and restriction of function may be highly applicable to syntaxin function in higher eukaryotes.

2.3 *Arabidopsis* syntaxin

AtSyp122. The ortholog of Syntaxin 1 in *Arabidopsis* was found to be phosphorylated after the induction of microbial defense response in suspension-cultured cells (Nuhse et al. 2003). *In vitro* studies revealed that AtSyp122 phosphorylation is Ca^{2+} dependent and both serine-6 and -8 were identified as potential phosphorylation sites (Nuhse et al. 2003). The signaling cascade and identity of the kinase responsible for this phosphorylation event are not known, and though the consequence of AtSyp122 phosphorylation has not been studied, it was proposed that Ca^{2+}-mediated signaling stimulates the exocytosis of defense-related molecules.

2.4 SNAP-25 and SNAP-23

SNAP-25 is a t-SNARE that contains two SNARE motifs (Sutton et al. 1998), which are separated by a linker region containing four palmitoylated cysteines that tether the protein to the plasma membrane (Hess et al. 1992). It interacts with both syntaxin and VAMP to generate the exocytic SNARE complex and is also known to directly interact with synaptotagmin (Chen and Scheller 2001).

SNAP-25. SNAP-25 phosphorylation was demonstrated both *in vivo* and *in vitro* by PKC and PKA (Shimazaki et al. 1996; Risinger and Bennett 1999). PKA-

mediated phosphorylation on threonine-138 of SNAP-25 did not alter the ability of SNAP-25 to enter into ternary SNARE complexes nor did it affect the binding to synaptotagmin when examined *in vitro* (Risinger and Bennett 1999). SNAP-25 was also shown to be phosphorylated on threonine-138 *in vivo* by PKA using for-skolin, a known activator of PKA in mammalian cells. However, this phosphoryla-tion event did not correlate with the modulation of secretion in PC12 cells (Hepp et al. 2002). More recently, PKA was shown to phosphorylate threonine-138 of SNAP-25 in adrenal chromaffin cells, which is required for the readily releasable pool (RRP) of vesicles to be in a primed and releasable state (Nagy et al. 2004). By employing photolysable caged calcium in capacitance and amperometry stud-ies, it was found that PKA phosphorylation affects the slowly releasable pool (SRP) of vesicles, leading to their re-filling and, thus, increasing the size of the RRP pool. Increased SNAP-25 phosphorylation was detected upon PKA activa-tion, although mutations in threonine-138 (both phosphomimetic and non-mimetic) inhibited the sizes of the SRP and the RRP. This may indicate that a cy-cle of phosphorylation and dephosphorylation is required for the linear maturation and release of vesicles. However, no effect upon SNARE assembly was examined in this study.

Extensive evidence for SNAP-25 phosphorylation and its effect upon secretion comes from studies on PKC-mediated phosphorylation. SNAP-25 was found to be phosphorylated on serine-187 in PC12 cells in a phorbol 12-myristate 13-acetate (PMA)-dependent manner, an ester known to activate PKC (Shimazaki et al. 1996; Hepp et al. 2002). This resulted in reduced binding to Syntaxin 1 (Shima-zaki et al. 1996), again suggesting a role for phosphorylation in inhibiting SNARE assembly. SNAP-25 phosphorylation by PKC was shown *in vivo* in hippocampal organotypic cultures (Genoud et al. 1999); however, this phosphorylation event was found not to be responsible for phorbol ester-mediated enhancement of secre-tion in these neurons (Finley et al. 2003). Similar results were obtained in insulin secreting cells, where SNAP-25 was shown to be phosphorylated upon PMA treatment, but again this did not correlate with changes in insulin secretion (Gonelle-Gispert et al. 2002). As phorbol esters are also known to bind to the C1 domains of synaptic proteins, such as Munc13, the enhancement of exocytosis as mediated by phorbol ester treatment may be PKC-independent (Silinsky and Searl 2003). Since phorbol esters may have pleiotropic effects leading to the regulation of both PKC and C1 domain-containing proteins, it is difficult to assess their con-tribution to the phosphorylative control of SNARE assembly.

Interestingly, phosphorylated SNAP-25 (on serine-187) was readily detected in non-differentiated PC12 cells and a small increase in phosphorylation of serine-187 was noticed after long exposure to neuronal growth factor (NGF), which in-duces neuronal differentiation (Kataoka et al. 2000). Yet, there too the authors were unable to show any effect on secretion. However, in a later study that also employed membrane capacitance measurement studies in chromaffin cells, it was shown that the phosphorylation of SNAP-25 on serine-187 affects the sustained phase of secretion by increasing vesicle recruitment after the RRP had undergone exocytosis (Nagy et al. 2002). Overexpression of a SNAP-25 mutant that simu-lates the non-phosphorylated form, as well as the addition of PKC inhibitors, sig-

nificantly reduced the exocytosis of vesicles from the SRP but not vesicles from the RRP (Nagy et al. 2002). Although the molecular mechanism by which phosphorylation of SNAP-25 works is not yet known, it might involve the dissociation of the binary SNARE complex, as shown in PC12 cells (Shimazaki et al. 1996). This may allow for the eventual reassembly of productive ternary SNARE complexes after dephosphorylation (Nagy et al. 2002).

It can be concluded from these studies that phosphorylation on serine-187 is occurring on a small fraction of the cellular SNAP-25 pool and that somehow it affects the subcellular localization of the t-SNARE (Kataoka et al. 2000; Gonelle-Gispert et al. 2002; Nagy et al. 2002). The ability to modulate secretion was observed only when amperometry was employed, which enabled dissection of the kinetics of exocytosis (Nagy et al. 2002). Thus, it is quite possible that in cases where secretion was not affected by SNAP-25 phosphorylation (Kataoka et al. 2000; Gonelle-Gispert et al. 2002), it was also due to the low resolution of the methods in use. Given that serine-187 is located within the COOH-terminal coil that is required for SNARE complex formation (Chapman et al. 1994; Fasshauer et al. 1997), as well as for Synaptotagmin 1 (Gerona et al. 2000) and Snapin binding (Ilardi et al. 1999), it would be interesting to determine whether SNAP-25 phosphorylation affects interactions with these SNARE regulators.

It is also important to mention Snapin, a SNAP-25/23 interacting protein (Ilardi et al. 1999; Buxton et al. 2003) that is phosphorylated by PKA (Chheda et al. 2001) and modulates secretion in cultured neurons (Ilardi et al. 1999) and chromaffin cells (Chheda et al. 2001). Snapin associates with the SNARE complex through direct binding to SNAP-25 and cAMP-dependent phosphorylation of Snapin induces an increase in synaptotagmin association with the SNARE complex (Chheda et al. 2001). Exocytosis of large dense-core vesicles was increased when a phospho-mimetic mutant of Snapin was introduced in chromaffin cells, demonstrating the physiological significance of Snapin phosphorylation and its importance in promoting membrane fusion.

SNAP-23. SNAP-23 is a ubiquitously expressed non-neuronal homolog of SNAP-25 (Ravichandran et al. 1996; Polgar et al. 2003). A possible role for SNAP-23 in Ca^{2+}-triggered exocytosis in platelets was demonstrated when anti-SNAP-23 antibodies were shown to block secretion from α, dense and lysosome granules in permeabilized cells (Polgar et al. 2003). Similar to Syntaxin 4, SNAP-23 is phosphorylated in activated platelets in a PKC-dependent manner (Polgar et al. 2003) and the phosphorylation sites were mapped to residues serine-23/threonine-24 and serine-161, which reside on the first and second SNARE motifs of SNAP-23, respectively. Phospho-mimetic mutations in the first two NH$_2$-terminal sites reduced Syntaxin 4 binding up to 50% *in vitro* (Polgar et al. 2003). This result is similar to that observed for Syntaxin 4 phosphorylation by PKC (Chung et al. 2000). The kinetics of dense granule (but not α-granule) exocytosis after thrombin induction were relatively fast and resembled those of thrombin-induced SNAP-23 phosphorylation, indicating that SNAP-23 phosphorylation may be an early event in activated platelets (Polgar et al. 2003). Yet, it is not clear how a reduction in the Syntaxin 4-SNAP-23 interaction stimulates secretion from platelets. Since both t-SNAREs are phosphorylated apparently by the same PKC-

dependent mechanism, one could speculate that phosphorylation may act to limit the secretion of coagulation factors after the initial release, similar to that shown for SNAP-25 in chromaffin cells (Nagy et al. 2002).

Interestingly, a novel serine/threonine kinase, SNAK was shown to phosphorylate SNAP-23 (Cabaniols et al. 1999). SNAK is expressed in various tissues and was isolated in the yeast two-hybrid screen as a Syntaxin 4-interacting protein (Cabaniols et al. 1999). It is not known which signaling pathways SNAK is involved in and, thus, the physiological significance of this phosphorylation event still needs further investigation. However, SNAK-mediated phosphorylation of SNAP-23 was also found to result in decreased binding to Syntaxin 4 when these proteins were overexpressed in HeLa cells (Cabaniols et al. 1999), indicating that SNAK may have a role in the regulation of exocytosis.

2.5 VAMP/Synaptobrevins

VAMP/Synaptobrevin family members are v-SNAREs that have been shown to participate in all intracellular trafficking and secretion events. VAMP proteins on isolated synaptic vesicles were shown to be phosphorylated by exogenous CKII and PKC, and by endogenous CaMKII (Nielander et al. 1995). However, mutations in the putative CaMKII phosphorylation sites on VAMP2 did not alter Ca^{2+}-mediated secretion in the HIT-T15 β-cell line (Regazzi et al. 1996). As of yet, no precise role for v-SNARE phosphorylation and SNARE assembly has been demonstrated. Interestingly, the yeast exo- and endocytic VAMP ortholog, Snc1 (Protopopov et al. 1993; Gurunathan et al. 2000), has been shown to undergo phosphorylation when expressed as a chimera with green fluorescent protein (GFP) (Galan et al. 2001). The phosphorylation of GFP-Snc1 appeared to decrease in yeast lacking two CKI isoforms, while phosphorylated GFP-Snc1 was suggested to be retained in an intracellular compartment and not sorted to the plasma membrane (Galan et al. 2001). Thus, v-SNARE phosphorylation may play a role in its ability to be recycled along the endosomal sorting pathway. However, no evidence for Snc v-SNARE phosphorylation alone has been put forth, thus, it is unclear whether GFP or Snc1 is the phosphorylated substrate in yeast.

2.6 SM Proteins

While SNAREs are likely to be the fusogens that mediate membrane fusion, several regulatory proteins act in physical and temporal proximity to the events leading to SNARE assembly and fusion (Gerst 2003). The Sec1/Munc18 (SM) protein family plays a central role in membrane fusion through their interaction with syntaxins (Gallwitz and Jahn 2003). Munc18 binds tightly to Syntaxin 1 holding it in a closed conformation that prevents assembly into a SNARE complex (Dulobova et al. 1999; Misura et al. 2000). Moreover, the Munc18-Syntaxin 1 complex is required in order for fusion to occur, as neurotransmitter release is eliminated in neurons from Munc18 knockout mice (Verhage et al. 2000). However, mainte-

nance of a closed conformation is not the only model by which SM proteins might operate. For example, some syntaxin family members do not adopt this conformation (Dulobova et al. 2001) and, yet, interact with SM proteins to promote SNARE complex formation (Carr et al. 1999; Peng and Gallwitz 2002). It is possible that the phosphorylation of SM proteins may take place in different sites and, perhaps, by different kinases. Unfortunately, the only information available on SM protein phosphorylation concerns Munc18, which functions in stimulus-coupled exocytosis in higher eukaryotes. The phosphorylation of Munc18 isoforms was shown in chromaffin cells (Barclay et al. 2003; Craig et al. 2003), synaptosomes and extracts from rat brain (Craig et al. 2003; Fletcher et al. 1999), and human activated platelets (Schraw et al. 2003). In addition, Munc18 was also shown to be a substrate of PKA and CKII *in vitro* (Dubois et al. 2002). In all cases, the phosphorylation of Munc18 was found to inhibit the binding to syntaxin.

PKC was shown to phosphorylate Munc18 and the phosphorylation site was mapped to serine residues 306 and 313 (Fujita et al. 1996). Importantly, serine-313 phosphorylation was confirmed in adrenal chromaffin cells following treatment with PMA (Barclay et al. 2003; Craig et al. 2003), which also induces an increase in exocytosis in a PKC-dependent manner. Thus, there is a correlation between PKC activation, Munc18 phosphorylation, and an increase in exocytosis. Munc18 binding to Syntaxin 1 was dramatically reduced following PKC-mediated phosphorylation (Fujita et al. 1996; Barclay et al. 2003) and, conversely, Munc18 phosphorylation was abrogated when pre-bound to Syntaxin 1 (Fujita et al. 1996). Phospho-mimetic mutations in serine-313, but not in serine-306, induced changes in the kinetics of single vesicle release from chromaffin cells that were similar to the effect obtained with phorbol ester (Barclay et al. 2003). In permeabilized chromaffin cells, an increase in serine-313 phosphorylation was observed in response to both Ca^{2+} and histamine, which are known to evoke exocytosis (Craig et al. 2003). Changes in the phosphorylation state of serine-313 were also observed in synaptosomes from rat brain that were triggered with a variety of stimuli to generate action potentials (Craig et al. 2003). X-ray crystallographic analysis of Munc18 in a complex with Syntaxin 1 shows that serine-313 is directly located in a positively charged region that contacts the Habc/H3 linker helix of Syntaxin 1 (Misura et al. 2000), which is the domain known to bind autophosphorylated CaMKII (Ohyama et al. 2002). Thus, it would be important to test whether CaM-KII also phosphorylates Munc18 on serine-313. Phosphorylation of serine-313 is likely to change the overall positive charge of this binding surface and, thus, reduce the affinity for Syntaxin 1 (Misura et al. 2000). Thus, the phosphorylation of Munc18 may release Syntaxin 1 from an inhibitory interaction in order to promote fusion.

Further evidence for enhanced Munc18 phosphorylation upon exocytosis comes from activated platelets, wherein phosphorylation of the Munc18c isoform was found to increase upon thrombin activation (Schraw et al. 2003), and was accompanied by reduced binding to both Syntaxin 4 and Syntaxin 2. These results are in accordance with the data obtained for catecholamine secretion from chromaffin cells (Barclay et al. 2003), indicating that regulation by PKC-mediated phosphorylation may be shared by all Munc18 isoforms involved in stimulated

exocytosis. Identification of the phosphorylation site on Munc18c that is responsible for altering the affinity for Syntaxins 4 and 2 should contribute to the understanding of this mechanism.

Interestingly, Munc18 co-purifies with (Shetty et al. 1995), and is phosphorylated by, cyclin-dependent kinase 5 (Cdk5) (Fletcher et al. 1999). Though its name implies involvement in cell cycle regulation, Cdk5 is found primarily in post-mitotic neuronal precursors and mature neurons (Smith and Tsai 2002). Cdk5 activity is involved in the regulation of the actin- and microtubule-based cytoskeleton, synaptic function, dopamine signaling, and brain development (Smith and Tsai 2002). Similar to the effect exerted by PKC-mediated phosphorylation of Munc18, it was demonstrated that when Cdk5 is bound to p35 (a neuronal specific activator) it phosphorylates Munc18a in a fashion that decreases the affinity for Syntaxin 1a (Shuang et al. 1998; Fletcher et al. 1999). Identified first in the yeast Sec1 ortholog (Aalto et al. 1991), Munc18a has two putative Cdk5 phosphorylation sites of which only threonine-574 was confirmed using site-directed mutagenesis (Fletcher et al. 1999). Importantly, mutation of threonine-574 to alanine blocked Cdk5-mediated dissociation of the Munc18-Syntaxin 1a heterodimer *in vitro* (Fletcher et al. 1999). An analogous mutation made in yeast Sec1 had no apparent effect, based upon its ability to complement temperature-sensitive mutations in *SEC1* (Brummer et al. 2001). However, phosphoproteomic analysis suggests that Sec1 is phosphorylated in yeast cells, but at a different location (Ficarro et al. 2002).

Interestingly, Cdk5 activation enhances exocytosis from chromaffin cells (Fletcher et al. 1999) and has been shown to phosphorylate neuronal transport proteins, such as amphiphysin (Floyd et al. 2001) and synapsin 1 (Matsubara et al. 1996). Moreover, a yeast Cdk5 ortholog, Pho85, phosphorylates Rvs167, a yeast amphiphysin ortholog that confers endocytosis (Huang et al. 1999). As mutations in Pho85 show severe abnormalities in the actin cytoskeleton (Tennyson et al. 1998), it may indicate that this conserved kinase mediates cross-talk between membrane trafficking events and the cytoskeleton. Since Cdk5 and Munc18 were recently shown to co-localize on neurofilaments (Bhaskar et al. 2004), it supports the idea that Munc18 may be involved in cytoskeleton dynamics.

2.7 Synaptotagmin

Synaptotagmins constitute a large family of proteins having an NH_2-terminal trans-membrane domain and two C2 Ca^{2+}-binding domains at the COOH-terminus, termed C2A and C2B. The vertebrate synaptotagmins constitute a family of 13 members with additional splice variants (Sudhof 2002). According to the current view, synaptotagmin is the Ca^{2+} sensor that helps trigger synaptic vesicle fusion upon the rapid increase in Ca^{2+} influx during membrane depolarization (Jahn et al. 2003). Synaptotagmin binds to both syntaxin and SNAP-25 individually, to the heterodimeric t-SNARE complex, as well as to the fully assembled v-t SNARE complex (Chapman 2002). Binding occurs near the membrane anchor of the t-SNAREs and, thus, places synaptotagmin in a position to regulate SNARE

assembly (Chapman 2002). Synaptotagmin 1, which resides on synaptic vesicles, was shown to be phosphorylated *in vitro* by several kinases such as CaMKII, CKII, and PKC (Popoli et al. 1993; Bennett at el. 1993, Davletov et al. 1993; Hilfiker et al. 1999). Endogenous Synaptotagmin I was shown to be phosphorylated on synaptosomes from rat brain and in intact PC12 cells in response to either K^+-evoked depolarization or phorbol-ester treatment (Hilfiker et al. 1999). Two CKII phosphorylation sites at threonines 125 and 128, and one PKC phosphorylation site at threonine-112 were identified (Davletov et al. 1993; Hilfiker et al. 1999). All three sites are positioned within the linker region that connects the two C2 domains and the trans-membrane domain, which is conserved within Synaptotagmin I orthologs (Nakhost et al. 2003), but is not required for the Ca^{2+}-dependent binding of syntaxin (Kee and Scheller 1996). As the linker region is not well characterized it is hard to determine the full significance of these phosphorylation sites in SNARE assembly. However, the PKC phosphorylation site is also conserved in evolution and was mutated in the *Aplysia* ortholog of Synaptotagmin I (Nakhost et al. 2003). Mutation of this site had no effect upon serotonin-induced synaptic recovery, which is thought to be mediated by a PKC substrate (Nakhost et al. 2003). The phosphorylation of Synaptotagmin I in an isolated synaptosomal fraction was shown to occur under conditions that activate CaMKII (Verona et al. 2000). Therein, phosphorylation was found to result in a mild increase in binding of Syntaxin 1 and SNAP-25. That said, no conclusive role for the phosphorylation of Synaptotagmin in SNARE assembly has been demonstrated.

3 Protein phosphorylation in the endocytic pathway

3.1 Yeast Tlg1,2 and Vps45

In yeast, the Snc v-SNAREs mediate both exo- and endocytosis (Protopopov et al. 1993; Gurunathan et al. 2000), and it was shown that CAPP activation also confers normal endocytic functioning in *snc* null cells (Gurunathan et al. 2002). In this case, one set of targets for dephosphorylation is the Tlg1 and Tlg2 endocytic t-SNAREs (Holthuis et al. 1998). Both Tlg1 and Tlg2 are phosphorylated by PKA *in vitro* and point mutations in putative PKA phosphorylation sites restored endocytosis in the absence of CAPP activation (Gurunathan et al. 2002). The phosphorylation sites identified in both the Sso and Tlg t-SNAREs reside within the N-terminal autoinhibitory domain. Thus, exocytosis and endocytosis in yeast are likely to be under the control of the same signaling pathways.

The endocytic SNARE complex, comprising of Tlg1, Tlg2, Vti1, and Snc2, and its interaction with the SM protein, Vps45, appears to be regulated by another protein phosphatase - Glc7/PP1 (Bryant and James 2003). This SNARE complex accumulates in temperature-sensitive *glc7-10* cells at non-permissive temperatures and is accompanied by the release of Vps45 from the complex (Bryant and James 2003). This suggests that dephosphorylation of an unknown component must take

place in order for membrane fusion to proceed, although based upon the Munc18 phosphorylation studies Vps45 may be an attractive candidate.

Vacuoles are contiguous with the endocytic pathway in yeast and are analogous to mammalian lysosomes. The homotypic fusion of vacuoles is Ca^{2+}-calmodulin dependent and it was shown that the final stages of this fusion depend upon Glc7/PP1 complexed with Ca^{2+}-calmodulin (Peters et al. 1999). Inhibition of PP1 using either microcystin, a specific inhibitor, or a temperature-sensitive allele of GLC7 was found to block membrane fusion after docking (Peters et al. 1999). However, the target of PP1 dephosphorylation on the vacuole membrane is still unknown. Taken together, these two studies indicate that PP1-dependent dephosphorylation events are part of a regulatory mechanism which confers fusion independent of temporal signaling cascades. Thus, it is likely that a cycle of phosphorylation and dephosphorylation is prerequisite for intracellular membrane trafficking events.

3.2 VAMP4

VAMP4 is a v-SNARE implicated in endosome to trans-Golgi network vesicle trafficking (Mallard et al. 2002) and was shown to be phosphorylated in vitro by CKII (Hinners et al. 2003). Phosphorylation of VAMP4 was mapped to serine-30 and a mutation in this site (along with removal of a dileucine motif) resulted in the mislocalization of VAMP4. This study showed that VAMP4 phosphorylation is required, in part, for interaction with the AP-1 and PACS-1 adaptor proteins, which are needed for the correct endosomal sorting of VAMP4 (Hinners et al. 2003). Similar to the case of Syntaxin 1 phosphorylation by CKII in neurons (Foletti et al. 2000), the phosphorylation of SNAREs by CKII may influence membrane trafficking by regulating their subcellular localization.

3.3 Rabs 4 and 5

Rab4 is localized to the early endosomes in mammalian cells and functions in the recycling of receptors to the plasma membrane (van der Sluijs et al. 1992a). Rab4 was shown to be a substrate for the Cdc2 mitotic kinase in vitro and phosphorylation correlates with its dissociation from membrane during mitosis. Mutation of a putative Cdc2 phosphorylation site (serine-196) resulted in a protein that remains bound to the membrane during mitosis and does not translocate to the cytosol (van der Sluijs et al. 1992b). However, this mutation had no effect upon Rab4 prenylation. A chimeric Rab4 that is permanently anchored to endosomal membranes, and inhibits transcytosis, was found to be specifically phosphorylated during mitosis (Mohrmann et al. 2002). Phosphorylation of Rab4 by Cdc2 inhibits binding to a guanyl-nucleotide dissociation inhibitor (GDI) (Ayad et al. 1997) and, since GDI proteins extract Rabs from membranes after vesicle docking (Pfeffer 2001), it is likely that the phosphorylation perturbs the Rab cycle in order to prevent early endosomal transport during mitosis.

It is now widely acknowledged that receptor tyrosine kinase (RTK) signaling is regulated by receptor endocytosis and that Rab5 seems to participate in these processes (Clague and Urbe 2001). Interestingly, the phosphorylation of GDI by p38 MAPK led to the sequestration of Rab5 from endosomal membranes and is required for membrane transport on the endocytic pathway (Cavalli et al. 2001). As of yet, though, the phosphorylation of Rab5 has not been demonstrated *in vivo* and it is not known whether Rab phosphorylation affects SNARE assembly. However, signaling cascades that control Rab protein localization may, in turn, modulate SNARE localization as already shown for RTK recycling from endosomal compartments.

4 Protein phosphorylation in the early secretory pathway

Golgi membranes are the main sorting apparatus for the secretory pathway in eukaryotic cells. Newly synthesized proteins destined for export travel from the endoplasmic reticulum (ER) to the Golgi were they undergo post-translational modification, sorting, and transport. Transport in the early secretory pathway is bidirectional and comprises not only the forward transport of secreted cargo but also the retrograde transport of chaperones and cargo receptors back to the ER (Lippincott-Schwartz et al. 2000). Almost two decades ago, it was shown that transport through the Golgi is regulated in a cell cycle-dependent manner (Featherstone et al. 1985). Mitotic kinases phosphorylate structural proteins, such as GM130 and GRASP65, required for the maintenance of Golgi structure and function during the process of mitotic Golgi fragmentation in mammalian cells (Lowe et al. 2000; Lin et al. 2000; Sutterlin et al. 2001). However, up until now there is no report of Golgi SNAREs or SNARE regulators as being targets for mitotic kinases. As extensive membrane fusion events take place during mitotic fragmentation and reassembly of the Golgi apparatus, it is likely that mitotic kinases directly target SNAREs or SNARE-regulators.

The p115 tethering factor that mediates the tethering of coat complex I (COPI) vesicles to Golgi membranes is phosphorylated by CKII (Sohda et al. 1998; Dirac-Svejstrup et al. 2000). Phosphorylation of p115 at its COOH-terminus results in its release to the cytosol (Sohda et al. 1998) and is required for the reassembly of mitotic fragments *in vitro* (Dirac-Svejstrup et al. 2000). Thus, the role of CKII in the homotypic fusion of mitotic fragments may be to enhance the interaction between p115 and its receptor in the Golgi, GM130, upon p115 phosphorylation (Dirac-Svejstrup et al. 2000). However, it is possible that other targets of CKII exist on the Golgi membrane. CKII is known to phosphorylate Syntaxin 1 and Syntaxin 4, as described above, although the only syntaxin required for the transport of proteins through the Golgi apparatus is Syntaxin 5 (Nichols and Pelham 1998). The unique status of Syntaxin 5 makes it an attractive candidate for kinase regulation, perhaps, by CKII.

Not only is phosphorylation by mitotic kinases connected with Golgi transport, a study on cellular volume changes revealed that dramatic alterations occur in both

anterograde and retrograde transport between the ER and the Golgi upon osmotic stress (Lee and Linstedt 1999). Osmotic stress led to a block in ER export, while retrograde transport from the Golgi to the ER was stimulated, both leading to a collapse of the ER-Golgi-intermediate compartment into the ER (Lee and Linstedt 1999). An inhibitor of serine/threonine kinases, H89, was found to inhibit hypotonically-induced, but not constitutive, Golgi to ER retrograde transport (Lee and Linstedt 2000). In addition, H89 inhibited the recruitment of Sec13, a component of the COPII coat complex, to ER exit sites (Lee and Linstedt 2000). Sec13 recruitment is preceded by the attachment of Sar1 GTPase to ER exit sites (Barlowe 2002), a step which is also inhibited by H89 and leads to the abrogation of cargo export from ER microsomes (Aridor and Balch 2000). H89 has been used traditionally as a specific inhibitor of PKA activity, which, therefore, suggests a role for this kinase in COPII recruitment, a process that is also SNARE-catalyzed (Miller et al. 2002). However, H89 has also been shown to act as an inhibitor for vesicle formation at the trans-Golgi network (Muniz et al. 1997), as well as for PKD-mediated vesiculation of the Golgi (Jamora et al. 1999). PKD is a member of PKC protein family (Rykx et al. 2003) and was shown to regulate the fission of membranes at the trans-Golgi network (Liljedahl et al. 2001). Thus, the fact that H89 inhibits PKD-mediated vesiculation of the Golgi suggests that PKA may not necessarily be the sole mediator of stress responses at the level of the ER-Golgi (Lee and Linstedt 2000)

Intra-Golgi and Golgi-to-ER transport are integral to the early secretory pathway, but whether these transport routes are also under the direct regulation of protein kinases and phosphatases is still an open question. Despite the dearth of information, it is known that PP1 phosphatase activity is required for trafficking between the ER and the Golgi (Peters et al. 1999). This implies that protein phosphorylation is likely to play an integral role in protein transport early in the secretory pathway.

5 Conclusions

The secretory pathway is one of the central processes that allow eukaryotic cells to translate external and internal signals into cellular responses. These responses include, for example, polarized cell growth, the secretion of specific small molecules and proteins, the internalization of RTKs, fragmentation and reassembly of the Golgi during mitosis, and so on. Importantly, all these processes involve the SNARE-catalyzed fusion of vesicles and target membranes to lead to the specific trafficking events. Protein phosphorylation plays a pivotal role in the transduction of signals leading to control over the membrane fusion machinery and exocytosis. As presented in this review, a growing body of evidence shows the direct phosphorylation of SNAREs and SNARE-regulators as a potential means of regulating protein trafficking and secretion. Though it is premature to draw up a unifying model, it appears that in most cases of SNARE or SNARE regulator (i.e. SM protein) phosphorylation the result is a decrease in SNARE-SNARE or SNARE-

SNARE regulator interactions (Table I and Figure 1). Most prominently, the outcome of t-SNARE phosphorylation, either of the syntaxins or the SNAP-25 homologs, is a reduction in t-SNARE complex formation that may lead to a diminution in exocytosis. This effect by phosphorylation is well conserved and has been described in both yeast and mammals (Table I). Furthermore, dephosphorylation enhances SNARE interactions and leads to membrane fusion in yeast (Peters et al. 1999; Marash and Gerst 2001; Bryant and James 2003), and which may also be true for mammals.

On one hand, the phosphorylation of t-SNAREs results in a decrease in t-SNARE–t-SNARE interactions. However, this is somewhat paradoxical since many of the kinases involved therein are known to stimulate secretion, such as PKA and PKC in neurons, chromaffin cells, and platelets (Evans and Morgan 2003; Takahashi et al. 2003). It may be then, as shown for stimulus-coupled exocytosis from platelets and chromaffin cells, that phosphorylative control of the SNAREs is not necessarily to regulate the RRP, but rather, exocytosis of the SRP or the sustained component of secretion. On the other hand, phosphorylation may also dissociate SNARE regulators, such as Munc18, which may promote both rapid and slow exocytosis at least in the mammalian system. Thus, phosphorylation is likely to occur on many proteins of the exocytic apparatus simultaneously, although the combinatorial effect is to promote, if not fine-tune, secretion. In addition, there are likely to be temporal and spatial constraints to allow for phosphorylative control of SNARE (and SNARE regulator) localization, recycling, and assembly in order to coordinate all the different events leading to secretion. Although a direct connection between signal transduction pathways and membrane trafficking has been established, we see only the tip of the iceberg in regards to understanding its regulation of the secretory pathway.

Acknowledgments

This work was supported by grants to J.E.G from the Israel Science Foundation (374/02-1) and the M.D. Moross Institute for Cancer Research. J.E.G holds the Henry Kaplan Chair in Cancer Research.

References

Aalto MK, Ruohonen L, Hosono K, Keranen S (1991) Cloning and sequencing of the yeast *Saccharomyces cerevisiae SEC1* gene localized on chromosome IV. Yeast 7:643-650

Aridor M, Balch WE (2000) Kinase signaling initiates coat complex II (COPII) recruitment and export from the mammalian endoplasmic reticulum. J Biol Chem 275:35673-35676

Ayad N, Hull M, Mellman I (1997) Mitotic phosphorylation of rab4 prevents binding to a specific receptor on endosome membranes. EMBO J 16:4497-4507

Barclay JW, Craig TJ, Fisher RJ, Ciufo LF, Evans GJ, Morgan A, Burgoyne RD (2003) Phosphorylation of Munc18 by protein kinase C regulates the kinetics of exocytosis. J Biol Chem 278:10538-10545

Barlowe C (2002) COPII-dependent transport from the endoplasmic reticulum. Curr Opin Cell Biol 14:417-422

Bennett MK, Calakos N, Scheller RH (1992) Syntaxin: a synaptic protein implicated in docking of synaptic vesicles at presynaptic active zones. Science 257:255-259

Bennett MK, Miller KG, Scheller RH (1993) Casein kinase II phosphorylates the synaptic vesicle protein p65. J Neurosci 13:1701-1707

Bhaskar K, Shareef MM, Sharma VM, Shetty AP, Ramamohan Y, Pant HC, Raju TR, Shetty KT (2004) Co-purification and localization of Munc18-1 (p67) and Cdk5 with neuronal cytoskeletal proteins. Neurochem Int 44:35-44

Blockmans D, Deckmyn H, Vermylen J (1995) Platelet activation. Blood Rev 9:143-156

Brennwald P, Kearns B, Champion K, Keranen S, Bankaitis V, Novick P (1994) Sec9 is a SNAP-25-like component of a yeast SNARE complex that may be the effector of Sec4 function in exocytosis. Cell 79:245-258

Brummer MH, Kivinen KJ, Jantti J, Toikkanen J, Soderlund H, Keranen S (2001) Characterization of the sec1-1 and sec1-11 mutations. Yeast 18:1525-1536

Bryant NJ, James DE (2003) The Sec1p/Munc18 (SM) protein, Vps45p, cycles on and off membranes during vesicle transport. J Cell Biol 161:691-696

Buxton P, Zhang XM, Walsh B, Sriratana A, Schenberg I, Manickam E, Rowe T (2003) Identification and characterization of Snapin as a ubiquitously expressed SNARE-binding protein that interacts with SNAP23 in non-neuronal cells. Biochem J 375:433-440

Cabaniols JP, Ravichandran V, Roche PA (1999) Phosphorylation of SNAP-23 by the novel kinase SNAK regulates t-SNARE complex assembly. Mol Biol Cell 10:4033-4041

Clague MJ, Urbe S (2001) The interface of receptor trafficking and signaling. J Cell Sci 114:3075-3081

Carr CM, Grote E, Munson M, Hughson FM, Novick PJ (1999) Sec1p binds to SNARE complexes and concentrates at sites of secretion. J Cell Biol 146:333-344

Cavalli V, Vilbois F, Corti M, Marcote MJ, Tamura K, Karin M, Arkinstall S, Gruenberg J (2001) The stress-induced MAP kinase p38 regulates endocytic trafficking via the GDI:Rab5 complex. Mol Cell 7:421-432

Chapman ER (2002) Synaptotagmin: A Ca^{2+} sensor that triggers exocytosis? Nat Rev Mol Cell Biol 3:498-508

Chapman ER, An S, Barton N, Jahn R (1994) SNAP-25, a t-SNARE which binds to both syntaxin and synaptobrevin via domains that may form coiled coils. J Biol Chem 269:27427-24732

Chen A, Duvvuri V, Schulmani H, Scheller R (1999) Calmodulin and protein kinase C increase Ca^{2+}-stimulated secretion by modulating membrane-attached exocytic machinery. J Biol Chem 274:26469–26476

Chen YA, Scheller RH (2001) SNARE-mediated membrane fusion. Nat Rev Mol Cell Biol 2:98-106

Chheda MG, Ashery U, Thakur P, Rettig J, Sheng ZH (2001) Phosphorylation of Snapin by PKA modulates its interaction with the SNARE complex. Nat Cell Biol 3:331-338

Chung SH, Polgar J, Reed GL (2000) Protein kinase C phosphorylation of syntaxin 4 in thrombin-activated human platelets. J Biol Chem 275:25286-25291

Cohen O, Kimchi A (2001) DAP-kinase: from functional gene cloning to establishment of its role in apoptosis and cancer. Cell Death Differ 8:6-15

Cohen O, Feinsein E, Kimchi A (1997) DAP-kinase is a Ca^{2+}/calmodulin-dependent, cytoskeletal-associated protein kinase, with cell death-inducing functions that depend on its catalytic activity. EMBO J 16:998-1008

Craig TJ, Evans GJ, Morgan A (2003) Physiological regulation of Munc18/nSec1 phosphorylation on serine-313. J Neurochem 86:1450-1457

Davletov B, Sontag JM, Hata Y, Petrenko AG, Fykse EM, Jahn R, Sudhof TC (1993) Phosphorylation of synaptotagmin I by casein kinase II. J Biol Chem 268:6816-6822

Deiss L, Feinsein E, Berissi H, Cohen O, Kimchi A (1995) Identification of a novel serine/threonine kinase and a novel 15-kD protein as potential mediators of the gamma interferon-induced cell death. Genes Dev 8:15-30

Dirac-Svejstrup AB, Shorter J, Waters MG, Warren G (2000) Phosphorylation of the vesicle-tethering protein p115 by a casein kinase II-like enzyme is required for Golgi reassembly from isolated mitotic fragments. J Cell Biol 150:475-488

Dubois T, Kerai P, Learmonth M, Cronshaw A, Aitken A (2002) Identification of syntaxin-1A sites of phosphorylation by casein kinase I and casein kinase II. Eur J Biochem 269:909-914

Dulubova I, Sugita S, Hill S, Hosaka M, Fernandez I, Sudhof TC, Rizo J (1999) A conformational switch in syntaxin during exocytosis: role of munc18. EMBO J 18:4372-4382

Dulubova I, Yamaguchi T, Gao Y, Min SW, Huryeva I, Sudhof TC, Rizo J (2002) How Tlg2p/syntaxin 16 'snares' Vps45. EMBO J 21:3620-3631

Dulubova I, Yamaguchi T, Wang Y, Sudhof TC, Rizo J (2001) Vam3p structure reveals conserved and divergent properties of syntaxins. Nat Struct Biol 8:258-264

Evens GJ, Morgan A (2003) Regulation of the exocytotic machinery by cAMP-dependent protein kinase: implications for presynaptic plasticity. Biochem Soc Trans 31:824-827

Fasshauer D, Bruns D, Shen B, Jahn R, Brunger AT (1997) A structural change occurs upon binding of syntaxin to SNAP-25. J Biol Chem 272:4582-4590

Fasshauer D, Sutton RB, Brunger AT Jahn R (1998) Conserved structural features of the synaptic fusion complex: SNARE proteins reclassified as Q- and R-SNAREs. Proc Natl Acad Sci USA 95:15781–15786

Featherstone C, Griffiths G, Warren G (1985) Newly synthesized G protein of vesicular stomatitis virus is not transported to the Golgi complex in mitotic cells. J Cell Biol 101:2036-4206

Ficarro SB, McCleland ML, Stukenberg PT, Burke DJ, Ross MM, Shabanowitz J, Hunt DF, White FM (2002) Phosphoproteome analysis by mass spectrometry and its application to *Saccharomyces cerevisiae*. Nat Biotechnol 20:301-305

Finley MF, Scheller RH, Madison DV (2003) SNAP-25 Ser187 does not mediate phorbol ester enhancement of hippocampal synaptic transmission. Neuropharmacology 45:857-862

Fletcher AI, Shuang R, Giovannucci DR, Zhang L, Bittner MA, Stuenkel EL (1999) Regulation of exocytosis by cyclin-dependent kinase 5 via phosphorylation of Munc18. J Biol Chem 274:4027-4035

Floyd SR, Porro EB, Slepnev VI, Ochoa GC, Tsai LH, De Camilli P (2001) Amphiphysin 1 binds the cyclin-dependent kinase (cdk) 5 regulatory subunit p35 and is phosphorylated by cdk5 and cdc2. J Biol Chem 276:8104-8110

Foletti DL, Lin R, Finley MA, Scheller RH (2000) Phosphorylated syntaxin 1 is localized to discrete domains along a subset of axons. J Neurosci 20:4535-4544

Foster LJ, Yeung B, Mohtashami M, Ross K, Trimble WS, Klip A (1998) Binary interaction of SNARE proteins Syntaxin-4 SNAP23, and VAMP-2 and their regulation by phosphorylation. Biochem 37:11089-11096

Fujita Y, Sasaki T, Fukui K, Kotani H, Kimura T, Hata Y, Sudhof TC, Scheller RH, Takai Y (1996) Phosphorylation of Munc-18/n-Sec1/rbSec1 by protein kinase C: its implication in regulating the interaction of Munc-18/n-Sec1/rbSec1 with syntaxin. J Biol Chem 271:7265-7268

Galan JM, Wiederkehr A, Seol JH, Haguenauer-Tsapis R, Deshaies RJ, Riezman H, Peter M (2001) Skp1p and the F-box protein Rcy1p form a non-SCF complex involved in recycling of the SNARE Snc1p in yeast. Mol Cell Biol 21:3105-3117

Gallwitz D, Jahn R (2003) The riddle of the Sec1/Munc-18 proteins - new twists added to their interactions with SNAREs. Trends Biochem Sci 28:113-116

Genoud S, Pralong W, Riederer BM, Eder L, Catsicas S, Muller D (1999) Activity-dependent phosphorylation of SNAP-25 in hippocampal organotypic cultures. J Neurochem 72:1699-1706

Gerona RR, Larsen EC, Kowalchyk JA, Martin TF (2000) The C terminus of SNAP25 is essential for Ca^{2+}-dependent binding of synaptotagmin to SNARE complexes. J Biol Chem 275:6328-6336

Gerst JE (2003) SNARE regulators: matchmakers and matchbreakers. Biochim Biophys Acta 1641:99-110

Gerst JE, Rodgers L, Riggs M, Wigler M (1992) *SNC1*, a yeast homolog of the synaptic vesicle-associated membrane protein/synaptobrevin gene family: genetic interactions with the *RAS* and *CAP* genes. Proc Natl Acad Sci USA 89:4338-4342

Gillis KD, Mossner R, Neher E (1996) Protein kinase C enhances exocytosis from chromaffin cells by increasing the size of the readily releasable pool of secretory granules. Neuron 16:1209-1220

Gonelle-Gispert C, Costa M, Takahashi M, Sadoul K, Halban P (2002) Phosphorylation of SNAP-25 on serine-187 is induced by secretagogues in insulin-secreting cells, but is not correlated with insulin secretion. Biochem J 368:223-232

Griffioen G, Thevelein JM (2002) Molecular mechanisms controlling the localization of protein kinase A. Curr Genet 41:199–207

Gurunathan S, Chapman-Shimshoni D, Trajkovic S, Gerst JE (2000) Yeast exocytic v-SNAREs confer endocytosis. Mol Biol Cell 11:3629-3643

Gurunathan S, Marash M, Weinberger A, Gerst JE (2002) t-SNARE phosphorylation regulates endocytosis in yeast. Mol Biol Cell 13:1594-1607

Hepp R, Cabaniols JP, Roche PA (2002) Differential phosphorylation of SNAP-25 *in vivo* by protein kinase C and protein kinase A. FEBS Lett 532:52-56

Hess DT, Slater TM, Wilson MC, Skene JH (1992) The 25 kDa synaptosomal-associated protein SNAP-25 is the major methionine-rich polypeptide in rapid axonal transport and a major substrate for palmitoylation in adult CNS. J Neurosci 12:4634-4641

Hilfiker S, Pieribone VA, Nordstedt C, Greengard P, Czernik AJ (1999) Regulation of synaptotagmin I phosphorylation by multiple protein kinases. J Neurochem 73:921-932

Hinners I, Wendler F, Fei H, Thomas L, Thomas G, Tooze SA (2003) AP-1 recruitment to VAMP4 is modulated by phosphorylation-dependent binding of PACS-1. EMBO Rep 4:1182-1189

Hirling H, Scheller RH (1996) Phosphorylation of synaptic vesicle proteins: modulation of the alpha SNAP interaction with the core complex. Proc Natl Acad Sci USA 93:11945-11949

Holthuis JC, Nichols BJ, Dhruvakumar S, Pelham HR (1998) Two syntaxin homologues in the TGN/endosomal system of yeast. EMBO J 17:113-126

Hu C, Ahmed M, Melia TJ, Sollner TH, Mayer T, Rothman JE (2003) Fusion of cells by flipped SNAREs. Science 300:1745-1749

Huang D, Patrick G, Moffat J, Tsai LH, Andrews B (1999) Mammalian Cdk5 is a functional homologue of the budding yeast Pho85 cyclin-dependent protein kinase. Proc Natl Acad Sci USA 96:14445-14450

Ilardi JM, Mochida S, Sheng ZH (1999) Snapin: a SNARE-associated protein implicated in synaptic transmission. Nat Neurosci 2:119-124

Jahn R, Lang T, Sudhof TC (2003) Membrane fusion. Cell 112:5519-5533

Jamora C, Yamanouye N, Van Lint J, Laudenslager J, Vandenheede JR, Faulkner DJ, Malhotra V (1999) Gbetagamma-mediated regulation of Golgi organization is through the direct activation of protein kinase D. Cell 98:59-68

Kataoka M, Kuwahara R, Iwasaki S, Shoji-Kasai Y, Takahashi M (2000) Nerve growth factor-induced phosphorylation of SNAP-25 in PC12 cells: a possible involvement in the regulation of SNAP-25 localization. J Neurochem 74:2058-2066

Kee Y, Scheller R (1996) Localization of synaptotagmin-binding domains on syntaxin. J Neurosci 76:1975-1981

Korolchuk VI, Banting G (2002) CK2 and GAK/auxilin2 are major protein kinases in clathrin-coated vesicles. Traffic 3:428-439

Lee TH, Linstedt AD (1999) Polo-like kinase is required for the fragmentation of pericentriolar Golgi stacks during mitosis. Proc Natl Acad Sci USA 98:9128-9132

Lee TH, Linstedt AD (2000) Potential role for protein kinases in regulation of bidirectional endoplasmic reticulum-to-Golgi transport revealed by protein kinase inhibitor H89. Mol Biol Cell 11:2577-2590

Liljedahl M, Maeda Y, Colanzi A, Ayala I, Van Lint J, Malhotra V (2001) Protein kinase D regulates the fission of cell surface destined transport carriers from the trans-Golgi network. Cell 104:409-420

Lin CY, Madsen ML, Yarm FR, Jang YJ, Liu X, Erikson RL (2000) Peripheral Golgi protein GRASP65 is a target of mitotic polo-like kinase (Plk) and Cdc2. Proc Natl Acad Sci USA 97:12589-12594

Lippincott-Schwartz J, Roberts TH, Hirschberg K (2000) Secretory protein trafficking and organelle dynamics in living cells. Annu Rev Cell Dev Biol 16:557-589

Lowe M, Gonatas NK, Warren G (2000) The mitotic phosphorylation cycle of the cis-Golgi matrix protein GM130. J Cell Biol 149:341-356

Mallard F, Tang BL, Galli T, Tenza D, Saint-Pol A, Yue X, Antony C, Hong W, Goud B, Johannes L (2002) Early/recycling endosomes-to-TGN transport involves two SNARE complexes and a Rab6 isoform. J Cell Biol 156:653-664

Marash M, Gerst JE (2001) t-SNARE dephosphorylation promotes SNARE assembly and exocytosis in yeast. EMBO J 20:411-421

Marash M, Gerst JE (2003) Phosphorylation of the autoinhibitory domain of the Sso t-SNAREs promotes binding of the Vsm1 SNARE regulator in yeast. Mol Biol Cell 14:3114-3125

Martin F, Salinas E, Barahona F, Vazquez J, Soria B, Reig JA (1998) Engineered peptides corresponding to segments of the H3 domain of syntaxin inhibit insulin release both in intact and permeabilized mouse pancreatic beta cells. Biochem Biophys Res Commun 248:83-86

Matsubara M, Kusubata M, Ishiguro K, Uchida T, Titani K, Taniguchi H (1996) Site-specific phosphorylation of synapsin I by mitogen-activated protein kinase and Cdk5 and its effects on physiological functions. J Biol Chem 271:21108-21113

Mayer A (2001) Membrane fusion in eukaryotic cells. Annu Rev Cell Dev Biol 18:289-314

Mayer A, Wickner W, Haas A (1996) Sec18p (NSF)-driven release of Sec17p (alpha-SNAP) can precede docking and fusion of yeast vacuoles. Cell 85:83-94

Meggio F, Pinna LA (2003) One-thousand-and-one substrates of protein kinase CK2? FASEB J 17:349-368

Miller E, Antonny B, Hamamoto S, Schekman R (2002) Cargo selection into COPII vesicles is driven by the Sec24p subunit. EMBO J 21:6105-6113

Misura KM, Scheller RH, Weis WI (2000) Three-dimensional structure of the neuronal-Sec1-syntaxin 1a complex. Nature 404:355-362

Mohrmann K, Gerez L, Oorschot V, Klumperman J, van der Sluijs P (2002) Rab4 function in membrane recycling from early endosomes depends on a membrane to cytoplasm cycle. J Biol Chem 277:32029-32035

Motojiro Y, Adolfsen B Littleton T (2003) Is synaptotagmin the calcium sensor? Curr Opin Neuro 13:315–323

Muniz M, Martin ME, Hidalgo J, Velasco A (1997) Protein kinase A activity is required for the budding of constitutive transport vesicles from the trans-Golgi network. Proc Natl Acad Sci USA 94:14461-14466

Munson M, Chen X, Cocina AE, Schultz SM, Hughson FM (2000) Interactions within the yeast t-SNARE Sso1p that control SNARE complex assembly. Nat Struct Biol 7:894–902

Nagy G, Matti U, Nehring RB, Binz T, Rettig J, Neher E, Sorensen JB (2002) Protein kinase C-dependent phosphorylation of synaptosome-associated protein of 25 kDa at Ser187 potentiates vesicle recruitment. J Neurosci 22:9278-9286

Nagy G, Reim K, Matti U, Brose N, Binz T, Rettig J, Neher E, Sorensen JB (2004) Regulation of releasable vesicle pool sizes by protein kinase A-dependent phosphorylation of SNAP-25. Neuron 41:417-429

Nakhost A, Houeland G, Castellucci VF, Sossin WS (2003) Differential regulation of transmitter release by alternatively spliced forms of synaptotagmin I. J Neurosci 23:6238-6244

Nichols BJ, Pelham HR (1998) SNAREs and membrane fusion in the Golgi apparatus. Biochim Biophys Acta 1404:9-31

Nielander HB, Onofri F, Valtorta F, Schiavo G, Montecucco C, Greengard P, Benfenati F (1995) Phosphorylation of VAMP/synaptobrevin in synaptic vesicles by endogenous protein kinases. J Neurochem 65:1712-1720

Nuhse TS, Boller T, Peck S (2003) A plasma membrane syntaxin is phosphorylated in response to the bacterial elicitor flagellin. J Biol Chem 278:45248-45254

Ohyama A, Hosaka K, Komiya Y, Akagawa K, Yamauchi E, Taniguchi H, Sasagawa N, Kumakura K, Mochida S, Yamauchi T, Igarashi M (2002) Regulation of exocytosis through Ca^{2+}/ATP-dependent binding of autophosphorylated Ca^{2+}/calmodulin-activated protein kinase II to syntaxin 1A. J Neurosci 22:3342-3351

Peng R, Gallwitz D (2002) Sly1 protein bound to Golgi syntaxin Sed5p allows assembly and contributes to specificity of SNARE fusion complexes. J Cell Biol 157:645-655

Perry DK, Hannun YA (1998) The role of ceramide in cell signaling. Biochim Biophys Acta 1436:233-243

Peters C, Andrews PD, Stark MJ, Cesaro-Tadic S, Glatz A, Podtelejnikov A, Mann M, Mayer A (1999) Control of the terminal step of intracellular membrane fusion by protein phosphatase 1. Science 285:1084-1087

Pevsner J, Hsu SC, Braun JEA, Calakos N, Ting AE, Bennett MK, Scheller RH (1994) Specificity and regulation of a synaptic vesicle docking complex. Neuron 13:353-361

Pfeffer SR (1999) Transport-vesicle targeting: tethers before SNAREs. Nat Cell Biol 1:E17-E22

Pfeffer SR (2001) Rab GTPases: specifying and deciphering organelle identity and function. Trends Cell Biol 11:487-491

Polgar J, Lane WS, Chung SH, Houng AK, Reed GL (2003) Phosphorylation of SNAP-23 in activated human platelets. J Biol Chem 278:44369-44376

Pombo I, Martin-Verdeaux S, Iannascoli B, Le Mao J, Deriano L, Rivera J, Blank U (2001) IgE receptor type I-dependent regulation of a Rab3D-associated kinase: a possible link in the calcium-dependent assembly of SNARE complexes. J Biol Chem 276:42893-42900

Popoli M (1993) Synaptotagmin is endogenously phosphorylated by Ca^{2+}/calmodulin protein kinase II in synaptic vesicles. FEBS Lett 317:85-88

Popoli M, Venegoni A, Buffa L, Racagni G (1997) Ca^{2+}/phospholipid-binding and syntaxin-binding of native synaptotagmin I. Life Sci 61:711-721

Protopopov V, Govindan B, Novick P, Gerst JE (1993) Homologs of the synaptobrevin/VAMP family of synaptic vesicle proteins function on the late secretory pathway in S. cerevisiae. Cell 74:855-861

Ravichandran V, Chawla A, Roche PA (1996) Identification of a novel syntaxin- and synaptobrevin/VAMP-binding protein, SNAP-23, expressed in non-neuronal tissues. J Biol Chem 271:13300-13303

Regazzi R, Sadoul K, Meda P, Kelly RB, Halban PA, Wollheim CB (1996) Mutational analysis of VAMP domains implicated in Ca^{2+}-induced insulin exocytosis. EMBO J 15:6951-6959

Risinger C, Bennett MK (1999) Differential phosphorylation of syntaxin and synaptosome-associated protein of 25 kDa (SNAP-25) isoforms. J Neurochem 72:614-624

Rykx A, De Kimpe L, Mikhalap S, Vantus T, Seufferlein T, Vandenheede JR, Van Lint J (2003) Protein kinase D: a family affair. FEBS Lett 546:81-86

Sakagami H, Kondo H (1997) Molecular cloning and developmental expression of a rat homologue of death-associated protein kinase in the nervous system. Brain Res Mol Brain Res 52:249-256

Schraw TD, Lemons PP, Dean WL, Whiteheart SW (2003) A role for Sec1/Munc18 proteins in platelet exocytosis. Biochem J 374:207-217

Seton K, Hakansson L, Karawajczyk M, Venge P (2003) The stimulus-dependent release of eosinophil cationic protein and eosinophil protein x increases in apoptotic eosinophils. Scand J Immunol 58:312-320

Shetty KT, Kaech S, Link WT, Jaffe H, Flores CM, Wray S, Pant HC, Beushausen S (1995) Molecular characterization of a neuronal-specific protein that stimulates the activity of Cdk5. J Neurochem 64:1988-1995

Shimazaki Y, Nishiki T, Omori A, Sekiguchi M, Kamata Y, Kozaki S, Takahashi M (1996) Phosphorylation of 25-kDa synaptosome-associated protein. Possible involvement in protein kinase C-mediated regulation of neurotransmitter release. J Biol Chem 271:14548-14553

Shuang R, Zhang L, Fletcher A, Groblewski GE, Pevsner J, Stuenkel EL (1998) Regulation of Munc-18/syntaxin 1A interaction by cyclin-dependent kinase 5 in nerve endings. J Biol Chem 273:4957-4966

Silinsky EM, Searl TJ (2003) Phorbol esters and neurotransmitter release: more than just protein kinase C? Br J Pharmacol 138:1191-1201

Smith DS, Tsai LH (2002) Cdk5 behind the wheel: a role in trafficking and transport? Trends Cell Biol 12:28-36

Sohda M, Misumi Y, Yano A, Takami N, Ikehara Y (1998) Phosphorylation of the vesicle docking protein p115 regulates its association with the Golgi membrane. J Biol Chem 273:5385-5388

Sollner T, Whiteheart SW, Brunner M, Erdjument-Bromage H, Geromanos S, Tempst P, Rothman JE (1993) SNAP receptors implicated in vesicle targeting and fusion. Nature 362:318-324

Sudhof TC (2002) Synaptotagmins: why so many? J Biol Chem 277:7629-7632

Sutterlin C, Lin CY, Feng Y, Ferris DK, Erikson RL, Malhotra V (2001) Polo-like kinase is required for the fragmentation of pericentriolar Golgi stacks during mitosis. Proc Natl Acad Sci USA 98:9128-9132

Sutton RB, Fasshauer D, Jahn R, Brunger AT (1998) Crystal structure of a SNARE complex involved in synaptic exocytosis at 2.4 A resolution. Nature 1998 395:347-353

Takahashi M, Itakura M, Kataoka M (2003) New aspects of neurotransmitter release and exocytosis: regulation of neurotransmitter release by phosphorylation. J Pharmacol Sci 93:41-45

Tennyson CN, Lee J, Andrews BJ (1998) A role for the Pcl9-Pho85 cyclin-cdk complex at the M/G1 boundary in *Saccharomyces cerevisiae*. Mol Microbiol 28:69-79

Tian J, Das S, Sheng Z (2003) Ca^{2+}-dependent phosphorylation of Syntaxin1A by the death-associated protein (DAP) kinase regulates its interaction with munc-18. J Biol Chem 278:26265-26274

van der Sluijs P, Hull M, Huber LA, Male P, Goud B, Mellman I (1992 b) Reversible phosphorylation-dephosphorylation determines the localization of rab4 during the cell cycle. EMBO J 11:4379-4389

van der Sluijs P, Hull M, Webster P, Male P, Goud B, Mellman I (1992 a) The small GTP-binding protein rab4 controls an early sorting event on the endocytic pathway. Cell 70:729-740

Verhage M, Maia AS, Plomp JJ, Brussaard AB, Heeroma JH, Vermeer H, Toonen RF, Hammer RE, van den Berg TK, Missler M, Geuze HJ, Sudhof TC (2000) Synaptic assembly of the brain in the absence of neurotransmitter secretion. Science 287:864-869

Verona M, Zanotti S, Schafer T, Racagni G, Popoli M (2000) Changes of synaptotagmin interaction with t-SNARE proteins *in vitro* after calcium/calmodulin-dependent phosphorylation. J Neurochem 74:209-221

Weber T, Zemelman BV, McNew JA, Westermann B, Gmachl M, Parlati F, Sollner TH, Rothman JE (1998) SNAREpins: minimal machinery for membrane fusion. Cell 92:759-772

Yoshihara M, Adolfsen B, Littleton JT (2003) Is synaptotagmin the calcium sensor? Curr Opin Neurobiol 13:315-323

Gerst, Jeffrey E.
Department of Molecular Genetics, Weizmann Institute of Science, Rehovot 76100, Israel
jeffrey.gerst@weizmann.ac.il

Weinberger, Adina
Department of Molecular Genetics, Weizmann Institute of Science, Rehovot 76100, Israel

Phosphoinositides and membrane traffic in health and disease

Anna Godi, Antonella Di Campli, and Maria Antonietta De Matteis

Abstract

The phosphorylated derivatives of phosphatidylinositol (PtdIns) are collectively known as the polyphosphoinositides (PIs) and they were originally identified as precursors of second messengers. In particular, PtdIns 4,5-bisphosphate (PtdIns45P$_2$), initially the most studied of the PIs, was shown to be the substrate of a PI-specific phospholipase C (PI-PLC), which upon agonist stimulation generates the water-soluble inositol 1,4,5-trisphosphate and the membrane-bound diacylglycerol. Over the last decade or so, it has become increasingly clear that the PIs fulfil a wider variety of functions, including the control of cytoskeleton assembly and membrane traffic. This occurs via reversible phosphorylation of the inositol ring, which generates a series of products that can bind to cytosolic or membrane proteins with variable affinities and specificities. Indeed, the number of domains or modules that are known to specifically bind the various PIs, such as FYVE fingers, and PH, PX and ENTH domains, have increased dramatically over just the past few years. By interacting with these protein domains, membrane PIs can control actin cytoskeleton remodelling, vesicle coat assembly, and signalling-complex formation. Here, we give an overview of the functions of the PI-metabolizing enzymes in the secretory pathway, and of the human pathologies that are associated with mutations in these enzymes.

1 PIs metabolism

PtdIns is the basic building block for the inositol lipids of eukaryotic cells. It largely consists of D-*myo*-inositol 1-phosphate linked through its phosphate group to 1-stearoyl, 2-arachidonoyl diacylglycerol. The prevalence in mammalian cells of these two particular fatty acids is probably related to the correct insertion of the PIs into the membranes, which has to allow an efficient exposure of the inositol headgroup for interactions with cytosolic proteins.

The inositol ring contains five free hydroxyl groups (2'– 6'); all of these have the potential to become phosphorylated. Thus, a number of derivatives of PtdIns can exist in cells, many of which have unique functions. To date, the following PIs have been identified in cells: PtdIns 3-phosphate (PtdIns3P), PtdIns4P, PtdIns5P,

Topics in Current Genetics, Vol. 10
S. Keränen, J. Jäntti (Eds.): Regulatory Mechanisms of Intracellular Membrane Transport
DOI 10.1007/b98497 / Published online: 13 May 2004

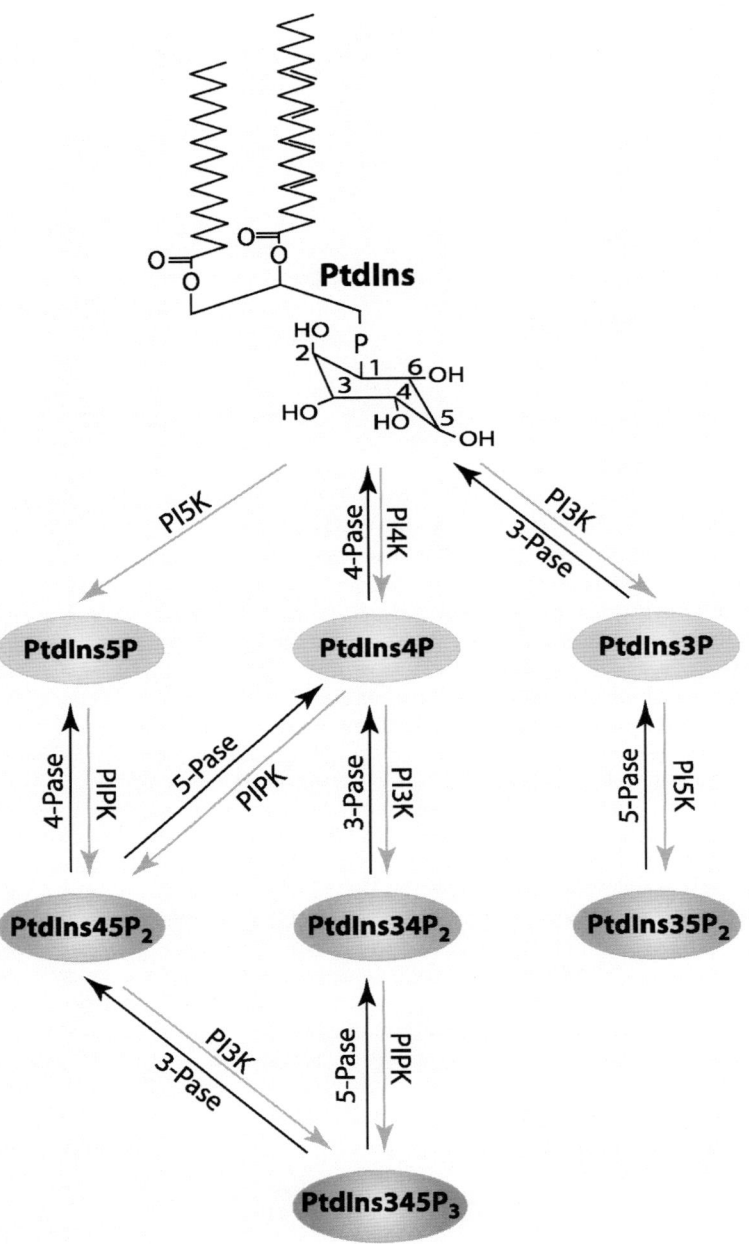

Fig. 1. The inositol ring of PtdIns can be phosphorylated in the 3, 4 and/or 5 positions, giving rise to seven different PIs. The most important phosphorylation and dephosphorylation pathways are indicated.

PtdIns 3,4-bisphosphate (PtdIns34P$_2$), PtdIns35P$_2$, PtdIns45P$_2$, and PtdIns 3,4,5-trisphosphate (PtdIns345P$_3$), as illustrated in Figure 1.

PtdIns is the most abundant of the inositol lipids in mammalian cells under basal conditions, and constitutes ~10% of the total membrane phospholipids. Of the monophosphorylated PIs in cells, PtdIns4P represents 90-94%, while PtdIns3P and PtdIns5P each make up 2-5%. PtdIns45P$_2$ is the most abundant of the bisphosphorylated PIs (up to 99%). The actual cellular levels of the single trisphosphorylated PI, PtdIns345P$_3$, can vary enormously in response to both external and internal stimulation, although its maximal levels of upregulation remain ~500-fold lower than those of PtdIns45P$_2$.

The PIs are mainly localized at the cytoplasmic face of cellular membranes (glycosyl-PtdIns-anchored proteins are an exception to this orientation), where they are substrates for different enzymes, including the PI kinases (PIKs), PI phosphatases and phospholipases. The synthesis of PtdIns occurs in the endoplasmic reticulum (ER), while the subsequent phosphorylation steps occur in post-ER compartments (Cockcroft and De Matteis 2001; Gehrmann and Heilmeyer 1998; Loijens et al. 1996). Indeed, the PIKs and the PI phosphatases have been localized to most intracellular membrane compartments, including the plasma membrane, the nucleus, secretory granules, endosomes, the ER, and the Golgi complex.

2 The PI kinases: their subcellular localization and role in the secretory pathway

Many of the PIKs were initially described as enzymatic activities capable of transferring a phosphate to a precise position on the inositol ring of PtdIns, or of its phosphorylated derivatives. Studies on the purified enzymes then led to their classification into three main families: the PtdIns 3-kinases (PI3Ks), PtdIns 4-kinases (PI4Ks) and PtdInsP 5-kinases (PIP5Ks). Over the last decade, the genes encoding several of these enzymes have been identified, and their protein sequence comparisons generally support this distinct classification of PI3Ks, PI4Ks and PIP5Ks, as summarised in Table 1 and described below.

2.1 The PtdIns 3-kinases

The PI3Ks have been subdivided into three main classes (I, II, III) based on sequence similarities, substrate specificities and physiological roles (Fruman et al. 1998). The class I PI3Ks are further subdivided into classes IA and IB. Class IA PI3Ks are heterodimeric enzymes consisting of a catalytic (p110α, p110β and p110γ) and a regulatory (p85α, p85β and P55γ) subunit. The single class 1B PI3K is made of a p110 catalytic subunit and a p101 regulatory subunit. The preferred substrate of the class I PI3Ks is PtdIns45P$_2$ *in vivo*, and hence their primary product is PtdIns345P$_3$. However, these kinases can also phosphorylate PtdIns and PtdIns4P *in vitro*. The class I PI3Ks have been a major focus of attention since

Table 1. Mammalian phosphoinositide kinases

Type	Subtype	Iso-forms	Molecular weight (kDa)	Major substrates	Yeast homo-logues
PI3K	Class IA	p110α	110	PtdIns45P$_2$ (PtdIns4P, PtdIns)	
		p110β	119	PtdIns45P$_2$ (PtdIns4P, PtdIns)	
		p110δ	115	PtdIns45P$_2$ (PtdIns4P, PtdIns)	
	Class IB	p110γ	120	PtdIns45P$_2$ (PtdIns4P, PtdIns)	
	Class II	PI3K-C2α	170	PtdIns, PtdIns4P	
		PI3K-C2β	180	PtdIns, PtdIns4P	
		PI3K-C2γ	170	PtdIns, PtdIns4P	
	Class III	Vps34	101	PtdIns	Vps34p
PI4K	Type II	PI4Kα	55	PtdIns	Lsb6p
		PI4Kβ	55	PtdIns	
	Type III	PI4Kα	100-230	PtdIns	Stt4p
		PI4Kβ	110	PtdIns	Pik1p
PI5K		PIKfyve	235	PtdIns3P, PtdIns	Fab1p
PIPK	Type I	PIP5Kα	68	PtdIns4P (PtdIns3P)	Mss4p
		PIP5Kβ	68	PtdIns4P (PtdIns3P)	
		PIP5Kγ	68	PtdIns4P (PtdIns3P)	
	Type II	PIP4Kα	53	PtdIns5P (PtdIns3P)	
		PIP4K β	53	PtdIns5P (PtdIns3P)	
		PIP4Kγ	47	PtdIns5P (PtdIns3P)	

All abbreviations as in text.

they are generally coupled to receptor activation by extracellular stimuli and have been implicated in a wide range of cellular processes, including cell-cycle progression and cell growth, motility, adhesion and survival.

The class II PI3Ks are large proteins that have catalytic domains that are 45-50% identical to those of the class I PI3Ks. The class II PI3Ks have a C-terminal region that has homology with the C2 domain of the classical protein kinase C isoforms that mediate Ca^{2+}/lipid binding. Although they preferentially phosphorylate PtdIns and PtdIns4P *in vitro*, their *in vivo* substrates and functions are less clear.

The class III PI3Ks are highly homologous to the yeast *VPS34* gene product, and as with the yeast enzyme, they are specific for PtdIns. Mutations in the Vps34 protein cause missorting of vacuolar proteins, changes in vacuole morphology, and defects in the endocytic pathway (Takegawa et al. 1995; Kihara et al. 2001; Stack et al. 1995).

In mammalian cells, wortmannin inhibition of the PI3Ks blocks transport of proteins from the Golgi complex to lysosomes (Brown et al. 1995; Davidson 1995), inhibits early endosomal trafficking, and causes the accumulation of pre-lysosomal vesicles. Thus, this indicates that the function of these kinases in secretory pathway is conserved from yeast to mammals.

Indeed, PtdIns3P is enriched in the endocytic/lysosomal compartment (Gillooly et al. 2000), where it is involved in the recruitment of different proteins that have domains known to bind this lipid with high affinity and specificity: the FYVE and PX domains. Among these proteins, the best characterized are the early endosomal GTPase Rab5, which regulates endocytic membrane fusion (Simonsen et al. 1998; Zerial and McBride 2001), and its effectors EEA1 and rabenosyn-5 (Christoforidis and Zerial 2001; Nielsen et al. 2000). Other proteins that localize to the endocytic compartment through FYVE domains are SARA (Smad anchor for receptor activation), which mediates TGFβ signalling (Tsukazaki et al. 1998; Miura et al. 2000; Hayes et al 2002), and Hrs (hepatocyte-growth-factor-regulated tyrosine kinase substrate), which is the homologue of the yeast Vps27p and is involved in endosome/lysosome trafficking and multivesicular body formation (Bache et al. 2003; Lloyd et al. 2002; Odorizzi et al. 1998).

Recently, it has been shown that two type II PI3Ks, PI3KIIα and γ associate with the Golgi complex (Domin et al. 2000; Ono et al. 1998). Moreover, PI3KIIα can bind clathrin and it is stimulated by this interaction. Overexpression of this kinase induces the redistribution of the mannose 6-phosphate receptor from the *trans*-Golgi network (TGN) to the cell periphery and inhibits endocytosis (Gaidarov et al. 2001). Thus, PI3KIIαmight control clathrin-dependent sorting events at the TGN through a localized generation of PtdIns3P at sites of clathrin-coated bud formation.

2.2 The PtdIns 4-kinases

The PI4Ks catalyse the phosphorylation of PtdIns on the D-4 position of the inositol ring. Unlike the PI3Ks and PIP5Ks, these enzymes appear to exclusively phosphorylate PtdIns, hence, leading to the production of PtdIns4P, which is the major precursor in the synthesis of the other PIs, including PtdIns45P$_2$, PtdIns34P$_2$ and PtdIns345P$_3$.

The PI4Ks were originally subdivided into two types (II and III) that were based on biochemical differences of the partially purified enzymes. Type II PI4K, was purified from a variety of mammalian sources and found to be strongly associated with membranes; it is believed to account for most of the PI4K activity in cells. It can be distinguished from type III PI4K by virtue of its lower K_m for ATP and PtdIns, its insensitivity to wortmannin, and its sensitivity to inhibition by adenosine (Pike 1992; Gehrmann and Heilmeyer 1998).

Over the last decade, several PI4Ks have been cloned, and these cloned enzymes do not strictly follow the classification described above. The first cloned PI4K was *PIK1* from *Saccharomyces cerevisiae* (Flanagan et al. 1993). This enzyme is mostly cytosolic, and has characteristics closer to those of the type III

mammalian enzyme (high K_m for ATP and low sensitivity to inhibition by adenosine) than to the type II enzymes, although it is insensitive to wortmannin. Initial characterization of Pik1p showed that it is localized in the nucleus and at the Golgi membranes. Although the nuclear function of Pik1p is still not clear, this enzyme is essential and has been shown to have an important role in secretion from the late-Golgi compartment (Hama et al. 1999; Walch-Solimena and Novick 1999; Audhya et al. 2000). The activity of Pik1p in yeast is stimulated by the homologue of the neuronal Ca^{2+}-binding protein frequenin, which binds to a conserved sequence motif in Pik1p (Hendricks et al. 1999).

Another PI4K, *STT4*, was cloned from yeast in a genetic screen for mutations defective in pathways upstream of the protein kinase C Pkc1p, (Yoshida et al. 1994). Stt4p shows 30% sequence identity in its C-terminal catalytic domain to the corresponding catalytic domains of Pik1p and Vps34p. Stt4p is the major target of wortmannin in yeast (Cutler et al. 1997) and has been suggested to function in a transport step required for phosphatidylserine metabolism (Trotter et al. 1998), in the maintenance of vacuole morphology, and in the organisation of the actin cytoskeleton (Audhya et al. 2000). However, the precise cellular roles of Stt4p in these pathways remain unclear.

The first mammalian PI4K to be cloned was human PI4Kα (Wong and Cantley 1994). This kinase has a molecular mass of 97 kDa, and shows a strong sequence similarity to Stt4p, although it is smaller. Interestingly, both Stt4p and PI4Kα contain putative PH domains, together with their catalytic kinase domain and the unique lipid kinase domain. From a biochemical point of view, PI4Kα appears to be a type II kinase, although its molecular mass is different from that of the purified enzyme. Two other PI4Ks have been identified: a 210-kDa enzyme closely related to Stt4p, and a 92-kDa protein related to Pik1p. Both of these are sensitive to the PI3K inhibitor wortmannin (although only at higher, μmolar, concentrations compared to PI3K) and insensitive to adenosine, consistent with type III kinases. The 210-kDa enzyme, known as PI4KIIIα, is identical to the shorter PI4Kα and, thus, appears to be a splice variant (Balla et al. 1997). PI4KIIIα contains an N-terminal Src-homology-3 (SH3) domain, a proline-rich region and a leucine zipper, all of which are missing in the shorter variant. The 92-kDa protein has been named PI4KIIIβ (Meyers and Cantley 1997). Both PI4PIIIα and PI4KIIIβ have been shown to associate with the Golgi complex, whereas, PI4Kα associates with the ER (Balla et al. 1997; Wong et al. 1997).

Recently, two new isoforms of PI4K have been cloned that have enzymatic characteristics and molecular masses of the type II PI4Ks (Barylko et al. 2001; Balla et al. 2002). They have, thus, been called PI4KIIα and PI4KIIβ. PI4KIIα was seen to be tightly associated to membranes and, thus, behaves as an integral membrane protein. Immunofluorescence studies have shown its localization both at the Golgi complex and in the endo-lysosomal compartments (Balla et al. 2002; Wei et al. 2002). PI4KIIβ is primarily cytosolic and associates with the plasma membrane, the ER and the Golgi complex. PI4KIIβ translocates to the plasma membrane upon growth factor stimulation, a process that is mediated by Rac1 activation (Wei et al. 2002).

The first evidence that the PI4Ks have roles in the secretory pathway in mammalian cells derived from the observation that expression of a dominant-negative dead-kinase form of PI4KIIIβ (PI4KD656A) induces alterations in the organization of the Golgi complex (Godi et al. 1999). PI4KIIIβ, however, is responsible for only a fraction of the PtdIns4P generated at the Golgi complex. Indeed, the Golgi complex possesses two PI4K activities: a basal type II activity due to PI4KIIα, and an ARF-induced type III activity, due to PI4KIIIβ (Godi et al. 1999; Jones et al. 2000; Wang et al. 2003).

The maintenance of the correct balance of PtdIns4P at the Golgi complex appears to be crucial for Golgi function since the overexpression of a functional enzyme (wt PI4KIIIβ) inhibits the rate of TGN-to-cell surface delivery of both influenza haemagglutinin (HA) and the temperature-sensitive variant of the G protein of vesicular stomatitis virus (VSVG) in MDCK cells. At the same time, the expression of kinase-dead PI4KIIIβ inhibits TGN-to-plasma membrane transport of VSVG while it stimulates TGN-to-plasma membrane delivery of the apical marker HA, possibly because of a defective incorporation of the apical cargo into membrane rafts (Bruns et al. 2002).

PI4KIIIβ associates to and is stimulated by the overexpression of recombinant frequenin, thus, mirroring the behaviour of its yeast homologue, Pik1p (Hendricks et al. 1999; Weisz et al. 2000; Zhao et al. 2001). The ability of frequenin to stimulate secretion in neuroendocrine cells is mediated by an enhanced PI4KIIIβ activity and a concomitant increase in PtdIns45P$_2$ levels at the plasma membrane (Rajebhosale et al. 2003). The same interaction between frequenin and PI4KIIIβ appears to negatively regulate GLUT4 translocation in 3T3L1 adipocytes (Mora et al. 2002).

Recently, an important role for PI4KIIα at the TGN has also been demonstrated using an RNA interference (RNAi) approach. Thus, RNAi of PI4KIIα decreases PtdIns4P levels at the Golgi, and blocks the recruitment of clathrin adaptor complex-1 (AP-1) to the Golgi, inhibiting AP-1-dependent clathrin coat assembly at the TGN (Wang et al. 2003). PI4KIIα is also responsible for the majority of the PI4K activity in the brain, and has been found to be concentrated both at the synapse and in the region of the Golgi complex in neuronal perikarya (Guo et al. 2003).

2.3 The PtdIns monophosphate kinases

Two classes of PIPKs that are capable of phosphorylating the PIPs to produce PtdIns45P$_2$ have been defined based on their sensitivity to phosphatidic acid (PA). Type I PIPKs are stimulated by PA, while type II PIPKs are not (Moritz et al. 1992; Jenkins et al. 1994). The genes encoding these two families of enzymes have been cloned, and shown to selectively phosphorylate different positions on the inositol ring. The type I enzymes phosphorylate PtdIns4P in the D-5 position to make PtdIns45P$_2$ and are, thus, PIP5Ks, whereas, the type II PIPKs phosphorylate PtdIns5P to make PtdIns45P$_2$, and are consequently PIP4Ks (Rameh et al.

1997). Furthermore, a broad *in vitro* substrate specificity of these PIPKs has been reported: type II PIP4K is able to phosphorylate PtdIns3P to produce PtdIns34P$_2$, and type I PIP5K produces PtdIns35P$_2$ and PtdIns345P$_3$ from PtdIns3P and PtdIns34P$_2$, respectively (Frumann et al. 1998; Tolias et al. 1998; Anderson et al. 1999).

To date, three isoforms (α, β and γ of both mammalian type I and type II PIPKs have been identified. A comparison of their primary sequences reveals a sequence identity of only 28-33%, whereas, the isoforms within each subtype are highly homologous (66-78% identical). The type I and type II kinases appear to be functionally non-redundant despite the fact that they synthesise the same product, PtdIns45P$_2$. They do, however, localize to different intracellular compartments: those of type I are found at the plasma membrane and in the nucleus, whereas, those of type II localize to the cytosol, the ER, the nucleus and the actin skeleton, but not at the plasma membrane.

The type I PIPKs have been implicated in the regulation of secretion, endocytosis and the actin cytoskeleton. The activity of this class of enzymes can be regulated by small GTP-binding proteins, such as Rho, Rac, and ARF, which have been found to interact directly with the type I, but not the type II, PIPK (Honda et al. 1999; Tolias et al. 2000; Aikawa and Martin 2003). The first evidence of a role for PtdIns45P$_2$ in exocytosis came from the identification of type I PIP5K as a cofactor required in the ATP-dependent priming step that precedes Ca^{2+}-triggered secretion of dense-core vesicles (DCVs; Hay et al. 1995). A more recent important advance is the finding that ARF6 has a role in DCV exocytosis through the control of the activity of PIP5K, and hence in synthesis of the required plasma-membrane pool of PtdIns45P$_2$. The physiological functions of the type II PIPKs are not yet well defined. A specific type II isoform, type II PIPKβ, interacts with the p55 subunit of the tumour necrosis factor-α (TNFα) receptor and may, thus, have a role in TNFα-mediated signalling (Castellino et al. 1997).

Although none of the known PIP5K isoforms have been visualized in the Golgi complex, there is evidence that indicates that PtdIns45P$_2$ has a role in ER-to-Golgi transport and in the formation/release of post-Golgi transport carriers. The former was deduced by the demonstration that the PtdIns45P$_2$-binding PH domains, such as those of the β-spectrins, inhibit ER-to-Golgi transport of VSVG in permeabilized NRK cells, potentially by inhibiting the association of spectrin with the Golgi complex and/or with pre-Golgi transport intermediates (Godi et al. 1998; De Matteis and Morrow 1998). Moreover, by inhibiting PtdIns45P$_2$ production with primary alcohols (which inhibit the synthesis of PA, an activator of PIP5K; Siddhanta et al. 2003; Sweeney et al. 2002), it has been shown that PtdIns45P$_2$ synthesis is required for the release of transport intermediates from the TGN and for maintaining the structural integrity and function of the Golgi apparatus, both in growth-hormone-secreting rat pituitary (GH3) cells and in isolated Golgi membranes (Siddhanta et al. 2000). In contrast, an increase in PtdIns45P$_2$ levels (induced by PIP5Kα transfection) triggers the formation of actin tails on vesicular structures, some of which derive from the Golgi complex (Rozelle et al. 2000;

Stamnes 2002). Thus, PtdIns45P$_2$ might be required both for the formation of post-Golgi transport carriers and for their actin-dependent movement.

PIKfyve is the mammalian homologue of yeast Fab1p (Sbrissa et al. 1999). It phosphorylates PtdIns and PtdIns3P in the D-5 position, generating PtdIns5P and PtdIns35P$_2$, respectively. PIKfyve has been localized in the late endocytic compartment, a localization seen due to the ability of its FYVE domain to bind PtdIns3P, which is enriched in this compartment. Recent studies using the expression of a kinase-dead PIKfyve point mutant and microinjection of specific antibodies have demonstrated the PIKfyve role in controlling fluid-phase endocytosis by the regulation of multivesicular body morphogenesis (Sbrissa et al. 2002; Ikonomov et al. 2002, 2003).

3 The PI phosphatases: localization and function in the secretory pathway

The PI phosphatases have been divided into three major categories, based on their abilities to hydrolyse the 3-, 4-, and 5-phosphorylated PIs (Table 2).

3.1 The PI 3-phosphatases

The PI 3-phosphatases include the tumour suppressor PTEN (Phosphatase and tensin homologue on chromosome ten), its related proteins TPIP (transmembrane phosphatase with tensin homology) and myotubularin (MTM), and the myotubularin-related proteins (MTMRs; Maehama et al. 2001; Wishart et al. 2001). The best-characterized substrate for PTEN *in vivo* is PtdIns345P$_3$, although it can dephosphorylate all of the 3-phosphorylated PIs *in vitro*. PTEN contains a phosphatase domain, a region with homology to tensin, and a PDZ protein-interaction domain. The crystal structure of PTEN shows that it also has a C2 domain (Lee et al. 1999). PTEN not only has a lipid phosphatase activity, but also a protein phosphatase activity, and it has been shown to be involved in many cellular functions, such as cell-cycle progression, apoptosis, and cell contact and migration. Furthermore, mutations in PTEN have been shown to be associated with a broad variety of human cancers (see below).

Recently, use of the PTEN catalytic core motif to screen the *Dictyostelium* genome using the BLAST programme has resulted in the identification of a new protein in the *Dictyostelium* genome that has been named phospholipid-inositol phosphatase (PLIP). PLIP has a preference for PtdIns5P as substrate (Merlot et al. 2003) and has been localized at the Golgi complex, where it is thought to be involved in membrane trafficking.

Table 2. . Mammalian phosphoinositide phosphatases

Type	Subtype	Isoforms	Major substrates	Yeast homo-logues
3-phosphatases				
		PTEN1, 2	PtdIns345P$_3$, PtdIns35P$_2$, PtdIns34P$_2$, PtdIns3P	Tep1p
		TPIPα, β	PtdIns345P$_3$	
		MTM1	PtdIns3P	Yjr110w
		MTMR1-8	PtdIns3P	
4-phosphatases				
		Type I, II	PtdIns34P$_2$	
		Sac1	PtdIns4P (PtdIns3P, PtdIns35P$_2$)	Sac1p
		Synaptojanin 1, 2	PtdIns4P (PtdIns3P, PtdIns35P$_2$)	Inp52p, Inp53p
5-phosphatases				
	Type I			
		5-phosphatase	InsPs	
	Type II			
		5-phospatase II	PtdIns45P$_2$, InsPs	
		Ocrl	PtdIns45P$_2$> InsPs	
		Synaptojanin 1, 2	PtdIns45P$_2$	Inp52p, Inp53p
		PIPP	InsPs > PtdIns45P$_2$	
		Sac2	PtdIns45P$_2$, PtdIns345P$_3$	
		72-kDa 5-phosphatase/pharbin	PtdIns345P$_3$, PtdIns34P$_2$, InsPs	
		SKIP	PtdIns45P$_2$, (PtdIns345P$_3$)> InsPs	Inp51p, Inp54p
	Type III			
		SHIP 1, 2	PtdIns345P$_3$, PtdIns35P$_2$, InsPs	
	Type IV			
		5-phosphatase	PtdIns345P$_3$> PtdIns45P$_2$	

Abbreviations as in text.
InsPs, soluble inositol phosphates of unspecified phosphorylation.

Other PTEN-related proteins have been shown to be localized at the Golgi complex: TPTE (Transmembrane Phosphatase with tensin homology; Guipponi et al. 2001), which is specifically expressed in testis, and TPIP (TPTE- and PTEN-homologous inositol lipid phosphatase; Walker et al. 2001), although their functions at the Golgi remain unknown.

The myotubularin-related genes define a large family of eukaryotic proteins that comprises 13 members. The MTMR proteins contain dual-specificity protein tyrosine phosphatase (PTP) activities, and MTM1, MTMR1, MTMR2, MTMR3, MTMR4, and MTMR6 have been shown to dephosphorylate the D-3 position on

the inositol ring of PtdIns3P as preferred substrate (Laporte et al. 2001; Wishart and Dixon 2002). Some of the MTM proteins are thought to be enzymatically inactive since they lack the conserved cysteine residue in their PTP motives, a prerequisite for this activity. Mutations in MTM1 result in the X-linked human disease myotubular myopathy, while mutations in MTMR2 and MTMR13 cause the Charcot-Marie-Tooth disease type 4B disorder (see below). The roles of the catalytically inactive MTMRs are unknown, but recently it has been proposed that they function as adapters for the active forms (Nandurkar et al. 2003; Kim et al. 2003). Recently, it has been shown that in the *Caenorhabditis elegans* worm, mutations in MTM6 and MTM9, which form a complex, block endocytosis in their coelomocytes (Dang et al. 2003).

3.2 The PI 4-phosphatases

Two PI 4-phosphatases have been cloned, type I and type II that use PtdIns34P$_2$ as their major substrate. Furthermore, a phosphatase activity is associated with the Sac phosphatase domain that has been shown to be present in yeast Sac1p and Inp53p, and in mammalian Sac1-3 and synaptojanins (Hughes et al. 2000). These proteins display activities that are principally towards PtdIns4P, PtdIns3P and PtdIns35P$_2$ *in vitro*. However, yeast sac1 mutants exhibit very high levels of PtdIns4P, thus, indicating that *in vivo* Sac1p has a preference for PtdIns4P versus the other two substrates, and suggesting that this should also be the case for other proteins possessing Sac domains. Yeast strains with mutations in the *SAC1* gene exhibit an array of phenotypes, including inositol auxotrophy (Whitters et al. 1993), secretory defects in chitin deposition, disorganization of the actin cytoskeleton, and impairment of ATP uptake and protein translocation to the ER (Mayinger et al. 1995). Moreover, mutations in *SAC1* are able to bypass the essential requirement for *SEC14* (responsible for the major yeast PI-transfer protein) in protein transport from the Golgi complex to the plasma membrane.

Recently, the human homologue of *SAC1* was cloned. hSac1 behaves like the yeast isoform in terms of its substrate specificity and its localization to the ER and Golgi complex (Rohde et al. 2003). Moreover, it has been shown that hSac1 interacts with the coatomer protein-I (COPI) complex; mutation of a putative COPI-binding motif (KXKXX) abolishes this interaction and results in the accumulation of hSac1 in the Golgi (Rohde et al. 2003).

3.3 The PI 5-phosphatases

The PI and inositol 5-phosphatases comprise a large family of phosphatases that are classified into four types according to their substrate specificities (Table 2). The type I enzymes hydrolyse only the water-soluble substrates (the inositol phosphates). The type II enzymes can utilize the lipid substrates, although not necessarily exclusively (Jefferson and Majerus 1995; Woscholski and Parker 1997). Indeed, the type II 5-phosphatases comprise a heterogeneous group of proteins that

can be further divided in four subgroups based on their sequence conservations: the GTPase-activating protein (GAP)-containing inositol 5-phosphatases, which have two members, OCRL and 5-phosphatase II; the Sac1p-containing inositol 5-phosphatases, which include synaptojanins 1 and 2 and hSac2 (Hughes et al. 2000); the proline-rich-domain-containing inositol 5-phosphatases, which include the 72-kDa 5-phosphatase/pharbin and PIPP (proline-rich inositol polyphosphate 5-phosphatase; Mochizuki and Takenawa 1999); and the skeletal muscle and kidney-enriched inositol 5-phosphatase (SKIP; Ijuin et al. 2000). The type III phosphatases, SHIP1 and SHIP2, contain SH2 domains and can hydrolyse the phosphate at the D-5 position of the PIs and inositol phosphates that also have a phosphate group at the D-3 position (Woscholski and Parker 1997). Finally, the type IV phosphatases hydrolyse only lipid substrates, such as $PtdIns345P_3$ and $PtdIns45P_2$ (Kisseleva et al. 2000).

The Golgi complex possesses three PI 5-phosphatases: hSac2, the 75-kDa 5-phosphatase pharbin and OCRL-1 (Dressman et al. 2000; Kong et al. 2000; Minagawa et al. 2001), although their functions remain unclear. OCRL-1 is known to interact with golgin 84, a coiled-coil type II Golgi membrane protein, and is localized at the TGN in fibroblasts and epithelial cells (Bascom et al. 1999; Dressman et al. 2000), with the presence of OCRL-1 in lysosomes in normal human kidney cells also having been reported (Zhang et al. 1998). Thus, OCRL-1 may function in the regulation of a specific pool of $PtdIns45P_2$ that is associated with the TGN and lysosomes, and that is important for lysosomal enzyme trafficking.

4 PI metabolism and disease

The importance of the fine regulation of the levels of the PIs is highlighted by the presence of several human diseases that are associated with mutations in the enzymes that are responsible for their metabolism. This is particularly true for the phosphatases acting on the D-3 (PTEN, SHIP and the MTM family members) or the D-5 (OCRL) positions of the inositol ring.

4.1 PTEN and human cancers

PTEN is frequently mutated in many cancers, including glioblastomas and endometrial, thyroid, and prostate cancers. These mutations lead to a complete inactivation of the phosphatase activity of PTEN as well as to either total or partial loss of expression of either its mRNA and/or the protein. Because of this inactivation, there is an accumulation of $PtdIns345P_3$ in cells, and a sustained activation of the PI3K signalling pathways that leads to increased cell survival, growth, and proliferation. Indeed, $PtdIns345P_3$ is a potent second messenger that can interact with and activate proteins that have PH domains, like the Akt family of proteins and their regulator, PI-dependent kinase 1 (PDK1), and it is involved in membrane traffic events, such as endocytosis of activated receptors and autophagy, the dis-

regulation of which has been seen to occur in many forms of human cancers (Arico et al. 2001; Naga Prasad et al. 2002).

Germ-line mutations of PTEN are also associated with rare autosomal-dominant syndromes that share the tendency for the development of hamartoma tumours: Cowden syndrome; Bannayan-Zonana syndrome; Proteus syndrome; and Lhermitte syndrome (Sulis and Parsons 2003; Zhou et al. 2003). These cancer syndromes have similar phenotypic characteristics, including mental retardation, gastrointestinal hamartomas, thyroid adenoma, breast fibroadenomas, macrocephaly, and mucocutanous lesions.

4.2 Myotubular myopathy and Charcot-Marie-Tooth disease

Myotubular myopathy is a severe X-linked congenital disorder that is characterized by hypotonia and respiratory insufficiency. Mutations in MTM1 cause this disorder, as has been demonstrated by positional cloning. This disease affects about one in 50,000 new-born males, most of which die within the first month of life due to respiratory failure, although a subset can survive for several years. More than 133 different mutations have been identified in patients from 328 families (Laporte et al. 2000). Most of these are point mutations that result in a truncated protein, while 26% are missense mutations that affect important residues that are conserved in the *Drosophila* orthologue and in the homologous MTMR1 gene. The truncating mutations cause often severe and early lethal phenotypes, while some missense mutations are associated with milder forms of the disease, and hence a prolonged survival. No significant clinical symptoms or signs of myopathy occur in female carriers.

Histopathological studies of skeletal muscle in myotubular myopathy patients has revealed the presence of small rounded muscle fibres containing centrally localized nuclei that resemble fetal myotubules, and it has, thus, been suggested that the disorder results from an arrest in normal development of muscle fibres (Sarnat 1990). Deletion of MTM1 in mice reproduces this muscle phenotype, and they develop a generalized and progressive myopathy starting from the first four weeks of age, with an accumulation of central nuclei in muscle fibres (Buj-Bello et al. 2002). However, and contrary to expectation, it has been shown that muscle differentiation in knockout mice occurs normally, and that the centralized myonuclei originate mainly from a structural maintenance defect that affects myotubularin-deficient muscle, rather than it being a regenerative process.

It is also important to note that although MTMR1 has been found localized in the cytoplasm and at the plasma membrane, its major substrate, PtdIns3P, is enriched in the endocytic compartment, where it actively participates in the subcellular localization and recruitment of FYVE- or PX-containing proteins, most of which are involved in membrane traffic (see above). Interestingly, results obtained in yeast have suggested a role for myotubularin in intracellular traffic regulation (Blondeau et al. 2000). In particular, it has been shown that in *S. pombe* overexpression of myotubularin reduces growth rate and induces the presence of large vacuoles. A similar phenotype has been observed through deletion of *VPS34*, the

yeast PI3K. In both of these cases, interfering with PtdIns3P metabolism results in alterations in vacuolar morphology and function. However, in the case of myotubularin, an indirect control of PtdIns3P levels through Vps34p regulation (i.e. by myotubularin-dependent phosphorylation) cannot be ruled out since the two proteins can co-immunoprecipitate (Blondeau et al. 2000) and myotubularin also has a protein kinase activity.

Mutations in the MTM-related proteins MTMR2 and MTMR13/SBF2 (Kim et al. 2002) cause severe demyelinating neuropathies, such as Charcot-Marie-Tooth disease type 4B1 and 4B2. Like myotubularin, MTMR2 is localized predominantly in the cytoplasm and at the plasma membrane, although it was recently shown that it preferentially dephosphorylates PtdIns35P$_2$ over PtdIns3P *in vitro* (Berger et al. 2002). The roles of PtdIns35P$_2$ and PtdIns5P remain poorly understood, although they appear to be important for endocytic traffic (Sbrissa et al. 2002; Ikonomov et al. 2003).

4.3 Lowe syndrome

The oculocerebrorenal syndrome of Lowe (OCRL) was first described in 1952 in a newborn with aminoaciduria, decreased renal ammonia production, hydrophthalmos and mental retardation. Descriptions of additional infants with similar features led to the identification of the cardinal features of the syndrome: congenital cataracts, Fanconi syndrome of the proximal renal tubules, and mental retardation inherited as an X-linked trait. OCRL is a rare disease with only a few hundred known patients in the United States. The main ophthalmologic manifestation of OCRL is congenital cataracts, seen as early as 20-24 weeks after gestation, and always present at birth. Female carriers of OCRL have no clinical symptoms, but do show lens abnormalities in the form of tiny punctuate opacities in the corte Mental retardation is commonly seen in OCRL patients. Neonatal and infantile hypotonia of central nervous system origin is also a frequent feature, although, this improves or disappears with age. The primary renal manifestations are Fanconi syndrome of renal tubular acidosis due to bicarbonate wasting, aminoaciduria, phosphaturia, proteinuria, and urinary concentrating defects. The kidney proximal tubule cells from Lowe syndrome patients have elevated levels of Golgi PtdIns45P$_2$, and the diagnostic test of the syndrome is based on measurement of the PtdIns45P$_2$ 5-phosphatase activity in skin fibroblasts. Ocrl-1p knockout mice do not develop the symptoms present in Lowe syndrome, presumably because of the functional overlap between Ocrl-1p and other 5-phosphatases (Janne et al. 1998).

Recently, a defect in actin cytoskeleton organization in patients with Lowe syndrome has been identified (Suchy and Nussbaum 2002). This defect consists of a decrease in long actin stress fibres, enhanced sensitivity to actin depolymerizing drugs, and an increase in punctate F-actin staining in a distinctly anomalous distribution in the centre of the cells. This defect also correlates with an abnormal distribution of two actin-binding proteins: α-actinin and gelsolin. Since actin polymerization is important for the function and maintenance of both adherent junctions,

which themselves are important for renal proximal tubule formation, and cell differentiation, taken together with the defect in secretion of lysosomal enzymes, this might explain the Lowe syndrome phenotype.

5 Conclusions

Here, we have summarized the evidence collected over the past decade demonstrating the key role of the PIs in almost every cellular membrane-traffic event. Different PIs serve as spatial recruiters or activators of protein complexes, performing distinct functions along the secretory pathway. To exert these functions, the PIs themselves have to be spatially and temporally regulated. This might be achieved by controlling the spatial distribution, regulation and activation of the enzymes, the kinases, and phosphatases involved in PI synthesis, and as a consequence, in the formation of membrane microdomains enriched in particular PI species. It is now clear that the activity of the secretory pathway and the regulation of lipid metabolic pathways, which generate and consume lipids and regulate exocytic events, are mutually interconnected. The understanding of how cells maintain and control these interconnections, as well as the identity/diversity and homeostasis of their organelles, remains an important challenge for the future.

Acknowledgements

The authors would like to thank Dr. C. P. Berrie for critical reading of the manuscript, and acknowledge the financial support of Italian Association for Cancer Research (AIRC, Milan, Italy), Telethon Italia, European TMR network, and MIUR.

References

Aikawa Y, Martin TF (2003) ARF6 regulates a plasma membrane pool of phosphatidylinositol(4,5)bisphosphate required for regulated exocytosis. J Cell Biol 162:647-659

Anderson RA, Boronenkov IV, Doughman SD, Kunz J, Loijens JC (1999) Phosphatidylinositol phosphate kinases, a multifaceted family of signaling enzymes. J Biol Chem 274:9907-9910

Arico S, Petiot A, Bauvy C, Dubbelhuis PF, Meijer AJ, Codogno P, Ogier-Denis E (2001) The tumor suppressor PTEN positively regulates macroautophagy by inhibiting the phosphatidylinositol 3-kinase/protein kinase B pathway. J Biol Chem 276:35243-35246

Audhya A, Foti M, Emr SD (2000) Distinct roles for the yeast phosphatidylinositol 4-kinases, Stt4p and Pik1p, in secretion, cell growth, and organelle membrane dynamics. Mol Biol Cell 11:2673-2689

Bache KG, Brech A, Mehlum A, Stenmark H (2003) Hrs regulates multivesicular body formation via ESCRT recruitment to endosomes. J Cell Biol 162:435-442

Balla T, Downing GJ, Jaffe H, Kim S, Zolyomi A, Catt KJ (1997) Isolation and molecular cloning of wortmannin-sensitive bovine type III phosphatidylinositol 4-kinases. J Biol Chem 272:18358-18366

Balla A, Tuymetova G, Barshishat M, Geiszt M, Balla T (2002) Characterization of type II phosphatidylinositol 4-kinase isoforms reveals association of the enzymes with endosomal vesicular compartments. J Biol Chem 277:20041-20050

Barylko B, Gerber SH, Binns DD, Grichine N, Khvotchev M, Sudhof TC Albanesi JP (2001) A novel family of phosphatidylinositol 4-kinases conserved from yeast to humans. J Biol Chem 276:7705-7708

Bascom RA, Srinivasan S, Nussbaum RL (1999) Identification and characterization of golgin-84, a novel Golgi integral membrane protein with a cytoplasmic coiled-coil domain. J Biol Chem 274:2953-2962

Berger P, Bonneick S, Willi S, Wymann M, Suter U (2002) Loss of phosphatase activity in myotubularin-related protein 2 is associated with Charcot-Marie-Tooth disease type 4B1. Hum Mol Genet 11:1569-1579

Blondeau F, Laporte J, Bodin S, Superti-Furga G, Payrastre B, Mandel JL (2000) Myotubularin, a phosphatase deficient in myotubular myopathy, acts on phosphatidylinositol 3-kinase and phosphatidylinositol 3-phosphate pathway. Hum Mol Genet 9:2223-2229

Brown WJ, DeWald DB, Emr SD, Plutner H, Balch WE (1995) Role for phosphatidylinositol 3-kinase in the sorting and transport of newly synthesized lysosomal enzymes in mammalian cells. J Cell Biol 130:781-796

Bruns JR, Ellis MA, Jeromin A, Weisz OA (2002) Multiple roles for phosphatidylinositol 4-kinase in biosynthetic transport in polarized Madin-Darby canine kidney cells. J Biol Chem 277:2012-2018

Buj-Bello A, Laugel V, Messaddeq N, Zahreddine H, Laporte J, Pellissier JF, Mandel JL (2002) The lipid phosphatase myotubularin is essential for skeletal muscle maintenance but not for myogenesis in mice. Proc Natl Acad Sci USA 99:15060-15065

Castellino AM, Parker GJ, Boronenkov IV, Anderson RA, Chao MV (1997) A novel interaction between the juxtamembrane region of the p55 tumor necrosis factor receptor and phosphatidylinositol-4-phosphate 5-kinase. J Biol Chem 272:5861-5870

Christoforidis S, Zerial M (2001) Purification of EEA1 from bovine brain cytosol using Rab5 affinity chromatography and activity assays. Methods Enzymol 329:120-132

Cockcroft S, De Matteis MA (2001) Inositol lipids as spatial regulators of membrane traffic. J Membr Biol 180:187-194

Cutler NS, Heitman J, Cardenas ME (1997) STT4 is an essential phosphatidylinositol 4-kinase that is a target of wortmannin in Saccharomyces cerevisiae. J Biol Chem 272:27671-27677

Dang H, Li Z, Skolnik EY, Fares H (2004) Disease-related myotubularins function in endocytic traffic in Caenorhabditis elegans. Mol Biol Cell 15:189-96

Davidson HW (1995) Wortmannin causes mistargeting of procathepsin D. evidence for the involvement of a phosphatidylinositol 3-kinase in vesicular transport to lysosomes. J Cell Biol 130:797-805

De Matteis MA, Morrow JS (1998) The role of ankyrin and spectrin in membrane transport and domain formation. Curr Opin Cell Biol 10:542-549

De Matteis M, Godi A, Corda D (2002) Phosphoinositides and the golgi complex. Curr Opin Cell Biol 14:434-447

Domin J, Gaidarov I, Smith ME, Keen JH, Waterfield MD (2000) The class II phospho-inositide 3-kinase PI3K-C2alpha is concentrated in the trans-Golgi network and present in clathrin-coated vesicles. J Biol Chem 275:11943-11950

Dressman MA, Olivos-Glander IM, Nussbaum RL, Suchy SF (2000) Ocrl1, a PtdIns(4,5)P(2) 5-phosphatase, is localized to the trans-Golgi network of fibroblasts and epithelial cells. J Histochem Cytochem 48:179-190

Flanagan CA, Schnieders EA, Emerick AW, Kunisawa R, Admon A, Thorner J (1993) Phosphatidylinositol 4-kinase: gene structure and requirement for yeast cell viability. Science 262:1444-1448

Fruman DA, Meyers RE, Cantley LC (1998) Phosphoinositide kinases. Annu Rev Biochem 67:481-507

Gaidarov I, Smith ME, Domin J, Keen JH (2001) The class II phosphoinositide 3-kinase C2alpha is activated by clathrin and regulates clathrin-mediated membrane trafficking. Mol Cell 7:443-449

Gehrmann T, Heilmeyer LM Jr (1998) Phosphatidylinositol 4-kinases. Eur J Biochem 253:357-370

Gillooly DJ, Morrow IC, Lindsay M, Gould R, Bryant NJ, Gaullier, JM, Parton RG, Stenmark H (2000) Localization of phosphatidylinositol 3-phosphate in yeast and mammalian cells. EMBO J 19:4577-4588

Godi A, Pertile P, Meyers R, Marra P, Di Tullio G, Iurisci C, Luini A, Corda D, De Matteis MA (1999) ARF mediates recruitment of PtdIns-4-OH kinase-beta and stimulates synthesis of PtdIns(4,5)P2 on the Golgi complex. Nat Cell Biol 1:280-287

Godi A, Santone I, Pertile P, Marra P, Di Tullio G, Luini A, Corda D, De Matteis MA (1999) ADP-ribosylation factor regulates spectrin skeleton assembly on the Golgi complex by stimulating phosphatidylinositol 4,5-bisphosphate synthesis. Biochem Soc Trans 27:638-642

Guipponi M, Tapparel C, Jousson O, Scamuffa N, Mas C, Rossier C, Hutter P, Meda P, Lyle R, Reymond A, Antonarakis SE (2001) The murine orthologue of the Golgi-localized TPTE protein provides clues to the evolutionary history of the human TPTE gene family. Hum Genet 109:569-575

Guo J, Wenk MR, Pellegrini L, Onofri F, Benfenati F, De Camilli P (2003) Phosphatidy-linositol 4-kinase type IIalpha is responsible for the phosphatidylinositol 4-kinase activity associated with synaptic vesicles. Proc Natl Acad Sci USA 100:3995-4000

Hama H, Schnieders EA, Thorner J, Takemoto JY, DeWald DB (1999) Direct involvement of phosphatidylinositol 4-phosphate in secretion in the yeast *Saccharomyces cerevisiae*. J Biol Chem 274:34294-34300

Hay JC, Fisette PL, Jenkins GH, Fukami K, Takenawa T, Anderson RA, Martin TF (1995) ATP-dependent inositide phosphorylation required for Ca(2+)-activated secretion. Nature 374:173-177

Hayes S, Chawla A, Corvera S (2002) TGF beta receptor internalization into EEA1-enriched early endosomes: role in signaling to Smad2. J Cell Biol 158:1239-1249

Hendricks KB, Wang BQ, Schnieders EA, Thorner J (1999) Yeast homologue of neuronal frequenin is a regulator of phosphatidylinositol-4-OH kinase. Nat Cell Biol 1:234-241

Honda A, Nogami M, Yokozeki T, Yamazaki M, Nakamura H, Watanabe H, Kawamoto K, Nakayama K, Morris AJ, Frohman MA, Kanaho Y (1999) Phosphatidylinositol 4-phosphate 5-kinase alpha is a downstream effector of the small G protein ARF6 in membrane ruffle formation. Cell 99:521-532

Hughes WE, Cooke FT, Parker PJ (2000) Sac phosphatase domain proteins. Biochem J 350:337-52

Ijuin T, Mochizuki Y, Fukami K, Funaki M, Asano T, Takenawa T (2000) Identification and characterization of a novel inositol polyphosphate 5-phosphatase. J Biol Chem 275:10870-5

Ikonomov OC, Sbrissa D, Mlak K, Kanzaki M, Pessin J, Shisheva A (2002) Functional dissection of lipid and protein kinase signals of PIKfyve reveals the role of PtdIns 3,5-P2 production for endomembrane integrity. J Biol Chem 277:9206-9211

Ikonomov OC, Sbrissa D, Foti M, Carpentier JL, Shisheva A (2003) PIKfyve controls fluid phase endocytosis but not recycling/degradation of endocytosed receptors or sorting of procathepsin D by regulating multivesicular body morphogenesis. Mol Biol Cell 14:4581-4591

Janne PA, Suchy SF, Bernard D, MacDonald M, Crawley J, Grinberg A, Wynshaw-Boris A, Westphal H, Nussbaum RL (1998) Functional overlap between murine Inpp5b and Ocrl1 may explain why deficiency of the murine ortholog for OCRL1 does not cause Lowe syndrome in mice. J Clin Invest 101:2042-2053

Jefferson AB, Majerus PW (1995) Properties of type II inositol polyphosphate   5-phosphatase. J Biol Chem 270:9370-9377

Jenkins GH, Fisette PL, Anderson RA (1994) Type I phosphatidylinositol 4-phosphate 5-kinase isoforms are specifically stimulated by phosphatidic acid. J Biol Chem 269:11547-11554

Jones DH, Morris JB, Morgan CP, Kondo H, Irvine RF, Cockcroft S (2000) Type I phosphatidylinositol 4-phosphate 5-kinase directly interacts with ADP-ribosylation factor 1 and is responsible for phosphatidylinositol 4,5-bisphosphate synthesis in the golgi compartment. J Biol Chem 275:13962-13966

Kihara A, Noda T, Ishihara N, Ohsumi Y (2001) Two distinct Vps34 phosphatidylinositol 3-kinase complexes function in autophagy and carboxypeptidase Y sorting in *Saccharomyces cerevisiae*. J Cell Biol 152:519-530

Kim SA, Taylor GS, Torgersen KM, Dixon JE (2002) Myotubularin and MTMR2, phosphatidylinositol 3-phosphatases mutated in myotubular myopathy and type 4B Charcot-Marie-Tooth disease. J Biol Chem 277:4526-4531

Kim SA, Vacratsis PO, Firestein R, Cleary ML, Dixon JE (2003) Regulation of myotubularin-related (MTMR)2 phosphatidylinositol phosphatase by MTMR5, a catalytically inactive phosphatase. Proc Natl Acad Sci USA 100:4492-4497

Kisseleva MV, Wilson MP, Majerus PW (2000) The isolation and characterization of a cDNA encoding phospholipid-specific inositol polyphosphate 5-phosphatase. J Biol Chem 275:20110-6

Kong AM, Speed CJ, O'Malley CJ, Layton MJ, Meehan T, Loveland KL, Cheema S, Ooms LM, Mitchell CA (2000) Cloning and characterization of a 72-kDa inositol-polyphosphate 5-phosphatase localized to the Golgi network. J Biol Chem 275:24052-24064

Laporte J, Biancalana V, Tanner SM, Kress W, Schneider V, Wallgren-Pettersson C, Herger F, Buj-Bello A, Blondeau F, Liechti-Gallati S, Mandel JL (2000) MTM1 mutations in X-linked myotubular myopathy. Hum Mutat 15:393-409

Laporte J, Blondeau F, Buj-Bello A, Mandel JL (2001) The myotubularin family: from genetic disease to phosphoinositide metabolism. Trends Genet 17:221-228

Laporte J, Blondeau F, Gansmuller A, Lutz Y, Vonesch JL, Mandel JL (2002) The PtdIns3P phosphatase myotubularin is a cytoplasmic protein that also localizes to Rac1-inducible plasma membrane ruffles. J Cell Sci 115:3105-3117

Lee JO, Yang H, Georgescu MM, Di Cristofano A, Maehama T, Shi Y, Dixon JE, Pandolfi P, Pavletich NP (1999) Crystal structure of the PTEN tumor suppressor: implications for its phosphoinositide phosphatase activity and membrane association. Cell 99:323-334

Lloyd TE, Atkinson R, Wu MN, Zhou Y, Pennetta G, Bellen HJ (2002) Hrs regulates endosome membrane invagination and tyrosine kinase receptor signaling in *Drosophila*. Cell 108:261-269

Loijens JC, Boronenkov IV, Parker GJ, Anderson RA (1996) The phosphatidylinositol 4-phosphate 5-kinase family. Adv Enzyme Regul 36:115-140

Maehama T, Taylor GS, Dixon JE (2001) PTEN and myotubularin: novel phosphoinositide phosphatases. Annu Rev Biochem 70:247-279

Mayinger P, Bankaitis VA, Meyer DI (1995) Sac1p mediates the adenosine triphosphate transport into yeast endoplasmic reticulum that is required for protein translocation. J Cell Biol 131:1377-1386

Merlot S, Meili R, Pagliarini DJ, Maehama T, Dixon JE, Firtel RA (2003) A PTEN-related 5-phosphatidylinositol phosphatase localized in the Golgi. J Biol Chem 278:39866-39873

Meyers R, Cantley LC (1997) Cloning and characterization of a wortmannin-sensitive human phosphatidylinositol 4-kinase. J Biol Chem 272:4384-4390

Minagawa T, Ijuin T, Mochizuki Y, Takenawa T (2001) Identification and characterization of a sac domain-containing phosphoinositide 5-phosphatase. J Biol Chem 276:22011-22015

Miura S, Takeshita T, Asao H, Kimura Y, Murata K, Sasaki Y, Hanai JI, Beppu H, Tsukazaki T, Wrana JL, Miyazono K, Sugamura K (2000) Hgs (Hrs), a FYVE domain protein, is involved in Smad signaling through cooperation with SARA. Mol Cell Biol 20:9346-9355

Mochizuki Y, Takenawa T (1999) Novel inositol polyphosphate 5-phosphatase localizes at membrane ruffles. J Biol Chem 274:36790-36795

Mora S, Durham PL, Smith JR, Russo AF, Jeromin A, Pessin JE (2002) NCS-1 inhibits insulin-stimulated GLUT4 translocation in 3T3L1 adipocytes through a phosphatidylinositol 4-kinase-dependent pathway. J Biol Chem 277:27494-27500

Moritz A, De Graan PN, Gispen WH, Wirtz KW (1992) Phosphatidic acid is a specific activator of phosphatidylinositol-4-phosphate kinase. J Biol Chem 267:7207-7210

Naga Prasad SV, Laporte SA, Chamberlain D, Caron MG, Barak L, Rockman HA (2002) Phosphoinositide 3-kinase regulates beta2-adrenergic receptor endocytosis by AP-2 recruitment to the receptor/beta-arrestin complex. J Cell Biol 158:563-575

Nandurkar HH, Layton M, Laporte J, Selan C, Corcoran L, Caldwell KK, Mochizuki Y, Majerus PW, Mitchell CA (2003) Identification of myotubularin as the lipid phosphatase catalytic subunit associated with the 3-phosphatase adapter protein, 3-PAP. Proc Natl Acad Sci USA 100:8660-8665

Nielsen E, Christoforidis S, Uttenweiler-Joseph S, Miaczynska M, Dewitte F, Wilm M, Hoflack B, Zerial M (2000) Rabenosyn-5, a novel Rab5 effector, is complexed with hVPS45 and recruited to endosomes through a FYVE finger domain. J Cell Biol 151:601-612

Odorizzi G, Babst M, Emr SD (1998) Fab1p PtdIns(3)P 5-kinase function essential for protein sorting in the multivesicular body. Cell 95:847-858

Ono F, Nakagawa T, Saito S, Owada Y, Sakagami H, Goto K, Suzuki M, Matsuno S, Kondo H (1998) A novel class II phosphoinositide 3-kinase predominantly expressed in the liver and its enhanced expression during liver regeneration. J Biol Chem 273:7731-7736

Pike LJ (1992) Phosphatidylinositol 4-kinases and the role of polyphosphoinositides in cellular regulation. Endocr Rev 13:692-706

Rajebhosale M, Greenwood S, Vidugiriene J, Jeromin A, Hilfiker S (2003) Phosphatidylinositol 4-OH kinase is a downstream target of neuronal calcium sensor-1 in enhancing exocytosis in neuroendocrine cells. J Biol Chem 278:6075-6084

Rameh LE, Tolias KF, Duckworth BC, Cantley LC (1997) A new pathway for synthesis of phosphatidylinositol-4,5-bisphosphate. Nature 390:192-196

Rohde HM, Cheong FY, Konrad G, Paiha K, Mayinger P, Boehmelt G (2003) The human phosphatidylinositol phosphatase SAC1 interacts with the coatomer I complex. J Biol Chem 278:52689-52699

Rozelle AL, Machesky LM, Yamamoto M, Driessens MH, Insall RH, Roth MG, Luby-Phelps K, Marriott G, Hall A, Yin HL (2000) Phosphatidylinositol 4,5-bisphosphate induces actin-based movement of raft-enriched vesicles through WASP-Arp2/3. Curr Biol 10:311-320

Sarnat HB (1990) Myotubular myopathy: arrest of morphogenesis of myofibres associated with persistence of fetal vimentin and desmin. Four cases compared with fetal and neonatal muscle. Can J Neurol Sci 17:109-123

Sbrissa D, Ikonomov OC, Shisheva A (1999) PIKfyve, a mammalian ortholog of yeast Fab1p lipid kinase, synthesizes 5-phosphoinositides. Effect of insulin. J Biol Chem 274:21589-21597

Sbrissa D, Ikonomov OC, Deeb R, Shisheva A (2002) Phosphatidylinositol 5-phosphate biosynthesis is linked to PIKfyve and is involved in osmotic response pathway in mammalian cells. J Biol Chem 277:47276-47284

Siddhanta A, Backer JM, Shields D (2000) Inhibition of phosphatidic acid synthesis alters the structure of the Golgi apparatus and inhibits secretion in endocrine cells. J Biol Chem 275:12023-12031

Siddhanta A, Radulescu A, Stankewich MC, Morrow JS, Shields D (2003) Fragmentation of the Golgi apparatus. A role for beta III spectrin and synthesis of phosphatidylinositol 4,5-bisphosphate. J Biol Chem 278:1957-1965

Simonsen A, Lippe R, Christoforidis S, Gaullier JM, Brech A, Callaghan J, Toh BH, Murphy C, Zerial M, Stenmark H (1998) EEA1 links PI(3)K function to Rab5 regulation of endosome fusion. Nature 394:494-498

Stack JH, DeWald DB, Takegawa K, Emr SD (1995) Vesicle-mediated protein transport: regulatory interactions between the Vps15 protein kinase and the Vps34 PtdIns 3-kinase essential for protein sorting to the vacuole in yeast. J Cell Biol 129:321-334

Stamnes M (2002) Regulating the actin cytoskeleton during vesicular transport. Curr Opin Cell Biol 14:428-433

Suchy SF, Nussbaum RL (2002) The deficiency of PIP2 5-phosphatase in Lowe syndrome affects actin polymerization. Am J Hum Genet 71:1420-1427

Sulis ML, Parsons R (2003) PTEN: from pathology to biology. Trends Cell Biol 13:478-483

Sweeney DA, Siddhanta A, Shields D (2002) Fragmentation and re-assembly of the Golgi apparatus *in vitro*. A requirement for phosphatidic acid and phosphatidylinositol 4,5-bisphosphate synthesis. J Biol Chem 277:3030-3039

Takegawa K, DeWald DB, Emr SD (1995) *Schizosaccharomyces pombe* Vps34p, a phosphatidylinositol-specific PI 3-kinase essential for normal cell growth and vacuole morphology. J Cell Sci 108:3745-3756

Tolias KF, Cantley LC, Carpenter CL (1995) Rho family GTPases bind to phosphoinositide kinases. J Biol Chem 270:17656-17659

Tolias KF, Rameh LE, Ishihara H, Shibasaki Y, Chen J, Prestwich GD, Cantley LC, Carpenter CL (1998) Type I phosphatidylinositol-4-phosphate 5-kinases synthesize the novel lipids phosphatidylinositol 3,5-bisphosphate and phosphatidylinositol 5-phosphate. J Biol Chem 273:18040-18046

Tolias K, Carpenter CL (2000) *In vitro* interaction of phosphoinositide-4-phosphate 5-kinases with Rac. Methods Enzymol 325:190-200

Trotter PJ, Wu WI, Pedretti J, Yates R, Voelker DR (1998) A genetic screen for amino-phospholipid transport mutants identifies the phosphatidylinositol 4-kinase, Stt4p, as an essential component in phosphatidylserine metabolism. J Biol Chem 273:13189-13196

Tsukazaki T, Chiang TA, Davison AF, Attisano L, Wrana JL (1998) SARA, a FYVE domain protein that recruits Smad2 to the TGFbeta receptor. Cell 95:779-791

Walch-Solimena C, Novick P (1999) The yeast phosphatidylinositol-4-OH kinase pik1 regulates secretion at the Golgi. Nat Cell Biol 1:523-525

Walker SM, Downes CP, Leslie NR (2001) TPIP: a novel phosphoinositide 3-phosphatase. Biochem J 360:277-283

Wang YJ, Wang J, Sun HQ, Martinez M, Sun YX, Macia E, Kirchhausen T, Albanesi JP, Roth MG, Yin HL (2003) Phosphatidylinositol 4 phosphate regulates targeting of clathrin adaptor AP-1 complexes to the Golgi. Cell 114:299-310

Wei YJ, Sun HQ, Yamamoto M, Wlodarski P, Kunii K, Martinez M, Barylko B, Albanesi JP, Yin HL (2002) Type II phosphatidylinositol 4-kinase beta is a cytosolic and peripheral membrane protein that is recruited to the plasma membrane and activated by Rac-GTP. J Biol Chem 277:46586-46593

Weisz OA, Gibson GA, Leung SM, Roder J, Jeromin A (2000) Overexpression of frequenin, a modulator of phosphatidylinositol 4-kinase, inhibits biosynthetic delivery of an apical protein in polarized madin-darby canine kidney cells. J Biol Chem 275:24341-24347

Whitters EA, Cleves AE, McGee TP, Skinner HB, Bankaitis VA (1993) SAC1p is an integral membrane protein that influences the cellular requirement for phospholipid transfer protein function and inositol in yeast. J Cell Biol 122:79-94

Wishart MJ, Taylor GS, Slama JT, Dixon JE (2001) PTEN and myotubularin phosphoinositide phosphatases: bringing bioinformatics to the lab bench. Curr Opin Cell Biol 13:172-181

Wishart MJ, Dixon JE (2002) PTEN and myotubularin phosphatases: from 3-phosphoinositide dephosphorylation to disease. Trends Cell Biol 12:579-585

Wong K, Cantley LC (1994) Cloning and characterization of a human phosphatidylinositol 4-kinase. J Biol Chem 269:28878-28884

Wong K, Meyers dd R, Cantley LC (1997) Subcellular locations of phosphatidylinositol 4-kinase isoforms. J Biol Chem 272:13236-13241

Woscholski R, Parker PJ (1997) Inositol lipid 5-phosphatases--traffic signals and signal traffic. Trends Biochem Sci 22:427-31

Yoshida S, Ohya Y, Goebl M, Nakano A, Anraku Y (1994) A novel gene, STT4, encodes a phosphatidylinositol 4-kinase in the PKC1 protein kinase pathway of *Saccharomyces cerevisiae*. J Biol Chem 269, 1166-1172

Zerial M, McBride H (2001) Rab proteins as membrane organizers. Nat Rev Mol Cell Biol 2:107-117

Zhang X, Hartz PA, Philip E, Racusen LC, Majerus PW (1998) Cell lines from kidney proximal tubules of a patient with Lowe syndrome lack OCRL inositol polyphosphate 5-phosphatase and accumulate phosphatidylinositol 4,5-bisphosphate. J Biol Chem 273:1574-1582

Zhao X, Varnai P, Tuymetova G, Balla A, Toth ZE, Oker-Blom C, Roder J, Jeromin A, Balla T (2001) Interaction of neuronal calcium sensor-1 (NCS-1) with phosphatidylinositol 4-kinase beta stimulates lipid kinase activity and affects membrane trafficking in COS-7 cells. J Biol Chem 276:40183-40189

Zhou XP, Waite KA, Pilarski R, Hampel H, Fernandez MJ, Bos C, Dasouki M, Feldman GL, Greenberg LA, Ivanovich J, Matloff E, Patterson A, Pierpont ME, Russo D, Nassif NT, Eng C (2003) Germline PTEN promoter mutations and deletions in Cowden/Bannayan-Riley-Ruvalcaba syndrome result in aberrant PTEN protein and dysregulation of the phosphoinositol-3-kinase/Akt pathway. Am J Hum Genet 73:404-411

De Matteis, Maria Antonietta
    Department of Cell Biology and Oncology, Consorzio Mario Negri Sud, Via Nazionale, 66030 Santa Maria Imbaro (Chieti), Italy
    demattei@negrisud.it

Di Campli, Antonella
    Department of Cell Biology and Oncology, Consorzio Mario Negri Sud, Via Nazionale, 66030 Santa Maria Imbaro (Chieti), Italy

Godi, Anna
    Department of Cell Biology and Oncology, Consorzio Mario Negri Sud, Via Nazionale, 66030 Santa Maria Imbaro (Chieti), Italy
    godi@negrisud.it

Regulation of exocytotic events by centrosome-analogous structures

Christof Taxis and Michael Knop

Abstract

Centrosomes, spindle pole bodies, and related structures, generally termed micro-tubule organising centres (MTOCs; Pickett-Heaps 1969), are best understood in the context of their function in the organisation of the mitotic spindle, spindle positioning via astral microtubule organisation, and regulation of cell cycle checkpoints (Doxsey 2001; Pereira and Schiebel 2001). However, cytological studies performed between 1960 and 1980 depict the involvement of MTOCs in membrane related events. These studies extend our definition of the processes that MTOCs regulate.

In this review, we summarise some of this cytological knowledge in light of our current understanding of how centrosomes work. The focus of this review is on centrosome related membrane biogenesis processes, where a new membrane is formed from clearly distinct precursors. Central to this discussion is our current understanding of one example that takes place during yeast meiosis.

1 MTOCs: a brief overview

Structurally, MTOCs are extremely diverse (Marshall 2001; Chapmann et al. 2000; Heywood and Magee 1976). The most prominent difference is the presence or absence of centrioles, tubulin-containing cylindrical structures that recruit the pericentriolar matrix, which contains the different functional components of centrosomes. Centrioles are not obligate components of MTOCs; lower eukaryotes such as yeasts, instead, contain spindle pole bodies (SPBs), plaque-like structures, which are inserted in the nuclear envelope. In protozoa, both, SPBs (often termed spindle plaques), as well as centrioles have been reported, in addition to other highly specialized MTOC structures. During the protist life cycle, these MTOCs appear to fulfil specific functions for discrete steps in cellular division and differentiation. *S. pombe* also contains MTOCs in addition to the SPBs, which nucleate various types of cytoplasmic microtubules. The microtubule nucleating activity of MTOCs requires a protein complex containing γ-tubulin, GRIP domain proteins and additional proteins as well as receptors that recruit the γ-tubulin complex (Zheng et al. 1995; Knop et al. 1997) to microtubule nucleating sites on the MTOCs, for example, to the nuclear and the cytoplasmic plaque of the SPB, as

Topics in Current Genetics, Vol. 10
S. Keränen, J. Jäntti (Eds.): Regulatory Mechanisms of Intracellular Membrane Transport
DOI 10.1007/b98734 / Published online: 16 June 2004
© Springer-Verlag Berlin Heidelberg 2004

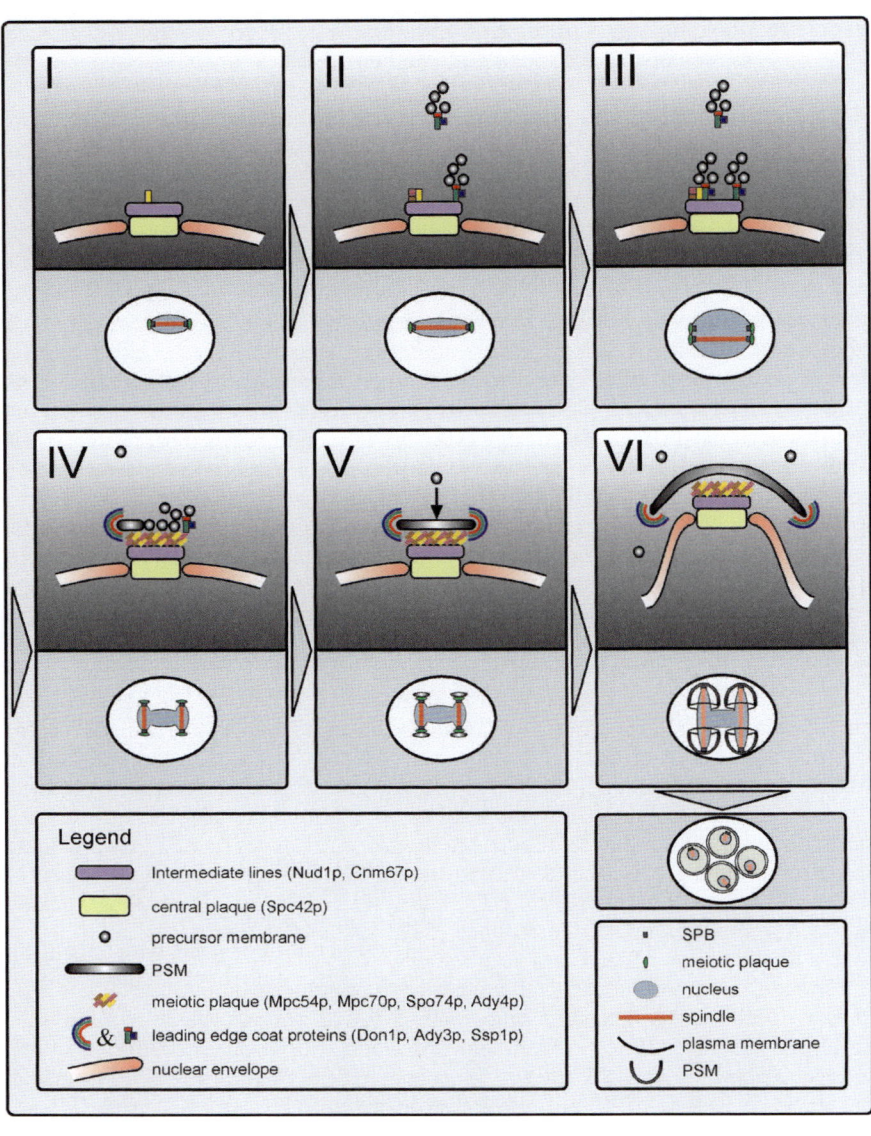

Fig. 1. PSM assembly in budding yeast. Function and localisation of proteins associated with initiation and shaping of the prospore membrane (PSM) in budding yeast during cellularization and spore formation in meiosis II. The cartoons illustrate the processes at individual spindle pole bodies (SPBs) (upper half), and an overview about the cell (lower half). The legend lists the molecules and structures that are shown. Usually, only one molecule is depicted as a representative for many others of the same kind. The cartoons put an emphasis on a specific set of proteins uniquely required for PSM formation, molecules associated with more general processes such as exocytosis and vesicle fusion are not depicted. A listing of the already known proteins and their localisation (if known) as well as a discussion of their function in this process can be found in the review by Shimoda (2004). Processes that lead to the formation of the PSM at the SPB start with the synthesis of specific components. (I) Some meiotic plaque components can already be detected during meiosis I at the SPBs, although, at a very low concentration (unpublished observation). (II) During later stages of meiosis I, also LEP coat markers become localised to the SPBs. (III) With exit of meiosis I and entry of meiosis II, the SPBs duplicate, and meiotic plaque components as well as LEP coat components become more abundant at the SPBs. Precursor membranes, that are decorated with LEP coat components are at this stage visible in the cytoplasm, as discrete dots. (IV) With entry into meiosis II, the precursor membranes assemble at the SPBs, and the meiotic plaque components become abundant at the SPBs. (V) PSM assembly starts during metaphase or with the beginning of anaphase of meiosis II. (VI) During PSM growth, LEP coats are visible as doughnut shaped structures inside the cell.

shown in budding yeast (Knop and Schiebel 1998; Knop et al. 1999). Mammalian centrosomes appear to have different receptors for the γ-tubulin complex (Takahashi et al. 2002). Binding of the γ-tubulin complex may be one way the microtubule activity of an MTOC is regulated; other modes of regulation are thought to occur through kinases (Pereira et al. 1998; Vogel et al. 2001). In order to understand MTOC function, one needs to know the molecular composition of these structures and in particular the dynamic localisation, activity, and sequestration of the component proteins. Many regulators of cell cycle progression and/or cellular differentiation become enriched at distinct intracellular sites, often on specific MTOCs. Currently, proteomics approaches have begun to define the molecules found at MTOCs (Wigge et al. 1998; Lange et al. 2000; Andersen et al. 2003), and functional analysis of individual proteins broadens our understanding of the different processes that constitute the centrosome cycle and define centrosome/SPB functions.

2 Spindle pole bodies and plasma membrane biogenesis in yeast meiosis

The yeasts, S. pombe and S. cerevisiae provide a classic system to study the intracellular assembly of a new plasma membrane. In fact, one of the earliest systematic genetic screens of S. cerevisiae addressed morphogenetic differentiation during sporulation (Esposito and Esposito 1969). De novo plasma membrane biogenesis occurs at the spindle pole bodies (SPBs) with the onset of meiosis II.

During formation, these membranes are termed fore spore membrane (FSM; *S. pombe*) and prospore membrane (PSM; *S. cerevisiae*). Two recent reviews by Shimoda (2004) and Moreno-Borchart and Knop (2003) summarize the current knowledge on FSM/PSM formation, so we will not discuss this in detail. Rather, we will focus on comparing PSM formation with mitotic plasma membrane biogenesis and creation of the *apicomplexan* inner membrane complex (IMC). We will emphasize general principles and the function of specific molecules.

Figure 1 illustrates the process of PSM generation in *S. cerevisiae*. We can distinguish three different steps of PSM formation, each will be discussed in the following chapters.

2.1 PSM initiation

During PSM formation, the membrane appears to arise *de novo*. There is no evidence at present time that an existing organellar structure, such as an endosome or a Golgi stack, acts as an acceptor membrane during the formation of the PSMs at the SPBs. Studies of mutants or conditions that specifically impair the ability of the SPBs to assemble the PSMs (Davidow et al. 1980; Okamoto and Iino 1982; Knop and Strasser 2000; Bajgier et al. 2001; Nickas et al. 2003), show vesicles of homogenous size (approx. 65-70 nm) as the earliest visible PSM structure in *S. cerevisiae*. This suggests that PSM formation occurs through homotypic fusion of these vesicles, which have the classical size of secretory vesicles. Similar vesicles accumulate in the bud of mitotic cells in mutants that affect exocytosis via impairing vesicle fusion with the plasma membrane (Aalto et al. 1997; Guo et al. 2000). Homotypic membrane fusion may underlie formation of an initial acceptor compartment, the starting structure for the PSM, while subsequent vesicle delivery then leads to growth of the PSM. The shape of the involved membranes suggests that initiation and subsequent growth of the membranes are two different processes. The first step is quite unusual, although homotypic membrane fusion events in the late secretory pathway have been proposed to occur in several systems, for example, melanosome biogenesis (Wendler et al. 2001), but never from such small vesicles and never as the initiating event for a new and completely different membrane system. The origin and targeting of these vesicles to the SPBs, and initiation of vesicle fusion locally at the SPBs are processes that raise a number of questions. *Where do the vesicles come from?* The involvement of late acting *SEC*-genes on PSM formation (Neiman 1998) suggests the Golgi as the origin of the membranes. However, it is unclear whether they are specific vesicles with a specific sorted cargo or, alternatively, a redirection of the entire pathway to the SPBs. *How do the vesicles find the SPBs?* If one would draw analogy to mitotic vesicle delivery to the bud, polar SPB nucleated and anchored actin filaments would be the answer. However, there is no evidence for this yet. *How do the vesicles bind to the SPBs?* In *S. cerevisiae*, binding of vesicles to the SPBs appears to occur via the interaction of Ady3p with Ssp1p on the vesicles and with Cnm67p and Nud1p of the SPBs. However, *ady3* mutants are still able to initiate PSM formation, suggesting additional interactions between the vesicles and the SPBs (see Fig. 2).

How is the localised vesicle fusion at the SPBs regulated? Initiation of PSM must be regulated in relation to the meiotic cell cycle, it occurs after separation of the SPBs in meiosis II, around metaphase. The significance of this is illustrated by localisation of the three known markers for these vesicles (Don1p, Ady3p, and Ssp1p) to the SPBs already prior to metaphase of meiosis II, during the later phases of meiosis I. Thus, tight control applied by cell cycle regulatory machinery must be considered. The SPBs, but also centrosomes are sites where such regulatory machinery, for example Cdc28 (Cdc2), cyclins as well as many other kinases can be found (Huh et al. 2003; Jackman et al. 2003), thus, vesicle fusion might be directly regulated. On the vesicle side, no molecules specifically involved in initiation of vesicle fusion have been identified. Even in the absence of Don1p, Ady3p, or Ssp1p the initiation of PSM biogenesis is not affected (Knop and Strasser 2000; Moreno-Borchart et al. 2001). Several *sec*-mutants affecting proteins involved in the last step of secretion, fusion of vesicles with the plasma membrane, show also impaired prospore membrane assembly (Neiman 1998). This suggests that prospore membrane biogenesis relies on general exocytotic machinery for vesicle fusion. However, the question how vesicle fusion is regulated remains unanswered. On the SPB side, proteins necessary for the triggering of vesicle fusion have been identified. These proteins form a meiosis specific structure on the SPB, the meiotic plaque (MP), which forms a protein aqueous scaffold that the vesicles appear to align on. It consists of four proteins, Mpc54p, Mpc70p, Spo74p, and Ady4p. The first three proteins are essential for membrane initiation and are apparently structural components. The fourth member, Ady4p exhibits an auxiliary function, it is not essential for membrane initiation, but promotes subsequent growth and anchorage of the membrane (Nickas et al. 2003). Two of the three essential proteins, Mpc54p and Mpc70p are coiled-coil proteins. Interestingly, Mpc70p is a distant homolog of Spc72p, which is required for anchorage of the microtubule nucleating γ-tubulin complex to the SPB in mitotic cells (Bajgier et al. 2001; Knop and Schiebel 1998). Mpc70p appears to function as a hetero dimer with Spo74p (Nickas et al. 2003), a non coiled-coil protein. Interestingly, the homology of Mpc70p and Spc72p lies within the central coiled-coil domain, whereas the N-terminal domain of Spc72p that directly interacts with microtubule nucleating machinery has been swapped for a novel domain with a yet unidentified function. Figure 2 depicts the interactions among the known components that are discussed in this review. Binding of the vesicles to the SPB is independent of the MP, but the MP appears to be necessary for membrane fusion. Perhaps the alignment of the vesicles along the scaffold or recruitment of other factors via the MP to the SPB is essential for this process. After initiation of membrane fusion a flat cysternae as large as the area of the MP is formed. Upon assembly, the PSM stays in direct contact with the MP, probably directly anchored to the MP proteins via a direct protein-membrane interaction until spore formation has been completed. At this point, the meiotic plaque disassembles and the nuclei are released into the spore cytoplasm (Knop and Strasser 2000).

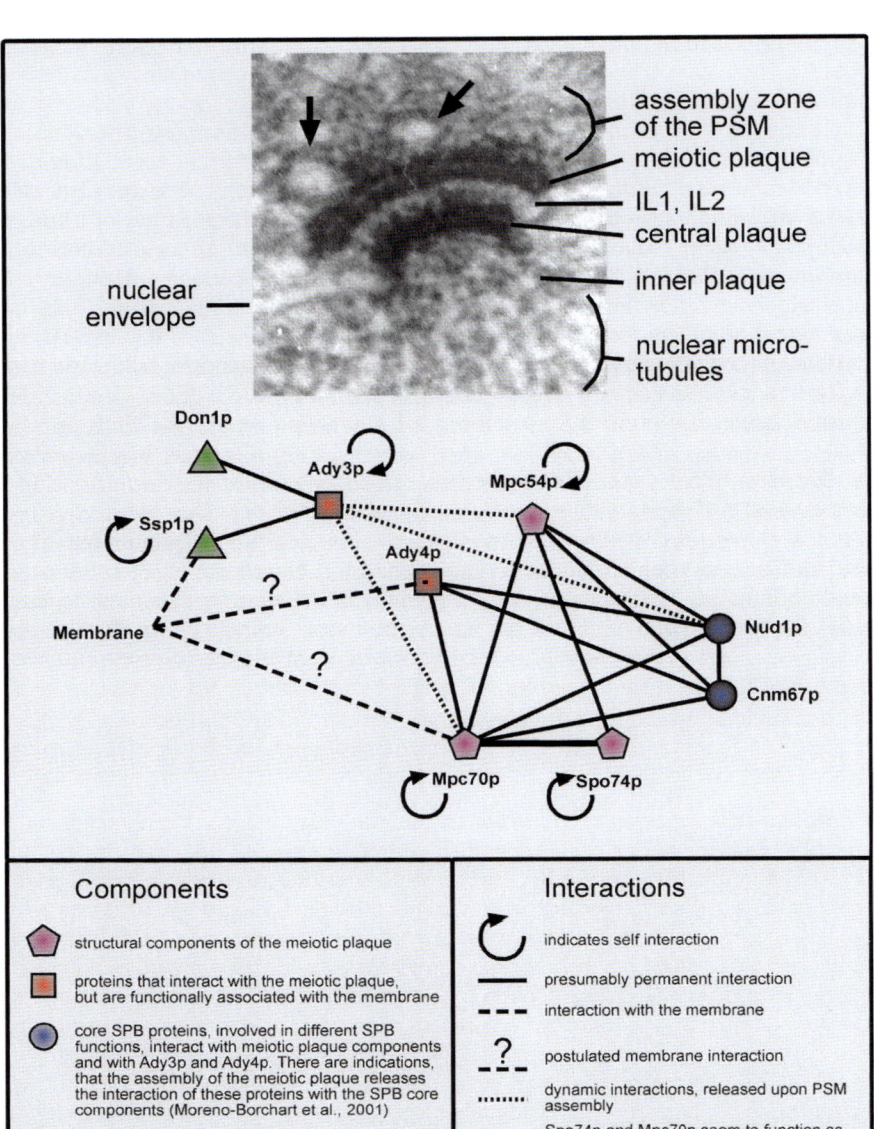

Fig. 2. Structure and proteins of the meiotic SPB of budding yeast. (I) The electron micrograph depicts an SPB from a cell in early stages of meiosis II with fully assembled but not yet elongated spindles. The meiotic plaque is assembled, while vesicular structures (arrows) that have not yet fused to a continuous PSM are detected bound to the meiotic plaque. Around the membranes, a diffuse mass of proteins can be seen as well. The nomenclature of the different SPB structures is in accordance with Wigge et al. (1998) and Knop and Strasser (2000). (II) The cartoon depicts the known interactions between the core SPB components, the proteins of the meiotic plaque and the LEP coat components. The legend explains the various types of interactions. Interactions of proteins with the membrane are indicated and are based on unpublished results (Knop et al. submitted; P. Maier and M. Knop unpublished data). The information given in the figure is based on the following publications: Knop and Strasser 2000; Moreno-Borchart et al. 2001; Bajgier et al. 2001; Nickas et al. 2003.

2.2 Growth and shaping of the PSM

Upon initiation of the PSM, the membrane becomes extended and starts to grow rapidly away from the SPB, roughly along the nuclear envelope. Obviously, membrane fusion now occurs between two structurally different membranes, with the PSM being clearly the acceptor membrane. Secretory vesicles, seen by EM in the vicinity of the PSM, appear of the same size as the vesicles responsible for the initiation of the membrane; however, they do not carry recognizable amounts of the three marker proteins Don1p, Ady3p, and Ssp1p that are characteristic for the vesicles that initiate the PSM. Thus, it might be plausible to assume that membrane initiation and membrane extension are different type of vesicle fusion steps, which might require different machinery for membrane fusion. Alternatively, membrane fusion may be subject to specific type of regulation. Another interesting point is the behaviour of the three marker proteins Don1p, Ady3p, and Ssp1p. During growth of the membrane, they form a protein aqueous coat, which covers the leading edge of the membrane. This coat can be seen by electron microscopy. Using GFP fusions to Don1p, the coat can be visualized during the entire growth phase of the PSM. Closure of the membrane occurs concomitantly with the disassembly of the coat. Don1p, Ady3p and Ssp1p are not essential for the formation of the membrane *per se*, however, in the absence of Ssp1p, which is required to recruit the other proteins to the membrane, the membrane appears to be missshaped. The membrane exhibits an exaggerated curvature, which leads to tight enclosure of the nuclear membrane, so that cytoplasmic contents are excluded from the forming PSM. This phenotype suggests that the function of the coat at the PSM is to keep the membrane open, while another, as of yet unidentified mechanism generates curvature of the membrane. This leads to the interesting question of how membrane curvature and shaping is achieved. The inner membrane complex (IMC) of *apicomplexan* protozoa is an analogous structure that requires specific microtubules but not actin filaments to form (Shaw et al. 2000). However, in *S. cerevisiae*, microtubules have not been found in association with the forming

membrane. In analogy to bud shaping in mitosis, one possibility would be that polarisation of secretion during PSM growth determines the place of membrane insertion and, thereby, the shape of the membrane. However, this would require a functional actin cytoskeleton, but depolymerization of actin using Latrunculin A does not impair the shape and growth of the PSM at all (Knop et al. submitted).

2.3 Closure of the PSM

The last step during formation of the PSM is closure of the membrane after nuclear fission at the end of telophase of meiosis II has taken place. This creates two membrane layers on top of each other. Obviously, the closure event of the PSM during yeast meiosis can be considered to be equivalent to cytokinesis in mitosis, because during PSM closure, the cytoplasm of the spore cell and the cytoplasm of the mother cell become topologically separated. Subsequent maturation processes then occur that lead to the formation of the spore wall between the two membranes (Briza et al. 1988). These processes are highly yeast specific and involve the formation of different spore wall layers that provide the necessary robustness towards environmental stress.

PSM closure has to be regulated in accordance with the meiotic cell cycle in order to occur with the correct timing during spore formation. We believe that regulated removal of the coat structure is the most obvious mechanism to account for closure of the membrane. This should allow the cell to decrease the hole size until only a small pinhole is left. This model does not explain the final closure, which leads to the physical separation of the inner and the outer membrane. This event may require specific machinery or it may occur spontaneously.

3 The IMC in *Apicomplexa*

The *Apicomplexa* are a diverse phylum of intracellular parasites that includes a number of important human and animal pathogens, such as *Toxoplasma, Plasmodium, Eimeria, Theileria*, and *Cryptosporidium*. All of them are obligate intracellular parasites that require host cells to grow. The *Apicomplexa* share a unique replication process that creates daughter parasites within a mother parasite cell by intracellular budding. In the case where one mitotic division leads to the formation of two new daughters, this process is termed endodyogeny. The process is termed schizogeny when many daughters are assembled concomitantly inside one mother cell from a corresponding number of previously formed nuclei. The formation of a specific new endomembrane system, the inner membrane complex (IMC), drives cellularization after serving to segregate daughter cell contents. The IMC also plays a pivotal role in parasite motility and host cell invasion. This latter process relies on a unique form of motility termed gliding motility. Thereby, actin-myosin motor driven translocation of apically secreted cell surface adhesions along the

long axis of the parasite promotes movement and host cell invasion (reviewed by Sibley 2003).

From a structural point of view, there is an amazing resemblance of the formation of IMCs during endodyogeny or schizogeny with PSM and FSM formation in yeasts. *Apicomplexan* IMCs are formed adjacent to nuclear lobes and they elongate through the cytoplasm around the underlying nuclear membrane and cytoplasmic content. Initiation of IMC formation occurs in close proximity to centrosome-like structures. The *Apicomplexa* have three kinds of MTOCs. First, many species have spindle plaques which are embedded in the nuclear envelope. There are also centrioles, consisting of singlet microtubules, associated with the spindle plaques (Morrissette and Sibley 2002). Moreover, *apicomplexans* have yet another MTOC called the polar ring (Russell and Burns 1984). The polar ring is a two-part structure; the posterior polar ring serves as a MTOC for the subpellicular microtubules that emanate underneath the surface of the IMC towards the posterior of the parasite. The anterior polar ring organizes an unusual type of 9 protofilament containing comma shaped microtubules that form the conoid, a cylindrical structure at the apex of the IMC (Hu et al. 2002). At the very apical region the IMC is open so that only the plasma membrane encloses the parasites. The anterior polar ring and the conoid are inserted in the IMC.

IMC formation is best studied in the apicomplexan *Toxoplasma gondii*. Live cell imaging and structural studies suggest an intimate relationship between the centrosome, Golgi and apicoplast (an essential plastid organelle) during IMC biogenesis (Pelletier et al. 2002; Striepen et al. 2000). The simplest explanation for this association is that biogenesis of the daughter IMCs must be closely controlled by the number of daughter structures to be formed. Since centrosomes and/or spindle plaques are subject to cell cycle controlled duplication processes, they would be an ideal site to regulate the process of daughter IMC formation. In yeast, this principle is exploited such that membrane formation occurs in direct association with the spindle pole bodies. *Apicomplexans* might make use of the specific properties of the cytoplasmic space around the centrosomes to restrict formation of one IMC per centrosome. For example, gradients of activities generated by centrosome-localised regulators could restrict the formation of the IMC to nearby areas of the cell. As centrosomes and related structures contain sequestered kinases and phosphatases, this mechanism might underlay regulated IMC assembly (Palazzo 2003). Alternatively, centrosome bound activities, may be localised on extensions of the centrosome or the related structures, trigger the initial steps. The precise mechanism of initiation of membrane formation, the nature and origin of the initial membranes and the exocytotic machinery involved are currently not known. The close proximity of the Golgi to the site of membrane formation suggests that either maturation of a late Golgi cisternae or exocytotic membranes from the Golgi are the donors of the membranes. However, defining the source of these membranes will require future studies.

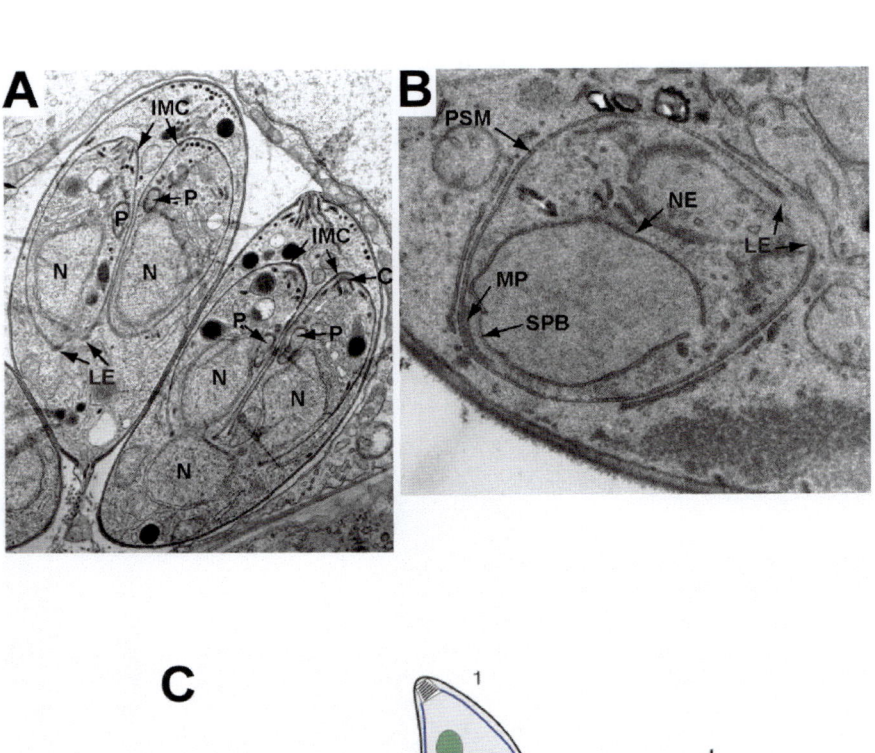

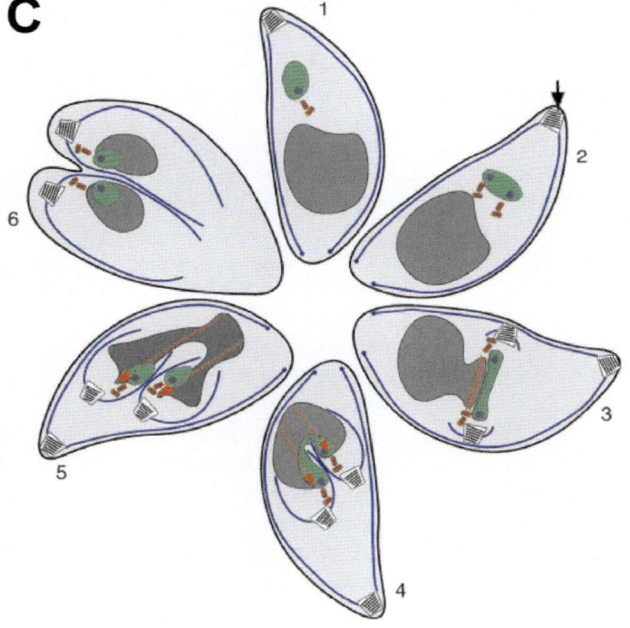

Fig. 3. Electron micrographs of IMC during *Toxoplasma gondii* daughter cell assembly and PSM formation in *S. cerevisiae*. (A) The electron micrograph shows a *Toxoplasma gondii* cell in process of intracellular budding. Some relevant structures are indicated. N, nucleus; C, centrioles; Con, conoid; IMC, inner membrane complex; LEP, leading edge protein coat; P, apicoplast. Picture courtesy of B. Striepen (Striepen et al. 2000). (B) Electron micrograph of a prospore membrane in budding yeast. Permanganese fixation allows the visualisation of all involved membranes while only the meiotic plaque from the SPB can be seen. The LEP coat cannot be seen using this fixation technique. SPB, position of the spindle pole body; NE, nuclear envelope; MP, meiotic plaque; LE, leading edge of the PSM. (C) *Toxoplasma gondii* cell cycle stages. The formation of the IMCs (blue) and new conoids (spiral structures, arrow) starts during mitosis after centrosome duplication (red), before nuclear fission (grey), and plastid (green) division (3). Upon completion of IMC assembly and probably disassembly of the mother conoid and IMC, the new cells acquire plasma membrane from the mother plasma membrane (5, 6, 1). Picture modified according to Striepen et al. (2000).

Interestingly, guidance, shaping and extension of the IMC depends on the subpellicular microtubules that extend in a slightly spiralled manner from the apical polar ring, suggesting intimate links between the membrane extending machinery and the microtubule cytoskeleton (Stokkermans et al. 1996; Morrissette et al. 1997; Morrissette and Sibley 2002).

Another interesting feature of IMC formation is that its formation is completely independent of the actin cytoskeleton, suggesting that polarized secretion is not the mechanism that generates shape of the IMC. Remarkably, PSM membrane shape and growth during meiosis is also completely unaffected by treatments with actin depolymerizing drugs (Knop et al. submitted). This is in contrast to bud growth and shape determination in mitosis that requires a polarized actin cytoskeleton (Karpova et al. 2000).

The PSM serves as the precursor structure for the plasma membrane of the spores, which arise at the end of meiosis. This means that it has to be completely closed at the end of the meiotic divisions. Closure occurs during cytokinesis and appears to involve the function of the leading edge protein (LEP) coat at the PSM. Electron micrographs show that the leading edges of the IMC in *Apicomplexa* are also covered by a coat-like structure (Fig. 3a), which strikingly resembles the LEP coat in yeast. Currently, there are no specific markers for this structure in the *Apicomplexa* known, therefore, we can only speculate about its composition. Additionally, electron micrographs have not elucidated whether the *Apicomplexan* IMC closes completely at the posterior end of the parasites. In either case, the subpellicular microtubules end before the posterior end of the parasite IMC, suggesting that something else is necessary in this region for shaping of the IMC, may be the structure that is akin to the yeast LEP coat.

4 Centrioles, basal bodies and plasma membrane

Basal bodies are centriole-related structures that contain the characteristic cylindrical bundle of 9 microtubule triplets. Basal bodies are formed by the addition of transition zone components to the plus end of centrioles and function in the formation of flagellar or ciliary axonemes. Basal body formation can be regulated by individual cells, for example, to control the number of ciliae. Basal bodies have a variety of attached structures, termed appendages, satellites and fibres that can be detected by electron microscopy. The intimate relationship between basal bodies and centrioles is best revealed in karyomastogont containing phyla, where the centriole-kinetosome fulfils both functions, as centrosome constituent and in ciliagenesis, in accordance with the cell cycle and differentiation stages; this functional relation is also preserved in motile gametes of other phyla (Chapman et al. 2000). Centrioles and basal bodies have been studied in diverse cells by electron microscopy, defining their intimate association with a variety of cellular structures. In *Drosophila* spermatogenesis, the centrosome is found at the plasma membrane in spermatocytes, where it appears to acquire a membrane, which associates with the distal part of the centriole throughout the meiotic divisions. After the divisions are complete, the membrane appears to vesiculate and the centriole transforms to a basal body and starts to template the flagellar axoneme (Tates 1971). Sorokin (1968) showed that new basal bodies formed inside the lung epithelia cells appear to recruit membranes, visible as vesicular structures. Upon migration to the plasma membrane, these membranes fuse and cilia formation occurs. In other cases, membrane related processes only occur after basal bodies are localized beneath the plasma membrane. In these cases, it is clear that specific membrane growth must occur in order to allow the formation of cilia. Most of the proteins or regulatory components of these membrane-associated processes are not known. In the green flagellate *Spermatozopsis similes*, the p210 protein is a component of the centriole-plasma membrane connector (Lechtreck et al. 1999). The essential role of centrioles in completion of cytokinesis depends upon the movement of the mother centriole to the midbody of the dividing cells. Again, this process requires centrioles and results in the physical separation of two daughter cells by membrane closure (Piel et al. 2001).

5 Conclusion

Membrane biogenesis, regulation of membrane growth and shape of PSMs and IMC and other specialized membranes are essential processes that are closely regulated by centrioles/basal bodies or spindle pole bodies. However, with the exception of the process of SPM/FSM generation in yeast, we have not defined any of the proteins that regulate these processes. Future studies that define the key players in these events in diverse cell types are critically important to understand the role of centrosome-membrane interactions in cell division, shape generation and differentiation, and to be able to derive a generalized understanding.

Acknowledgement

We would like to thank John M. Murray and, especially, Naomi S. Morrissette for help and discussions during writing of this review. Boris Striepen and the Journal of Cell Biology are kindly acknowledged for permission to reproduce Figure 3a and 3c. The specific focus of this review did not permit complete reviewing of all aspects of the field. Thus, we apologize to colleagues whose work has not been cited here.

References

Aalto MK, Jantti J, Östling J, Keränen S, Ronne H (1997) Mso1p: a yeast protein that functions in secretion and interacts physically and genetically with Sec1p. Proc Natl Acad Sci USA 94:7331-7336

Andersen JS, Wilkinson CJ, Mayor T, Mortensen P, Nigg EA, Mann M (2003) Proteomic characterization of the human centrosome by protein correlation profiling. Nature 426:570-574

Bajgier BK, Malzone M, Nickas M, Neiman AM (2001) SPO21 is required for meiosis-specific modification of the spindle pole body in yeast. Mol Biol Cell 12:1612-1621

Briza P, Ellinger A, Winkler G, Breitenbach M (1988) Chemical composition of the yeast ascospore wall. The second outer layer consists of chitosan. J biol Chem 263:11569-11574

Chapman MJ, Dolan MF, Margulis L (2000) Centrioles and kinetosomes: form, function, and evolution. Q Rev Biol 75:409-429

Davidow LS, Goetsch L, Byers B (1980) Preferential occurrence of nonsister spores in two-spored asci of *Saccharomyces cerevisiae*: evidence for regulation of spore-wall formation by the spindle pole body. Genetics 94:581-595

Doxsey S (2001) Centrosomes as command centres for cellular control. Nat Cell Biol 3:E105-E108

Esposito MS, Esposito RE (1969) The genetic control of sporulation in *Saccharomyces*. I. The isolation of temperature-sensitive sporulation-deficient mutants. Genetics 61:79-89

Guo W, Sacher M, Barrowman J, Ferro-Novick S, Novick, P (2000) Protein complexes in transport vesicle targeting. Trends Cell Biol 10:251-255

Heywood P, Magee PT (1976) Meiosis in protists: some structural and physiological aspects of meiosis in algae, fungi and protozoa. Bact Rev 40:190-140

Hu K, Roos DS, Murray JM (2002) A novel polymer of tubulin forms the conoid of *Toxoplasma gondii*. J Cell Biol 156:1039-1050

Huh WK, Falvo JV, Gerke LC, Carroll AS, Howson RW, Weissman JS, O'Shea EK (2003) Global analysis of protein localization in budding yeast. Nature 425:686-691

Jackman M, Lindon C, Nigg EA, Pines J (2003) Active cyclin B1-Cdk1 first appears on centrosomes in prophase. Nat Cell Biol 5:143-148

Karpova TS, Reck-Peterson SL, Elkind NB, Mooseker MS, Novick PJ, Cooper JA (2000) Role of actin and Myo2p in polarized secretion and growth of *Saccharomyces cerevisiae*. Mol Biol Cell 11:1727-1737

Knop M, Pereira G, Geissler S, Grein K, Schiebel E (1997) The spindle pole body component Spc97p interacts with the gamma-tubulin of *Saccharomyces cerevisiae* and functions in microtubule organization and spindle pole body duplication. EMBO J 16:1550-1564

Knop M, Schiebel E (1998) Receptors determine the cellular localization of a gamma-tubulin complex and thereby the site of microtubule formation. EMBO J 17:3952-3967

Knop M, Pereira G, Schiebel E (1999) Microtubule organization by the budding yeast spindle pole body. Biol Cell 91:291-304

Knop M, Strasser K (2000) Role of the spindle pole body of yeast in mediating assembly of the prospore membrane during meiosis. EMBO J 19:3657-3667

Lange BM, Bachi A, Wilm M, Gonzalez C (2000) Hsp90 is a core centrosomal component and is required at different stages of the centrosome cycle in *Drosophila* and vertebrates. EMBO J 19:1252-1262

Lechtreck KF, Teltenkotter A, Grunow A (1999) A 210 kDa protein is located in a membrane-microtubule linker at the distal end of mature and nascent basal bodies. J Cell Sci 112:1633-1644

Marshall WF (2001) Centrioles take center stage. Curr Biol 11:R487-R496

Moreno-Borchart AC, Strasser K, Finkbeiner MG, Shevchenko A, Shevchenko A, Knop M (2001) Prospore membrane formation linked to the leading edge protein (LEP) coat assembly. EMBO J 20:6946-6957

Moreno-Borchart AC, Knop M (2003) Prospore membrane formation: how budding yeast gets shaped in meiosis. Microbiol Res 158:83-90

Morrissette NS, Murray JM, Roos DS (1997) Subpellicular microtubules associate with an intramembranous particle lattice in the protozoan parasite *Toxoplasma gondii*. J Cell Sci 110:35-42

Morrissette NS, Sibley LD (2002) Disruption of microtubules uncouples budding and nuclear division in *Toxoplasma gondii*. J Cell Sci 115:1017-1025

Neiman AM (1998) Prospore membrane formation defines a developmentally regulated branch of the secretory pathway in yeast. J Cell Biol 140:29-37

Nickas ME, Schwartz C, Neiman AM (2003) Ady4p and Spo74p are components of the meiotic spindle pole body that promote growth of the prospore membrane in *Saccharomyces cerevisiae*. Eukaryot Cell 2:431-445

Okamoto S, Iino T (1982) Genetic block of outer plaque morphogenesis at the second meiotic division in an *hfd1-1* mutant of *Saccharomyces cerevisiae*. J Gen Microbiol 128:1309-1317

Palazzo RE (2003) Centrosome and spindle pole body dynamics: a review of the EMBO/EMBL Conference on Centrosomes and Spindle Pole Bodies, Heidelberg, September 13-17. Cell Motil Cyto 54:148-154

Pelletier L, Stern CA, Pypaert M, Sheff D, Ngo HM, Roper N, He CY, Hu K, Toomre D, Coppens I, Roos DS, Joiner KA, Warren G (2002) Golgi biogenesis in *Toxoplasma gondii*. Nature 418:548-552

Pereira G, Knop M, Schiebel E (1998) Spc98p directs the yeast gamma-tubulin complex into the nucleus and is subject to cell cycle-dependent phosphorylation on the nuclear side of the spindle pole body. Mol Biol Cell 9:775-793

Pereira G, Schiebel E (2001) The role of the yeast spindle pole body and the mammalian centrosome in regulating late mitotic events. Curr Opin Cell Biol 13:762-769

Pickett-Heaps JD (1969) The evolution of the mitotic apparatus: an attempt at comparative ultrastructural cytology in dividing plant cells. Cytobios 3:257-280

Piel M, Nordberg J, Euteneuer U, Bornens M (2001) Centrosome-dependent exit of cytokinesis in animal cells. Science 291:1550-1553

Russell DG, Burns RG (1984) The polar ring of Coccidian sporozoites: a unique microtubule-organizing centre. J Cell Sci 65:193–207

Shaw MK, Compton HL, Roos DS, Tilney LG (2000) Microtubules, but not actin filaments, drive daughter cell budding and cell division in *Toxoplasma gondii*. J Cell Sci 113:1241-1254

Shimoda C (2004) Forespore membrane assembly in yeast: coordinating SPBs and membrane trafficking. J Cell Sci 26:389-396

Sorokin SP (1968) Reconstructions of centriole formation and ciliogenesis in mammalian lungs. J Cell Sci 3:207-230

Stokkermans TJ, Schwartzman JD, Keenan K, Morrissette NS, Tilney LG, Roos DS (1996) Inhibition of *Toxoplasma gondii* replication by dinitroaniline herbicides. Exp Parasitol 84:355-370

Striepen B, Crawford MJ, Shaw MK, Tilney LG, Seeber F, Roos DS (2000) The plastid of *Toxoplasma gondii* is divided by association with the centrosomes. J Cell Biol 151:1423-1434

Takahashi M, Yamagiwa A, Nishimura T, Mukai H, Ono Y (2002) Centrosomal proteins CG-NAP and kendrin provide microtubule nucleation sites by anchoring γ-tubulin ring complex. Mol Biol Cell 13:3235-3245

Tates AD (1971) Cytodifferentiation during spermatogenesis in *Drosophila Melanogaster*. PhD-thesis, Faculty of Natural Sciences, University of Leiden, Netherlands

Vogel J, Drapkin B, Oomen J, Beach D, Bloom K, Snyder M (2001) Phosphorylation of gamma-tubulin regulates microtubule organization in budding yeast. Dev Cell 1:621-631

Wendler F, Page L, Urbe S, Tooze SA (2001) Homotypic fusion of immature secretory granules during maturation requires syntaxin 6. Mol Biol Cell 12:1699-1709

Wigge PA, Jensen ON, Holmes S, Soues S, Mann M, Kilmartin JV (1998) Analysis of the *Saccharomyces* spindle pole by matrix-assisted laser desorption/ionization (MALDI) mass spectrometry. J Cell Biol 141:967-977

Zheng YX, Wong ML, Alberts B, Mitchison T (1995) Nucleation of microtubule assembly by a gamma-tubulin-containing ring complex. Nature 378:578-583

Knop, Michael
    EMBL, Cell Biology and Biophysics Programme, Meyerhofstr. 1, 69117 Heidelberg
    knop@embl.de

Taxis, Christof
    EMBL, Cell Biology and Biophysics Programme, Meyerhofstr. 1, 69117 Heidelberg

Index

β-amyloid precursor protein (APP), 133
α-ear, 23, 28
α-ear-binding motif, 24
α-SNAP, 99

14-3-3 proteins, 77

5-phosphatase II, 182

ABC transporters, 49
adaptin, 10, 15
adaptor proteins, 159
adaptor-associated kinase 1 (AAK1), 15
Adhesion to curved particles, 44
Ady3p, 197
Ady4p, 197
aftiphilin, 30
amphiphysin, 12, 45
Amphiphysin, 51
ANTH domain, 18
AP-1, 10, 26, 159, 177
AP180, 18, 20
AP2, 50
AP-2, 9, 23
AP-2 in cargo recruitment, 17
AP-2 in clathrin recruitment, 15
apicomplexan, 196, 199, 200
apicoplast, 201
Aplysia, 158
ARF, 178
Arf proteins, 71
ARF1, 26
Arf6, 98
ARH, 21
AtSyp122, 152
autosomal dominant polycystic kidney
    disease (ADPKD, 98
autosomal recessive
    hypercholesterolemia (ARH), 21
auxilin, 12

Bannayan-Zonana syndrome, 183
Basal bodies, 204
Bet1p, 94, 95

Bet3p, 91
Bet5p, 92
bf, 120
bicaudal-D proteins, 76
Bos1p, 94, 95

Ca^{2+}, 124, 125, 133, 50, 151, 152, 157,
    159, 178
CaMKII, 150, 155, 156, 158
cAMP, 154
cancers, 182
cargo selection, 12
Cargo sorting, 71
carnation (car), 120, 130
Casein kinase II (CKII), 147
CASK, 133
CCV formation, 14
CCV formation at the TGN, 26
Cdc2, 159
Cdc28, 197
Cdc42p, 97, 101
centriole, 193, 201
centriole-related, 204
centrosome, 204
ceramide-activated protein phosphatase
    (CAPP), 152
Charcot-Marie-Tooth disease, 184
choroideremia, 78
CKII, 151, 155, 156, 158, 159, 160
class C Vps, 75, 100
clathrin, 9, 44, 45, 50, 177
Clathrin coated vesicles (CCVs), 9
Clathrin trafficking at the TGN, 26
clathrin triskelions, 10
clathrin-coated pits (CCPs), 10
clathrin-mediated endocytosis (CME)., 9
Cnm67p, 196
Coat assembly, 44
coat proteins, 71
COG complex, 95
COPI, 44, 73, 93, 95, 160, 181
COPII, 73, 92, 93, 161
Cowden syndrome, 183
Cryptosporidium, 200
cyclin-dependent kinase 5 (Cdk5), 157
cytoskeleton, 104

death-associated protein kinase (DAPK),
    150
deep-orange, 130
demyelinating neuropathies, 184
dense-core vesicles (DCVs), 178
diacylglycerol (DAG), 54
diacylglycerol., 171
disease
    autosomal dominant polycystic
        kidney disease (ADPKD), 98
    Bannayan-Zonana syndrome, 183
    cancers, 182
    Charcot-Marie-Tooth disease, 184
    choroideremia, 78
    Cowden syndrome, 183
    demyelinating neuropathies, 184
    glioblastomas, 182
    Griscelli syndrome, 78
    Hermansky-Pudlak syndrome, 78
    Lhermitte syndrome, 183
    Lowe syndrome, 184
    Myotubular myopathy, 183
    OCRL, 184
    Proteus syndrome, 183
disease and development, 78
DNF1, 48
DNF2, 48
DOC2, 133
Don1p, 197
DRS2, 48
dynactin, 76
dynamin, 12, 54
dynein, 42, 76

early endosome, 98
EEA1, 74, 90, 98, 175
effectors, 66, 104
Eimeria, 200
endocytic accessory proteins, 11, 16, 18,
    20, 23
endophilin, 51
endosomal antigen-1 (EEA1), 130
endosomes, 75
Ent3p, 29
Ent5p, 29
ENTH domain, 18, 29, 51
Enthoprotin, 29
epsin, 19, 20, 45, 50
Exo70p, 96
Exo84p, 96

exocyst complex, 96
Exocyst complex, 145
exocytosis, 96

Fab1p, 179
flip-flop, 46
flippases, 50
Fore spore membrane (FSM), 196
frequenin, 176
FXDXF motif, 24
FYVE, 104
FYVE domain, 29, 74, 98
FYVE finger protein, 130

GAE domain, 22, 28
GARP/VFT complex, 101
GDI displacement factor (GDF), 71
GDP dissociation inhibitor (GDI), 66
GGTase II, 70
giantin, 93
Glc7/PP1, 158
glioblastomas, 182
GLUT4, 177
GM130, 73, 93, 160
GOS-28, 94
G-protein-coupled receptors (GPCRs, 21
Granuphilin, 131
GRASP65, 73, 93, 160
GRIP domain, 193
Griscelli syndrome, 78
GTPγS, 92
GTPase activating proteins (GAPs), 66
guanine nucleotide exchange factors
    (GEFs), 66
Gyp1, 74

Hermansky-Pudlak syndrome, 78
HIP1, 18, 21
HOPS, 123, 130, 75, 100, 103
Hrs, 175

in disease, 79
Inner membrane complex (IMC), 199,
    200
    Formation, 203
    Structure, 203
inositol 5-phosphatases, 181
inositol lipids, 171
inositol phospholipids, 26
Inp53p, 181

KEULE, 120

Leading edge protein (LEP) coat, 203
Lhermitte syndrome, 183
lipid acyltransferases, 53
Lipid domain-induced, 56
Lipid hydrolases, 53
lipid translocases, 48
lipids
    flip-flop, 46
    flippases, 50
    lipid translocases, 48
    polymorphism, 52
    remodelling, 51
Lowe syndrome, 184
Lysobisphosphatidic acid (LBPA), 54

meiosis II, 195
Meiotic plaque (MP), 197
melanophilin, 76
membrane budding, 9, 39
    inward budding, 40
    outward budding, 40
membrane curvature, 12, 19, 39
    Coat assembly, 44
    Cytoskeletal elements, 42
    Lipid domain-induced, 56
    regulation, 42
membrane curvtaure
    Adhesion to curved particles, 44
Membrane deformation, 39
membrane docking, 89
membrane domains, 75
membrane fusion, 73, 89,104
membrane microdomains, 75
membrane recycling, 98
membrane tethering, 73
membrane tubulation, 39
metabolism, 171
Microtubule organizing centre (MTOC),
    193
microtubules, 42
Mint (Munc18 interacting), 133
Mitotic kinases, 160
motor proteins, 42, 76, 104
Mpc54p, 197
Mpc70p, 197
MSO1, 129
Mso1p, 129
MTM1, 183

MTMR, 179
MTMR1, 183
multivescicular endosome, 54
multivesicular bodies, 29
Munc18, 156
Munc18-1, 120
Munc18-2, 124
Munc18a, 120, 126, 133
Munc18b, 124
Myo2, 77
Myosin Va, 78
MyosinVa, 76
Myotubular myopathy, 183
myotubularin (MTM, 179

NECAP, 25
n-Sec1, 120
NSF, 99, 100
Nud1p, 196

OCRL, 182, 184
Organelle transport, 76
outward budding, 40

p115, 73, 90, 93, 160
PA, 177
PACS-1, 159
PDZ domain, 133
Pep12, 147
Pep12p, 118
pharbin, 182
Pho85, 157
phosphoinositides
    PI kinases, 173
phosphatidylinositol 3-kinases, 104
phosphatidylinositol(4,5)P$_2$, 12
phosphoinositide phosphatases, 180
phosphoinositides, 171
    metabolism, 171
    PI phosphatases, 179
    PtdIns45P2, 171
phospholipids, 133
phosphorylation, 147
phosphotyrosine binding domain (PTB),
    133
PI 4-phosphatases, 181
PI kinases, 173
PI phosphatases, 179
PI3K, 174
PI4K, 175
PI4Kα, 176

PI-dependent kinase 1 (PDK1), 182
PIK1, 175
PIKfyve, 179
PIPP, 182
PKA, 151, 152, 156
PKC, 151, 153, 155, 158
Plasmodium, 200
PLIP, 179
polar ring, 201
polarized growth, 97
polymorphism, 52
Prenylation and localization, 70
Prospore membrane (PSM), 196
    Assembly, 195
    Closure, 200
    Growth, 199
    Initiation, 196
Proteus syndrome, 183
PTB domain, 21
PtdIns, 171
PtdIns 3-kinases, 173
PtdIns monophosphate kinases, 177
PtdIns(3)P, 74
PtdIns(4)P, 29
PtdIns(4,5)P$_2$, 19
PtdIns45P$_2$, 171, 178, 184
PTEN, 179, 182
P-type ATPases, 48

Rab 38, 76
Rab GTPase, 145
Rab GTPase cycle, 66
Rab GTPases, 65, 90, 159
    disease and development, 78
    effectors, 66, 104
    GDI displacement factor (GDF), 71
    GDP dissociation inhibitor (GDI), 66
    GGTase II, 70
    GTPase activating proteins (GAPS),
      66
    guanine nucleotide exchange factors
      (GEFs), 66
    in disease, 79
    membrane fusion, 73
    membrane microdomains, 75
    Organelle transport, 76
    Prenylation and localization, 70
    Rab GTPase cycle, 66
    REP (Rab Escort Protein), 70
    signaling pathways, 77
    Structural features, 68

vesicle tethering, 72, 73
Rab1, 73
Rab11, 75
Rab27, 76, 78
Rab32, 77
Rab3b, 71
Rab4, 75, 159
Rab5, 131, 73, 74, 98, 99, 160, 175
Rab6, 76
Rab7, 74
Rab9, 73
Rabaptin5, 98
rabenosyn-5, 175
Rabenosyn-5, 130, 74, 103
Rabex5, 98
Rabkinesin-6, 76
Rac, 178
Ral protein, 97
Ras-related GTPases, 68
rbSec1, 120
real-time imaging, 45
remodelling, 51
REP (Rab Escort Protein), 70
retrograde transport, 101
Rho family GTPases, 97
Rho, 178
Rho1p, 97, 101
RILP, 76
RIN1, 77
Rop, 120
ROP, 124
Rvs167, 157

Sac1-3, 181
Sac1p, 181
Saccharomyces cerevisiae, 118
Sar proteins, 71
Seb1p, 98
SEC1, 118, 126, 129, 157
Sec1/Munc18 (SM), 119, 145, 155
    function, 134
    Model of function, 133
    mutant mice, 125
    Phosphorylation, 126, 128, 156, 157
    Structure, 116, 120
    Syntaxin binding, 122
Sec10p, 96
Sec13, 161
SEC14, 181
Sec15p, 74, 96
Sec18p, 100

Sec1p, 118, 97
Sec2,, 74
Sec22p, 95
Sec34p, 95
Sec35p, 130, 95
Sec3p, 96
Sec4p, 74, 129, 90, 96, 98
Sec5p, 96
Sec6p, 96
Sec8p, 96
Sec9p, 123, 102, 152
second messenger, 182
Sed5p, 118, 126, 94, 95
SHIP, 182
signal transduction, 77
signaling pathways, 77
SKIP, 182
SLP1, 118
sly1, 125
SLY1, 129
SLY1-20, 118, 121, 129
Sly1p, 66, 118, 123, 126, 94
    Structure, 120
SNAK, 155
SNAP-23, 154
SNAP-25, 152
Snapin, 154
SNARE, 115, 73, 89, 97, 99, 101, 145,
    158
    assembly, 102
    cis SNARE, 146
    cis-SNARE complexes, 116
    phosphorylation, 147
    Q-SNARE, 116, 146
    R-SNARE, 116, 146
    trans SNARE, 146
    trans-SNARE complexes, 115
    t-SNARE, 115
    v-SNARE, 115
Snc1, 123
Snc2, 158
sorting, 105
Spc72p, 197
Spermatozopsis similes, 204
Spindle pole body (SPB), 193, 195
    Structure, 199
Spo74p, 197
Sso1, 123
SSO1, 118
Sso1, 2p, 103
Sso1/2, 151

SSO2, 118
Ssp1p, 197
Structural features, 68
Stt4p, 176
synaptojanin, 181
Synaptotagmin I, 154, 158
Syntaxin 1, 147
Syntaxin 2, 156
Syntaxin 3, 124
Syntaxin 4, 131, 151, 154, 156
Syntaxin 5, 94
Syntaxin 6, 99

targeting specificity, 102
TC10, 97
tethering, 145
tethering proteins, 89
    COG complex, 95
    cytoskeleton, 104
    EEA1, 90, 98
    exocyst complex, 96
    functions, 102
    GARP/VFT complex, 101
    HOPS, 100
    interactions, 103
    motor proteins, 104
    p115, 90, 93
    regulations, 105
    sorting, 105
    TRAPP, 91
    TRAPP complex. *See*
    Uso1p, 90, 94
Tethering proteins
    EEA1, 74
TGN, 101
Theileria, 200
tight junction, 96
Tlg1p, 118, 101,158
Tlg2p, 118, 127, 147, 158
TNFα, 178
Tomosyn, 131
Toxoplasma, 200
Toxoplasma gondii, 201
TPIP, 179
Transbilayer area asymmetry, 46
transcytosis, 71
transport of mitochondria, 77
TRAPP complex, 74, 91, 90
Trs120p, 92
Trs130p, 92
Trs20p, 92

Trs23p, 92
Trs31p, 92
Trs33p, 92
Trs65p, 92
Trs85p, 92

Ufe1p, 118
UNC-13, 131
unc-18, 118, 125
UNC-18, 131
USO1, 129
Uso1p, 90, 94

Vac1p, 130, 74, 100, 103
Vam3, 147
Vam3p, 118, 130, 100
VAMP/Synaptobrevin, 155
VAMP4, 159
Veli1, 133
vesicle budding, 46, 71
vesicle tethering, 72, 130
Vps10p, 101
Vps11p, 130, 100

Vps16p, 130, 100
Vps18p, 130
Vps21, 74
Vps21p, 130
Vps27p, 29, 175
Vps33p, 118, 123, 125, 130, 100, 103
VPS34, 174, 183
Vps34p, 176
Vps39p, 100
Vps41p, 100
Vps45p, 118, 123, 125, 127, 103,158
Vps51p, 101
Vps52p, 101
Vps53p, 101
Vps54p, 101
Vti1p, 118,158

Ykt6p, 95
Ypt11, 77
Ypt1p, 66, 90, 92, 94
Ypt31/32, 92
Ypt32, 74
Ypt7p, 130, 100

Printed in Poland
by Amazon Fulfillment
Poland Sp. z o.o., Wrocław

87951392R00127